METHODS IN PORPHYRIN PHOTOSENSITIZATION

ADVANCES IN EXPERIMENTAL MEDICINE AND BIOLOGY

METHODS IN PORPHYRIN PHOTOSENSITIZATION

Edited by

David Kessel

Wayne State University School of Medicine
Detroit, Michigan

PLENUM PRESS • NEW YORK AND LONDON

Library of Congress Cataloging in Publication Data

Porphyrin Photosensitization Workshop (1984: Wayne State University)
  Methods in porphyrin photosensitization.

  "Proceedings of the Porphyrin Photosensitization Workshop, held July 6–7, 1984, in
Philadelphia, Pennsylvania" — T.p. verso.
  Bibliography: p.
  Includes index.
  1. Photochemotherapy — Congresses. 2. Cancer — Radiotherapy — Congresses. 3. Por-
phyrin and porphyrin compounds — Therapeutic use — Congresses. 4. Porphyrin and
porphyrin compounds — Diagnostic use — Congresses. 5. Photosensitization, Biological
— Congresses. I. Kessel, David. II. Title. III. Series. [DNLM: 1. Photochemotherapy —
methods — congresses. 2. Porphyrins — pharmacodynamics — congresses. 3. Porphyrins
— therapeutic use — congresses. QU 110 P8372m 1984]
RC271.P43P67   1984   615.5'8   85-25627
ISBN 0-306-42210-7

Proceedings of the Porphyrin Photosensitization workshop, held July 6–7, 1984,
in Philadelphia, Pennsylvania

© 1985 Plenum Press, New York
A Division of Plenum Publishing Corporation
233 Spring Street, New York, N.Y. 10013

Printed in the United States of America

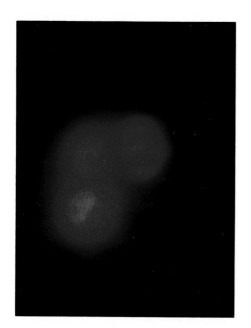 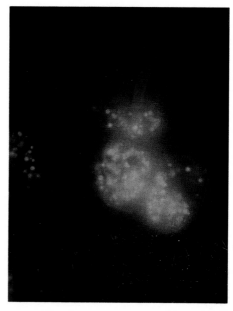

Porphyrin localization *in vitro*

Photographs of L1210 cells obtained by fluorescence microscopy. Left: after 10 min incubation with HPD, showing the diffuse peripheral fluorescence associated with drug accumulation at membrane loci. Right: after 24 hr incubation with HPD followed by washing to remove readily-diffusible porphyrins, showing intracellular sites of sensitizer binding. For further details of this experiment, see p. 223.

PREFACE

The use of porphyrins for localization and photodynamic therapy of neoplastic disease has been the topic of several international symposia, reviews and books during the preceeding five years. The literature on this topic has continued to grow, as have numbers of presentations at national and international meetings relating to photobiology, chemistry, and lasers. In this volume, it is the intention of the editor to provide both information on current research projects, and detailed methodology used in such investigations. A bibliography on the subjects of porphyrin localization and therapy is included.

The manuscripts contained in this volume are based on reports given at a Porphyrin Photosensitization Workshop which was held in Philadelphia, July 6-7, 1984. This Workshop was supported by NIH grant CA 36746, together with funds from the Fogarty International Center. Authors were requested to update their contributions to provide a summary of progress to mid 1985. Manuscripts containing material not presented in Philadelphia are also included, notably a series of articles describing current clinical and pre-clinical results from China. Since the Philadelphia Workshop, a meeting was held in Alghero, Sardinia (May, 1985), and additional conferences are now being planned; this attests to the continued interest in photodynamic therapy involving porphyrin photosensitization.

The successful development of this modality has required communication among a variety of workers in diverse fields. These include porphyrin chemistry, photochemistry, and photobiology, analytical chemistry, biophysics, pharmacology, surgery and medicine. While the clinical progress in photodynamic therapy was a major objective, a byproduct has been the mingling of workers from many different disciplines, with resulting collaborations sometimes extending to unrelated topics.

The use of porphyrins for tumor-localization had its beginnings in work reported by Lipson 25 years ago. This involved a collaborative effort with Dr. Samuel Schwartz, at the Mayo Clinic, which yielded a porphyrin product termed HPD (hematoporphyrin derivative) which was then, and still remains, the most effective tumor photosensitizer known. This work was of little initial interest to the biomedical field, since appropriate instrumentation for its exploitation was not available until the development of lasers and optical fibers for light distribution. But

a small group of workers continued to examine porphyrin-tumor affinity
relationships, generally with little outside support.

The current cascade of research projects, literature reports and
conferences was initiated by a series of investigations by Dr. T.J.
Dougherty, at the Roswell Park Memorial Institute in Buffalo. Dougherty
provided a reliable source of drug, together with information on light
sources, clinical protocols, and other information to interested investigators. A series of pioneering physicians, notably James Kennedy in
Ontario, Ian Forbes in Australia, Oscar Balchum in Los Angeles, Yoshihiro
Hayata in Japan, and Denis Cortese at the Mayo Clinic, provided the early
results which stimulated additional research. Because of this concerted
effort, many of the fundamentals of porphyrin photosensitization have
been reported, and guidelines for further work are clearly visible. This
volume is intended to provide a reference for further studies in the
field of porphyrin phototherapy, as well as an description of the current
status of selected fields of research.

David Kessel

CONTENTS

METHODS IN PORPHYRIN PHOTOSENSITIZATION

HEMATOPORPHYRIN PHOTODYNAMIC THERAPY FOR TRANSITIONAL CELL CARCINOMA

OF THE BLADDER - AN UPDATE

Ralph C. Benson, Jr.

Associate Professor of Urology
Mayo Clinic, Rochester, Minnesota

Carcinoma in situ (Tis) of the urinary bladder is usually treated with conservative measures such as fulguration and intravesical chemotherapy. The majority of patients, however, will experience disease recurrence and the only effective treatment for recurrent resistant carcinoma in situ of the bladder has been cystectomy and urinary diversion. The necessity for such a radical approach has been demonstrated, yet the use of surgery of such magnitude and consequence for superficial cancer is unfortunate.

Previous studies have convincingly demonstrated that hematoporphyrin derivative (HpD) is concentrated or retained selectively in severely dysplastic and neoplastic transitional cells.[1] This fact, plus the knowledge that activation of intracellular HpD with light of appropriate wavelength results in cytotoxicity, prompted a trial of HpD photodynamic therapy in patients with focal and diffuse resistant (Tis) of the urinary bladder.[2-11]

PATIENT SELECTION

All patients had transitional cell carcinoma in situ that was resistant to conventional modalities of treatment. Most patients had undergone frequent fulgurations and also intravesical instillations of chemotherapeutic agents such as thiotepa, mitomycin-C and doxorubicin and had, nonetheless, experienced disease recurrence. Carcinoma in situ was confirmed by bladder biopsy and also by urinary cytologic examination. All patients had been advised to undergo cystectomy and urinary diversion and had refused. The patients were treated under an experimental protocol previously approved by our institutional review board. There were 15 patients with focal Tis, and 12 patients with diffuse Tis (Table 1). Of the 12 patients with diffuse Tis, two also had focal T_2 disease (pt. no. 1 and 8) and two had focal T_a lesions (pt. no. 9 and 10) coexistent with their Tis. One patient (no. 2) had diffuse carcinoma in situ and also carcinoma in situ within a large bladder diverticulum.

PREPARATION OF HEMATOPORPHYRIN DERIVATIVE

Hematoporphyrin derivative was prepared by a modification of the technique of Lipson and associates.[2] Ten grams of hematoporphyrin

3

Table 1. Whole Bladder HpD Photodynamic Therapy, Patient History and Treatment results.

Pt.	(Age (yr)) Sex	Pathologic findings	Previous therapy	Bladder capacity (cc)	Response at 3 months	Follow-up time and status
1	70,M	Multifocal Tis, gr. 2; focal T_a	Multiple fulgurations thiotepa, Adriamycin mitomycin-C, focal HpD-PT	150	Partial (persistent) T_a)	8 mo; repeat Rx: neg.
2	75,M	Multifocal Tis, gr. 3	Fulgurations, thiotepa	200	Complete	6 mo; focal recurrence
3	72,M	Multifocal Tis, gr. 3	Fulgurations, thiotepa, mitomycin-C, Adriamycin	250	Complete	9 mo; two low-grade papillary tumors fulgurated
4	76,M	Multifocal Tis, gr. 2	Fulgurations, thiotepa, mitomycin-C, focal HpD-PT	250	Complete	6 mo; neg.
5	65,M	Multifocal Tis, gr. 2	Fulguration, thiotepa Adriamycin	150	Complete	10 mo; neg.
6	62,M	Multifocal Tis, gr. 2	Fulgurations, thiotepa, mitomycin-C, Adriamycin, focal HpD-PT	180	Complete	5 mo; neg.
7	60,M	Multifocal Tis, gr. 3	Fulgurations, partial cystectomy, thiotepa, mitomycin-C	250	Complete	7 mo; neg.
8	60,M	Multifocal Tis, gr. 2; focal T_a	Fulgurations, mitomycin-C	180	Partial (persistent T_a)	
9	49,M	Multifocal Tis,gr.3; focal T_2, gr. 3	Fulgurations, thiotepa, mitomycin-C, Adriamycin	150	Partial	8 mo; persistent T_2 disease
10	77,F	Multifocal Tis,gr.3; focal T_2, gr. 3	Fulguration, thiotepa, mitomycin-C, cis-platinum	200	Partial	8 mo; cystectomy for recurrent disease
11	73,F	Multifocal Tis, gr. 3	Fulguration, thiotepa	150	Complete	5 mo; neg.
12	72,M	Multifocal Tis, gr. 3	Fulguration, thiotepa mitomycin-C, Adriamycin	200	Complete	6 mo; neg.

gr., grade; HpD-PT, hematoporphyrin derivative photodynamic therapy; neg., negative; Tis, carcinoma in situ.

dihydrochloride were dissolved in 100 ml. glacial acetic acid and concentrated sulfuric acid, 19:1. This was done in a brown bottle, with stirring for 15 minutes at room temperature. The solution was then filtered through a coarse glass filter into a 500 ml. vacuum flask wrapped 2,000 ml. beaker, and 1.5 l. of 3 per cent sodium acetate were added and stirred. The resulting solution was allowed to stand overnight at room temperature. The hematoporphyrin derivative was a thick precipitate and was filtered by vacuum filtration through a 2.1 porous glass filter wrapped in foil. The filtered hematoporphyrin derivative was washed with 2.5 l. of cold distilled water. The washed hematoporphyrin derivative was then transferred to a vacuum bottle and lyophilized. The lyophilized hematoporphyrin derivative yield was 6.42 gm.

The hematoporphyrin derivative was prepared for patient use by dissolving it in 0.9 per cent sterile sodium chloride and stirring for approximately 15 minutes. The pH was then increased to between 6.4 and 7.0 by the addition of 0.1 normal sodium hydroxide. The solution was stirred for an additional hour in the dark. More 0.1 normal sodium hydroxide solution was added to increase the pH to 7.4, and the product was diluted to a final concentration of 5 mg./ml. by the addition of 0.9 per cent saline. The final hematoporphyrin derivative was sterilized by millipore filtration and placed in sterile amber glass 100 ml. vials for injection. The solution was tested for sterility and pyrogenicity.

For focal treatment the optical guide used was a medical grade quartz fiber 5 m. in length and having a core diameter of 400 μm. The exit beam divergence was approximately 20 degrees. The fiber was passed through the open channel of the endoscope so that the distal end protruded from the tip of the instrument. During treatment the distal end of the fiber was positioned approximately 2 to 2.5 cm. from the surface of the tumor whenever the light was directed on the external surface of the tumor. In this case the diameter of the beam was approximately 1 cm. In an attempt to increase the size of the intense spot the distal tip of the fiber was formed into a microlens by heating in a flame. In this way the intense spot could be spread over a diameter of 2 cm. when the fiber was held 2 to 2.5 cm. from the surface of the tumor.

The optical guide employed for whole bladder therapy was a medical-grade quartz optical fiber with a core diameter of 400 μm. For light divergence, a spherical bulb made of clear polycarbonate was attached to the end of this fiber (Fig. 1).

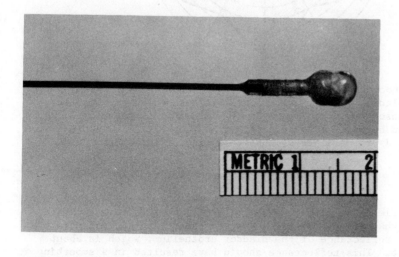

Fig. 1. Quartz fiber with a clear polycarbonate "light bulb" tip. From Benson, RC, Jr.: Treatment of diffuse transitional cell carcinoma in situ by whole bladder hematoporphyrin derivative photodynamic therapy. J Urol [in press]. By permission of the Journal of Urology)

With the fiber positioned approximately in the center of the bladder, the output of light in relation to the angle of impingement on the urothelium was that shown in Figure 2. The overall coupling efficiency of the system was approximately 70% from laser to bulb, and the bulb efficiency was 96%; that is, 96% of the power in the fiber emerged from the bulb. The average output was 81% of the maximum recorded at any one point, and the rear octant was 50% of the average output (Fig. 2).

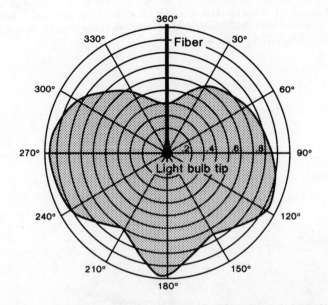

Fig. 2. Polar coordinate plot of relation of output to angle of light impingement on bladder urothelium. Bladder is assumed to be a perfect sphere. (From Benson, RC, Jr: Treatment of diffuse transitional cell carcinoma in situ by whole bladder hematoporphyrin derivative photo-dynamic therapy. J Urol [in press]. By permission of the Journal of Urology)

The rest of the field was within about 25% of the average value. These figures assumed that the bladder is a uniform sphere and did not take into account the reflectance of the bladder urothelium, which is about 50% for red light. This reflectance should have resulted in a smoothing out of the light distribution.

The proximal end of the quartz fiber was attached to an optical positioning device, which was coupled to the output of a continuous wave dye laser. Adjustments were made so that the dye laser beam was focused directly into the optical fiber. Rhodamine B dye was circulated in the dye laser. The argon laser had a maximal power output of 20 watts and was capable of producing an output of as much as 4 watts at 630 mm. from the dye laser.

TREATMENT PROCEDURE

Focal:

    For initial treatment all patients received HpD intravenously at a
dose of 2.5 mg./kg. body weight approximately 3 hours before light
irradiation. Several patients were also given a second treatment 48
hours after drug administration. Standard rigid cystoscopes equipped
with the Storz laser fiber guide were used to deliver the light when
possible. When lesions were located in the dome or at the bladder neck
a flexible fiberoptic bronchoscope or a flexible nephroscope-choledocho-
scope was used. Most procedures were done with the patient under
general anesthesia. However, if the lesions were easily accessible and
the patient requested, treatment was performed with the patient under
local anesthesia. All areas of in situ carcinoma treated were $\leq$ 3 cm.
in diameter, although in a few cases the disease was multifocal. Even
though the in situ carcinomas were not > 3 cm. in diameter the diameter
of the light irradiation spot often was too small to cover the entire
tumor when the probe was held in one position. However, all tumors
seemed to be covered adequately by irradiating in two overlapping locations or by moving the probe slowly back and forth over the involved
area. Light dose approximated 150 joules per cm. for each treatment
session and all treatment was performed at a wavelength of 630 nm.
Patients were released from the hospital either the evening of the
treatment day or the next morning.

Diffuse:

    All patients received 4 to 5 mg. of HpD per kg. of body weight
3 hours (initial 5 patients) to 48 hours (subsequent patients) before
exposure of the entire bladder to red light (630 nm.).

    Figure 3 shows the number of joules (Watt-seconds) delivered per
square centimeter per minute for various bladder volumes at an output
of 1 watt from the dye laser (Fig. 3).

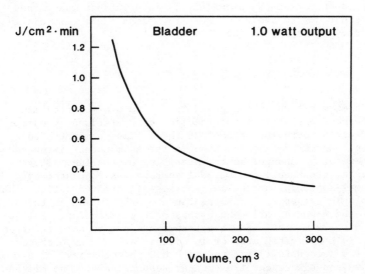

Figure 3. Plot of bladder volume vs. joules/cm^2/min. Power output
from argon pumped dye laser is assumed to be 1.0 watt, and bladder to be
a perfect sphere. From Benson, RC, Jr. J Urol. (in press). By permission
of the Journal of Urology.

Using this graph, one can calculate the time needed to deliver the desired number of joules per square centimeter. Most of the patients treated had measured bladder capacities of 150 to 250 ml.; this volume was kept constant during treatment (Table 1). Illumination times of approximately 60 minutes with power outputs of 1 to 1.5 watts resulted in the delivery of 25 to 45 joules/cm^2 to the entire urothelium. All treatments were performed with general anesthesia and standard rigid cystoscopes equipped with a Storz laser fiber guide; slightly cool saline was used as the irrigant. The fiber usually was placed as best possible in the center of the bladder. The patient with both diffuse Tis and Tis in a bladder diverticulum (pt. no. 2) was treated with the fiber in the center of the bladder for 60 minutes and in the center of the diverticulum for 30 minutes. On occasion, the fiber was located closer to the hemisphere of the bladder in which the disease was most prevalent in an attempt to increase the incident dose in the area of known tumor. This resulted in a diminished, but probably adequate, dose to the opposite, uninvolved hemisphere because of bladder reflectance.

PATIENT FOLLOW-UP AND INTERPRETATION OF RESULTS

All patients were informed of the need for close follow-up, including multiple cystoscopic and biopsy studies, to assess therapeutic response. After the initial treatment all patients were examined by cystoscopy at least once within the first 7 days after treatment. With experience we found that it takes at least 4 weeks for the initial edema and exudative reaction to subside before it is possible to obtain adequate revisualization, photographic documentation and biopsy proof of the therapeutic result. Accordingly, the efficacy of HpD photodynamic therapy was determined by repeat cystoscopic examination and urinary cytologic examination and biopsy 3 months after treatment and every 3 months thereafter (biopsies were performed at these examinations if urine cytologic findings were positive or if a visual abnormality was present).

Since the skin absorbs a quantity of hematoporphyrin derivative sufficient to produce marked photosensitivity each patient was warned carefully and repeatedly to avoid exposure to bright light, especially direct and indirect sunlight, for up to 4 weeks.

RESULTS

Focal Therapy

Cystoscopic examination in all patients 4-8 days after treatment revealed a pronounced exuberant local reaction characterized by edema, hemorrhage, and early necrosis. Biopsy at these times demonstrated a marked hemorrhagic and exudative reaction in the submucosal tissue and significant degenerative changes in the overlying urothelium. Thirty to sixty days after treatment, the mucosal surface was populated by essentially normal-appearing urothelial cells. All treated patients had biopsy and cytologic evidence of disappearance of the lesion in the treated area at the 3-month follow-up cystoscopic examination. Follow-up ranged from 6 to 32 months. Eight patients had one or more recurrent lesions, either in the treated area or in distant areas. In each case, the recurrence was successfully treated with HpD phototherapy. Two patients had five and six repeat treatments, respectively. One of the 15 patients underwent cystectomy and urinary diversion because of progression of T_1 disease after multiple recurrences.

The only adverse effect encountered was severe sunburn in two

patients who did not heed warnings to remain out of direct or indirect
sunlight.

Whole Bladder Therapy

In the 8 patients with Tis alone, biopsy and urinary cytologic
examination at the 3-month follow-up cystoscopic examination showed
complete disappearance of tumor (Table 1). Two patients who had both
Tis and T_a lesions were noted to have disappearance of Tis but persistence of one (pt. no. 1) or two (pt. no. 8) T_a lesions at the 3-month
follow-up examination. Of these eight patients, two were later noted
to have focal recurrent disease (no. 2 at 6 months and no. 3 at 9
months after treatment). Patient 8 had not yet returned for his
6-month follow-up cystoscopic examination. The two patients who had
diffuse Tis and also focal invasive disease (pt. no. 9 and 10) were
noted to have persistence of invasive disease and disappearance of
carcinoma in situ. Patient 10 underwent cystectomy and urinary diversion. At the time of this report, patient 9, who had persistently
refused cystectomy, is receiving BCG intravesical therapy.

Adverse effects of treatment were minimal. All patients were
warned to wear protective clothing and to use sunscreen preparation if
they were exposed to sunlight. Cutaneous photosensitivity (sunburn),
although a significant nuisance, was not severe in any of these patients.
Specifically, the two patients who had undergone previous focal HpD
photodynamic therapy did not experience any increased photosensitivity.
Most patients noted bladder irritability characterized by frequency,
urgency, and diminished bladder capacity shortly after therapy. Most
patients had reduced bladder capacities before HpD photodynamic therapy
(Table 1) due to their underlying disease and the types of previous
therapy that they had received. However, all patients reported that
their subjective bladder capacity had returned to its pretreatment
volume by the time of the first 3-month follow-up visit. Treatment
results did not appear to be affected by pretreatment bladder volumes
or by the timing of light delivery after administration of HpD (that is,
whether at 3 hours or at 48 hours).

DISCUSSION

Several investigators have shown that HpD phototherapy has a
significant antitumor effect in patients with transitional cell carcinoma in situ of the bladder.[12-14] In our early treatment attempts, a
quartz fiber with a flat or slightly convex end was used, and the result
was a spot of light only 2 to 2.5 cm. in diameter. This small spot size
greatly limited the area of tumor that could be treated at any one
sitting. Our expanded series of 15 patients has confused the efficacy
of focal HpD photodynamic therapy in patients with resistant Tis of the
bladder, but the recurrence rates for patients with a field-change
disease such as carcinoma in situ of the bladder have remained significant. For this reason, our group and several others have investigated
ways to deliver an ample amount of light to the entire bladder in hopes
of reducing recurrence. Jocham et al introduced the possibility of the
use of a lipid diffusing medium, but no clinical trials in humans have
been published.[15] Hisazumi et al devised a laser light-scattering
optic coupled with a driving instrument that irradiates the entire
bladder in sequential segmental strips.[16] They used an argon ion,
pumped dye laser as a light source in the treatment of two patients,
delivering a dose of approximately 10 joules/cm^2 to the entire bladder.
Early results in these two patients appeared encouraging, urinary
cytologic findings having returned to normal after treatment. This
system, however, is cumbersome and costly.

We approached the problem of light delivery to the entire bladder by modifying the distal end of the delivery fiber. Our early attempts at forming a bulbous tip at the end of the fiber resulted in inadequate illumination at the bladder neck, but more recently we have used the previously described fiber with a "light bulb" tip that delivers fairly uniform light to the entire bladder (Fig. 1 and 2). This fiber is simple and relatively inexpensive, and it allows significant light energy to be delivered. Although it has been suggested that as little as 10 joules/cm^2 may be effective in the treatment of carcinoma in situ, we prefer to deliver as close to 50 joules/cm^2 as technically possible. However, formal dose-response relationships for HpD photodynamic therapy of transitional cell carcinoma in situ are not yet available.

The optimal dose of HpD and timing of light delivery for this disease are likewise unknown. Most authors have suggested the administration of 2.5 to 5.0 mg. of HpD per kg. of body weight. More recently, Dougherty et al separated HpD by gel exclusion chromatography and identified a new component of the mixture.[17] The component was found to be responsible for the tumor-photosensitizing ability of HpD. It is possible that this new active component will be effective at lower doses and perhaps with less morbidity. The timing of light delivery after HpD administration has varied in the few clinical studies reported dealing with the bladder. Most investigators treat between 48 and 72 hours after administration of the drug. We occasionally treat as early as 3 hours after giving the HpD and again at 48 or 72 hours, if necessary. Early treatment was based on the appearance of bright fluorescence at 3 hours in our bladder tumor localization studies and also in other tumor systems.[1,18,19] Recent studies in mice indicate that treatment before 3 hours is ineffective and that the therapeutic effect reaches its maximum at 3 hours and then levels off. Treatment at either 3 hours or 48 hours in the present study was arbitrary and occasionally was based on patient convenience. None of the patients with diffuse Tis underwent fluorescence localization of the malignancy, because of its obviously diffuse nature and because of our conviction, based on previous studies, that HpD localizes in dysplastic and malignant transitional cells by 3 hours after administration.

All the patients to whom we administered whole bladder HpD photodynamic therapy had had multiple previous treatments of various types (Table 1) and, therefore, represented a highly select group of patients with resistant disease. We continue to believe that resistant carcinoma in situ is the stage of disease best treated by HpD photodynamic therapy, and the present study confirms that it is possible to effectively treat not only diffuse disease but also Tis in a bladder diverticulum. Traditionally, electrofulguration of lesions in a diverticulum has been somewhat worrisome because of the thinness of the wall and the fear of perforation or damage to adjunctive viscera. The light bulb fiber is easily inserted into a diverticulum, and appropriate light doses can be delivered without difficulty.

Comparison of treatment results with other therapeutic modalities is difficult, but it seems clear that our initial suspicion that patients with invasive disease are poor candidates for HpD photodynamic therapy as it is currently delivered is true (pt. no. 9 and 10). Although we noted recurrence of tumor in two of eight patients with diffuse resistant carcinoma in situ, we believe at this time that whole bladder HpD photodynamic therapy holds promise in the treatment of this most difficult disease entity. Until more is known about the natural history and underlying causes of this problem, it may be naive to consider that any one treatment short of cystectomy will be curative. Perhaps consideration should be given to some form of adjuvant continuing therapy in combination with HpD photodynamic therapy who refuse surgical therapy.

REFERENCES

1. Benson, R. C., Jr., Farrow, G. M., Kinsey, J. H., Cortese, D. A., Zincke, H. and Utz, D. C.: Detection and localization of in situ carcinoma of the bladder with hematoporphyrin derivative. Mayo Clin. Proc., 57:548, 1982.
2. Lipson, R. L., Baldes, E. J. and Olsen, A. M.: The use of a derivative of hematoporphyrin in tumor detection. J. Natl. Cancer Inst., 26:1, 1961.
3. Lipson, R. L. and Baldes, E. J.: The photodynamic properties of a particular hematoporphyrin derivative. Arch. Dermatol., 82:508, 1960.
4. Gregorie, H. B., Jr., Horger, E. O., Ward, J. L., Green, J. F., Richards, T., Robertson, H. C., Jr. and Stevenson, T. B.: Hematoporphyrin-derivative fluorescence in malignant neoplasms. Ann. Surg., 167:820, 1968.
5. Kelly, J. F. and Snell, M. E.: Hematoporphyrin derivative: a possible aid in the diagnosis and therapy of carcinoma of the bladder. J. Urol., 115:150, 1976.
6. Kinsey, J. H., Cortese, D. A., Moses, H. L., Ryan, R. J. and Branum, E. L.: Photodynamic effect of hematoporphyrin derivative as a function of optical spectrum and incident energy density. Cancer Res., 41:5020, 1981.
7. Gomer, C. J. and Dougherty, T. J.: Determination of [^{3}H]- and [^{14}C] hematoporphyrin derivative distribution in malignant and normal tissue. Cancer Res., 39:146, 1979.
8. Diamond, I., Granelli, S. G., McDonagh, A. F., Nielsen, S., Wilson, C. B. and Jaenicke, R.: Photodynamic therapy of malignant tumours. Lancet, 2:1175, 1972.
9. Dougherty, T. J., Grindey, G. B., Fiel, R., Weishaupt, K. R. and Boyle, D. G.: Photoradiation therapy. II. Cure of animal tumors with hematoporphyrin and light. J. Natl. Cancer Inst., 55:115, 1975.
10. Granelli, S. G., Diamond, I., McDonagh, A. F., Wilson, C. B. and Nielsen, S. L.: Photochemotherapy of glioma cells by visible light and hematoporphyrin. Cancer Res., 35:2567, 1975.
11. Dougherty, T. J., Gomer, C. J. and Weishaupt, K. R.: Energetics and efficiency of photoinactivation of murine tumor cells containing hematoporphyrin. Cancer Res., 36:2330, 1976.
12. Benson, R. C., Jr., Kinsey, J. H., Cortese, D. A., Farrow, G. M. and Utz, D. C.: Treatment of transitional cell carcinoma of the bladder with hematoporphyrin derivative phototherapy. J. Urol., 130:1090, 1983.
13. Tsuchiya, A., Obara, N., Miwa, M., Ohi, T., Kato, H. and Hayata, Y.: Hematoporphyrin derivative and laser photoradiation in the diagnosis and treatment of bladder cancer. J. Urol., 130:79, 1983.
14. Hisazumi, H., Misaki, T. and Miyoshi, N.: Photoradiation therapy of bladder tumors. J. Urol., 130:685, 1983.
15. Jocham, D., Staehler, G., Chaussy, C., Schmiedt, E., Unsold, E., Hammer, C., Lohrs, U., Brendel, W. and Gorisch, W.: Integrated dye laser therapy of bladder tumors after photosensitization with hematoporphyrin-derivative (HpD) - The use of a light-dispersion medium (abstract). Presented at the Seventy-Eighth Annual Meeting of the American Urological Association, Las Vegas, Nevada, April 17-21, 1983, p. 155.
16. Hisazumi, H., Miyoshi, N., Naito, K. and Misaki, T.: Whole bladder wall photoradiation therapy for carcinoma in situ of the bladder: a preliminary report. J. Urol., 131:884-1984.
17. Dougherty, T. J., Boyle, D. G., Weishaupt, K. R., Henderson, B. A., Potter, W. R., Bellnier, D. A. and Wityk, K. E.: Photoradiation therapy - clinical and drug advances. In: Porphyrin Photosensi-

tization. Edited by D. Kessel and T. J. Dougherty. New York: Plenum Press, 1983, p. 3.

18. Lipson, R. L., Baldes, E. J. and Olsen, A. M.: Hematoporphyrin derivative: a new aid for endoscopic detection of malignant disease. J. Thorac. Cardiovasc. Surg., 42:623, 1961.

19. Gregorie, H. B., Jr., Horger, E. O., Ward, J. L., Green, J. F., Richards, T., Robertson, H. C., Jr. and Stevenson, T. B.: Hematoporphyrin-derivative fluorescence in malignant neoplasms. Ann. Surg. 167:820, 1968.

PHOTODYNAMIC THERAPY IN THE MANAGEMENT OF CANCER:

AN ANALYSIS OF 114 CASES

Wei-Ming Cai, Yue Yang, Nai-Wu Zhang
Jian-Guo Xie, Xiu Zhang, Xian-Jun Fan
Hong Shen, Xiao-Jiang Hu and Xian-Wen Ha

Section of Lasers, Cancer Institute, Chinese
Academy of Medical Sciences, Beijing, China

This article reports on results of treatment of 114 cancer patients at the Cancer Institute, Chinese Academy of Medical Sciences, between September 1981 and December 1984. In September 1981, a study group of clinicians, physicians and basic scientists was set up at the Cancer Institute for the study of photodynamic therapy (PDT) with HPD. During the following 39 months, 114 patients were treated by PDT alone, 8 by PDT + hyperthermia and 3 by PDT + radiotherapy. The analysis in this presentation is based on PDT alone, since the numbers of patients in the other groups is insufficient for statistical analysis.

MATERIAL AND METHODS

All patients were examined physically, and lesions biopsied for histological confirmation. Among our patients, 63 were previously untreated; the others had recurrent disease, or tumors nonresponsive to other modalities. The distribution of cancer sites is shown in Table 1; histology in Table 2.

The photosensitizer HPD was prepared by the Institute of Pharmaceutical Industry, Beijing. It was tested for sterility, and stored at -20 to -30°C. Fluorescence studies in human serum indicated an excitation maximum at 405 nm with the emission maximum at 630 nm.

The light source was a 2-4 watt argon pumped CW dye laser, from the Institute of Electronology, Chinese Academy of Sciences. This provides 625-630 nm light at 250-400 mW, using Rhodamine 640 as the dye medium. At the beginning of each treatment, the light output was measured with a broad power meter with wavelength checked with a monochromator.

Patients received a routine physical examination with histological diagnosis. Each received 5 mg/kg of HPD in 100 ml isotonic saline over 30 min (i.v.). Phototherapy was begun 48-72 hr later. The estimated energy was 150-400 J/cm². All patients were informed of the hazards of photosensitization. If a tumor residue remained after one PDT treatment, or if local recurrence was observed, more courses were given. In this series, 62 patients had one course of PDT; 38 had 2 courses, 10 had

Table 1. Distribution of Cancer Sites

Site	Number of patients
Skin	51
Primary	42
Metastatic	9
Head & Neck	27
Urinary	32
Cervix	3
Esophagus	1
Total	114

3 courses; 3 had 4 courses, and 1 had 5 courses. Of these, 54 involved
direct irradiation, 129 involved use of a quartz fibre, and 2 were
interstitial.

RESULTS

Tumor response was classified into four grades:
1. Complete remission (CR): no evidence of tumor macroscopically and
    microscopically for one month.
2. Significant remission (SR): >50% of original tumor disappears one
    month after PDT.
3. Minor remission (MR): <50% of tumor disappears after one month.
4. No remission (NR): no change or continued growth after PDT, or tumor
    size initially diminishes but returns within one month.

The results of PDT are summarized in Table 3. Three patients (2
basal cell carcinomas, 1 metastatic breast carcinoma) lost to follow-up
are counted as no remission. The effectiveness rate (CR + SR + MR) is
85.1%. The longest duration of tumor control was 34 months.

Fourteen patients showed no response. PDT failure may be related
to the extent of disease or poor blood supply to the tumor because of
marked fibrosis. Tumors not previously treated showed the best response
rate, but even in residual or recurrent tumors, the response rate was
76% (Table 7).

Illustrative Cases

A male, 68, formerly a pilot, presented with a 1.5 x 1.5 cm basal
cell carcinoma located in his left nasal ala for >13 months. This lesion was treated with PDT (200 J) and a total remission was obtained.
This patient has now been followed for 34 months without recurrence
(Fig. 1).

A male, 42, peasant, presented with a 3 x 5.5 cm squamous cell carcinoma of the lip observed 8 months before. The energy dose employed
was 300-400 J for each course. A complete response was obtained, and
biopsies showed no remaining disease (Fig. 2).

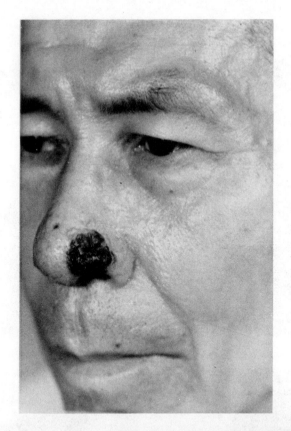

Figure 1. Case 1, basal cell carcinoma of the left nasal ala before (top) and after (bottom) PDT.

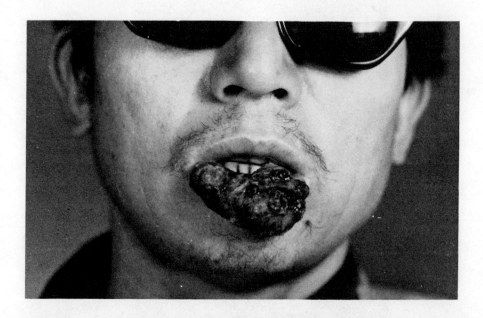

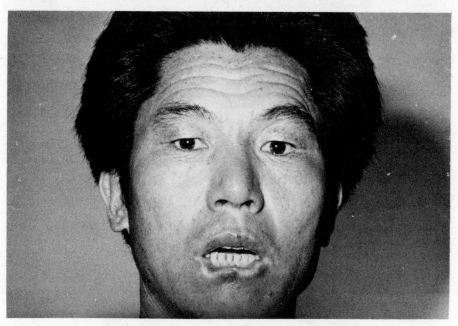

Figure 2. Case 2. Squamous cell carcinoma of the lower lip before (top) and after (bottom) PDT.

Table 2. Distribution of Histologic Types

Histology	Number of patients
Squamous cell carcinoma	40
Basal cell carcinoma	23
Transitional cell ca	31
Adenocarcinoma	9
Mixed tumor	2
Osteogenic sarcoma	2
Mucoepithelial ca	2
Malignant melanoma	1
Mycosis fungoides	1
Malignant papillary tumor	1
Retinoblastoma	1
Malignant fibrohisteocytoma	1
Total	114

DISCUSSION

This report describes results of a series begun in September 1981 using principles first described by Dougherty at Roswell Park, but with medical facilities, HPD and laser systems of Chinese manufacture. The immediate effective response to PDT was 85.1%: complete remission 68/114, significant remission in 15, minor remission in 14, no remission and unknown in 17.

Results of early experimental work by Dougherty's group were encouraging (1). Since then, other investigators have described their clinical experiences (2-5). The short-term clinical results are very promising. Hayata's group found complete tumor regression in 9 of 10 cases of early-stage lung cancer, with a significant remission in one (4). In a group of bladder tumors, the remission rate was 70% (5). Ha et al reported on a short-term study where 45 of 47 patients responded favorably (7). In an extensive review, Dougherty (8) pointed out that the overall response rate to PDT was 70%, with an 80% rate observed in early neoplasia. Better results are expected with improvements in light sources and the sensitizer.

In this study, the effective rate was 85.1% (97/114) which is lower than in an earlier report (7). This difference is attributed to the treatment of patients with more extensive cancer. We suggest multiple courses of PDT in extensively infiltrated lesions. In one case described here, the tumor size was initially 3 x 5.5 cm. The first PDT course affected only the superficial tumor, but repeated courses led to an eventual complete eradication. In such large tumors, interstitial irradiation would be more effective than surface illumination.

PDT may also be useful for therapy of local residual or recurrent tumor which has failed other modalities (Table 4). Because of the limited penetration of 630 nm light, deeply infiltrated lesions may be inherently untreatable. Curative results might therefore be expected for only early or superficial lesions. A combination of PDT + hyperthermia and/or radiotherapy would be indicated for such tumors. Such work is currently in progress in our institution.

Table 3. Results of PDT Treatment

Tumor and location	Cases	Immediate response				
		CR	SR	MR	NR	Unknown
Squamous Cell ca	40					
Skin		7	5	3	1	
Nasal/paranasal sinus		3		2	2	
Tongue		3	1	1		
Soft palate		1			1	
Buccal mucosa				1		
Lip		2				
Nasopharynx					2	
Cervix				3		
Esophagus				1		
Penis					1	
Basal Cell ca (skin)	23	17	3		1	2
Transitional Cell ca	31					
Urethral meatus		1				
Bladder		26	1	1	2	
Adenocarcinoma	9					
Breast (metast)		5	2			1
Liver (metast)				1		
Malignant mixed	2					
Parotid					1	
Nasopharynx		1				
Osteogenic sarcoma (maxilla)	2		1	1		
Mucoepithelial ca	2	1	1			
Malignant melanoma (skin)	1				1	
Mycosis fungoides (trunk)	1		1			
Malignant papillary (vagina)	1	1				
Retinoblastoma	1				1	
Fibrohisteocytoma	1				1	
Total	114	68	15	14	14	3

Table 4. Responses: Untreated vs. Recurrent Tumors

Lesion	Response					Rate
	CR	SP	MR	NR	Unknown	
Untreated	45	8	5	3	2	92%
Residual and recurrent	23	7	9	11	1	76%
Total	68	15	14	14	3	85.1%

Bladder tumors tend to occur at multiple foci. As an alternative to cystectomy, PDT is suggested. In the present series, 26/30 bladder tumors had complete responses, with one case disease-free for 20 months. The addition of external bladder radiotherapy or more extensive bladder PDT might increase the overall survival rate. Although ectopical recurrences are expected, repeated PDT courses will eradicate such recurrences, avoiding the need for surgery.

REFERENCES

1. TJ Dougherty et al. (1978) J. Natl. Cancer Inst. 55, 115.
2. JC Kennedy et al. (1981) in Cancer in Ontario, p. 118.
3. G Bruce et al. (1982) Am. J. Obstet. Gynecol., 2, 124.
4. H Hayata et al. (1984) Lasers Surg. Med. 4, 39.
5. IJ Forbes et al. (1984) Med. J. Aust. 2, 489.
6. Y Hayata et al. (1983) Abstracts, 13th International Cancer Congress.
7. XW Ha et al. (1983) Chin. Med. J. 96, 754.
8. TJ Dougherty (1984) Proc. Amer Assn. Cancer Res. 25, 408.

PRELIMINARY RESULTS OF HEMATOPORPHYRIN DERIVATIVE-LASER TREATMENT

FOR 13 CASES OF EARLY ESOPHAGEAL CARCINOMA

Mao-En Tian[1], Song-Ling Qui[2] and Qing Ji[1]

[1]Departments of Endoscopy and Photoradiation Therapy
Henan Tumor Institute and Hospital; [2]Department of
Pathology, Henan Institute of Medical Sciences
Zheng Zhou, Henan Province, China.

INTRODUCTION

As a result of the wide application of diagnostic fibroesophagoscopy and cytologic examination, the rate of detection of early
esophageal carcinoma (EEC) is increasing. But many patients would not
accept surgical treatment in the absence of compelling symptoms. The
task facing the medical profession is the identification of a non-
surgical, safe and effective therapy for this disease. Photodynamic
therapy meets this need because of its high selectivity for neoplastic
tissues, minimal damage to surrounding normal tissue, the 4-15 mm depth
of penetration feasible (1-3), and the usefulness of this procedure for
diagnosis preceeding therapy. Presented here are preliminary results of
13 cases of porphyrin photodynamic therapy of EEC between August 1982
and April 1985.

MATERIALS AND METHODS

Clinical Data

The 13 cases include 9 males and 4 females ranging in age from 42-65.
These patients had squamous cell cancer diagnosed through balloon
cytologic examination as EEC (clinical stage 0), with confirmation using
combined biopsy and brushing cytology via fibroesophagoscopy. Sites included upper third of the esophagus (1 case), middle third (7 cases) and
lower third (5 cases). Endoscopic typing (4) indicated 1 congestive, 4
erosive, 7 plaque-like and 1 polyp type cases.

Photosensitizer

HPD, batch 8112 K02, was provided by Oncology Research and Development, Cheektowaga, NY, USA. Each vial contained 30 ml of a 5 mg/ml
solution which was stored at 0-4°.

Light source

Two sources were employed: [1] Argon ion lasers, Model 360 (5.6 W)
and Model 362 (13 W) 457 and 514.8 nm wavelength, obtained from NETF,

Table 1. Results of Photodynamic Therapy of Early Esophageal Cancer

Sex	Age	Location	Endoscopic Typing	Follow-up (months)	Relapse
F	49	Middle	Erosive	3 (days)	+
M	53	Lower	Plaque-like	32	-
M	57	Middle	Erosive	29	-
M	42	Lower	Congestive	29	-
F	61	Middle	Plaque-like	29	-
M	52	Middle	Plaque-like	26	-
F	58	Lower	Erosive	23	-
M	43	Middle	Plaque-like	23	-
M	46	Middle	Plaque-like	21	-
M	51	Upper	Plaque-like	21	-
F	65	Lower	Erosive	21	-
M	47	Lower	Polyploid	21	-
M	55	Middle	Plaque-like	21	-

In all cases, the histology was that of squamous-cell cancer, clinical stage 0. Follow-up duration was in months unless otherwise indicated. Location indicates the upper, middle or lower third of the esophagus.

China. [2] Model BCI dye laser using Rhodamine B (emission 625-630 nm) obtained from BCI, China. The light-coupling assembly consisted of [1] a 10 X objective lens which focuses red light onto a quartz optical fibre (Olympus Optical Co., Ltd., Japan). [2] A 5-direction precise micro-adjuster, model SO-2A (Shanghai, China No. 4 MRP). [3] A quartz cylindrical optical fiber, 20 m long, 400 μ diameter with 95% light conductivity (Shanghai, China XHGM). [4] Fibroesophagoscope Model EF-B$_2$ (Olympus Optical Co., Japan).

HPD Administration and Irradiation

Only patients with a prior negative HPD scratch test were treated. HPD was administered at a level of 2.5-3 mg/kg via an IV 'push' within 8-10 min. Patients remained in a dark room after treatment. The light output at the fiber tip was adjusted to 100-400 mw/cm^2, and the tip was inserted through the biopsy channel of the endoscope. Irradiation was carried out with the fiber tip 0.5-2 cm from the lesion for 15-35 in. The lesions were examined 3 days and 1 month after irradiation using balloon cytological examination and biopsy and brushing cytology under direct-vision endoscopy. A follow-up examination was given every three months.

Table 2. Follow-Up Duration in 12 cases of Early Esophageal Cancer

Type	Number of cases	Months				
		32	29	26	23	21
Congestive	1		1			
Erosive	3		1		1	1
Plaque-like	7	1	1	1	1	3
Polyploid	1					1
Total	12	1	3	1	2	5

RESULTS

In one of the 13 cases, only a lesion at 29 cm was detected and treated. Three days post-irradiation, squamous tumor cells were found in the lower third of the esophagus under section-by-section balloon cytology, but no lesion was detected by fibroesophagoscopy. Surgical resection was therefore performed 21 days after irradiation. Macro-pathologic and histopathologic examination of the resected specimen showed that the lesion irradiated at 29 cm was now non-malignant, but a focus of in-situ multiple-origin cancer was detected at 36 cm.

In the remaining 12 cases, no relapse was found through balloon cytology, brushing cytology and biopsy under fibroesophagoscopy, with follow-up for 21-32 months (Tables 1,2; Fig. 1-9).

The thickness of the esophageal mucosa is 0.5-0.8 mm (15), and a thickness of 1.5 mm from the tumor surface to the tonica is rarely seen (except in the polyp type). Since the penetrating light depth is 4-15 mm, a cure of early cancer of the clinical 0-stage is feasible. of the 13 cases reported here, surgical resection was needed in only one. The other 12 showed complete remissions for 21-32 months. Photodynamic therapy of early esophageal cancer is therefore a promising modality.

DISCUSSION

Photodynamic therapy has been a rapidly-advancing field. In clinical centers in the US, Japan, Italy, Australia, Canada, China and elsewhere, treatment of 2500-3000 cases have been reported. Success has been reported in the fields of skin, lung, esophageal, gastric, gynecologic, brain and bladder cancer (5-13). As stated by Dougherty (14), one of the major obstacles to the study of this modality is the lack of suitable patients. It is difficult to find cases of early lung and bladder cancer. In Henan Province, there is a well-organized esophageal cancer prevention network, and a mass-screening program, since this a high-incidence region. Attention has been paid to early detection and timely treatment. So there is a good source of early esophageal cancer patients.

Macropathologically speaking, EEC is a polymorphous disease (4). proper localization, estimation of the size of the lesion, and correct diagnosis for occult and multiple-origin cancer will be the keys to curative therapy.

The output power at the tip of the optical fiber can be decreased by peristalsis and spasm of the esophagus, and contamination with blood and mucus during irradiation. Uneven irradiation can occur if the fiber is improperly aligned, and some damage to normal tissues can occur. The development of hemi-cylindrical, spherical, diffusive and water-cooled fibers will aid in the phototherapy of intaluminal tumors.

Of the 13 cases treated, only one patient observed erythema and edema from skin photosensitization. In another case, resection was carried out 21 days post irradiation. No phototoxic reactions were observed, nor was wound healing affected.

CONCLUSIONS

PDT provides an effective treatment modality for those who refuse surgery, and for elderly patients who are poor surgical risks. Safe and encouraging therapeutic effects are seen in early esophageal carcinoma. The key to increasing the cure rate is correct tumor localization and diagnosis, especially in occult and multiple origin cancer.

ACKNOWLEDGMENT

The authors wish to express their deep gratitude to Dr. TJ Dougherty of the Roswell Park Memorial Institute, and to K.R. Weishaupt of Photofrin Medical Inc., who supplied us with HPD and much helpful advice.

REFERENCES
1. JH Kinsey et al. Thermal considerations in murine tumor killing using hematoporphyrin derivative phototherapy. Cancer Res. 43, 562 (1983).
2. TJ Dougherty et al. Photoradiation therapy for the treatment of malignant tumors. Cancer Res. 38, 628 (1978).
3. H Kato et al., Hematoporphyrin derivative and laser photoradiation in the diagnosis and treatment of lung cancer. (日文) 37, 517 (1982).
4. Yang Guanrei et a. Endoscopic diagnosis of 115 cases of early esophageal carcinoma. Endoscopy 14, 489 (1980).
5. TJ Dougherty et al. Photodynamic therapy and cancer, in "Principles and Practice of Oncology" In Press (1985).
6. IJ Forbes et al. Phototherapy of human tumors using hematoporphyrin derivative. Med. J. Aust. 2, 489 (1980).
7. Tian Mao-en et al., Results of hematoporphyrin derivative-laser treatment for 30 cases of malignant skin tumors. Tumor Res. (中文) In press (1985).
8. Y Hayata et al. Evaluation of photoradiation therapy in 115 cases Abstracts, 13th International Cancer Congress, Seattle WA (1982).
9. Y Hayata et al. Hematoporphyrin derivative and laser photoradiation in the treatment of lung cancer. Chest 81, 269 (1982).
10. DA Cortese et al., Endoscopic management of lung cancer with hematoporphyrin derivative phototherapy. Mayo Clinic Proc. 57, 543 (1982).

11. OJ Balchum et al. Photoradiation therapy of endobronchial lung cancer employing the photodynamic action of hematoporphyrin derivative. Lasers in Surg. Med. In Press (1985).

12. A Tsuchiya et al. Hematoporphyrin derivative and laser photoradiation in the diagnosis and treatment of bladder cancer. J. Urol. 130, 79 (1983).

13. CJ Gomer et al. Hematoporphyrin derivative photoradiation therapy for the treatment of intraocular tumors: examination of acute normal ocular tissue toxicity. Cancer Res. 43, 721 (1983).

14. T Hager, Update: Photoradiation therapy for cancer. JAMA 248, 795 (1982).

15. G Vantrappen et al. Diseases of the esophagus. Springer-Verlag Berlin, Heidelberg & New York, p. 17 (1974).

OPTIMALIZATION OF THE DEPTH OF SELECTIVE RESPONSE

OF NEOPLASTIC TISSUE

Lars O. Svaasand

Division of Physical Electronics
University of Trondheim, Norway

INTRODUCTION

An optimal therapy of malignant tumors administers a lethal dose to the neoplastic cells simultaneously with a minimal damage to the surrounding normal cells. (1), (2)

The therapeutic dose in photodynamic therapy is dependent on the local optical fluence as well as the local concentration of the photosensitizer. The optical fluence rate in tissue decreases exponentially with distance from the irradiated surface. The optical penetration depth, i.e. the distance corresponding to a reduction by a factor e = 2.7, is typically in the range 0.5-1.5 mm for light at 514 nm wavelength. The corresponding value for light at 630 nm wavelength is 1-3 mm. The typical ratio between the retention of the tumor localizing photosensitizer hematoporphyrin derivative (Hpd) in neoplastic tissue and in normal tissue is in the range 3-10 (11). The maximum distance from the irradiated surface where a lethal dose may be administered to the neoplastic cells without simultaneously introducing permanent damage to normal cells in the surface layer, is therefore expected to be proportional to the optical penetration depth and to the logarithm of the retention ratio. This maximum distance is thus in the range 0.5-3.5 mm for 514 nm irradiation and in the range 1-7 mm for 630 nm irradiation.

The absorption of light in tissue will also initiate a local temperature rise. This temperature rise may, under certain conditions, result in a hyperthermic contribution to cell necrosis. (3), (4). There may also be synergistic effect between the thermotoxic and the phototoxic mechanism (6)(7). The thermotolerance of malignant tumors is usually less than that of normal tissue. An exposure of tumor cells to 43-44^{0}C for a period of 30 min. usually introduces a lethal damage (8). But normal cells typically require a two degree higher temperature for obtaining the same damage.

The finite depth of the selective necrosis of malignant tissue may represent a limitation for the therapy of certain subcutaneous or submucosal tumors. In many cases it will be necessary to minimize the damage to the normal skin or mucosa and simultaneously have a selective necrosis of malignant tissue to a depth of several millimeters. This condition may be difficult to satisfy exclusively with photodynamic therapy; the maximum depth for selective necrosis for the case with 3 mm optical penetration depth and a retention ratio of 10 between the concentration of Hpd in malignant- to normal cells is, as discussed above, about 7 mm. The corresponding depth is for the most usual condition with an optical penetration depth 2 mm and a retention ratio of 3, is only slightly more than 2 mm.

This depth of selective necrosis might be significantly enhanced by combining photodynamic and hyperthermic therapy. An interesting modality is to utilize forced cooling of the surface during irradiation. The temperature of the skin or the mucosa may then be maintained in the sublethal range simultaneously with a lethal temperature rise in the distal regions. Any infiltrating malignant cells in the region proximal to the surface may be destroyed by the photodynamic mechanism. The dominant mechanism for destruction of malignant cells will hence vary with distance from the surface. The photodynamic mechanism will be dominant in the layer proximal to the surface where the temperature is low and the optical fluence rate is high, and the hyperthermic mechanism will dominate in the distal layer with high temperature and low optical fluence rate. Both mechanisms will be of importance in an intermediate region, and the response might here be additive or synergistic.

This modality is characterized by a low administered Hpd-dose together with a high incident optical irradiation and forced cooling of the surface.

THEORY

The cumulated probability for tumor response may be expressed in terms of a Gaussian probability distribution (4)

$$
S = \frac{\mathrm{erf}(\frac{1}{\sqrt{2}\sigma}) + \mathrm{erf}(\frac{1}{\sqrt{2}\sigma}(D-1))}{\mathrm{erf}(\frac{1}{\sqrt{2}\sigma}) + 1} \tag{1}
$$

where S is the cumulated probability, σ is a dimension-less parameter characterizing the width of the distribution and D is a normalized dose.

The normalized photodynamic dose D_p may be expressed

$$
D_p = \frac{\varphi t}{\alpha} \tag{2}
$$

where φ is the local optical fluence rate, t is the exposure time and α is a photodynamic parameter dependent on the local concentration of the photosensitizer. The parameter α will, however, not be constant with respect to time if for example the concentration of the active drug is depleted during irradiation, or if the blockage of microvessels during the treatment affects the oxygen concentration (9), (10). These cases might be accounted for by expressing the normalized dose as a power series in the fluence. The expression in Eq. (2) with a constant α may be considered as the first order term in this power series expansion.

28

The fluence rate may be expressed (5)

$$\varphi = 2(1 + \gamma)Ie^{-\frac{x}{\delta}} \qquad (3)$$

where γ is the optical reflection coefficient, x is the distance from the irradiated surface, δ is the optical penetration depth and I is the incident irradiation.

The normalized hyperthermic dose D_H may be written

$$D_H = \xi_1 e^{\xi_2 T_L} t \qquad (4)$$

where T_L is the local temperature. The parameters ξ_1 and ξ_2 characterize the absolute thermosensitivity of the tissue and the temperature dependence of this sensitivity, respectively.

The steady state value of the local temperature may be expressed (4)

$$T_L = T_0 + \frac{(1-\gamma)I\delta}{\kappa\left(1-\left(\frac{\delta}{\delta_V}\right)^2\right)} \left[\frac{A + \frac{\kappa}{\delta_V}}{A + \frac{\kappa}{\delta_V}} e^{-\frac{x}{\delta}} - e^{-\frac{x}{\delta}} \right] \qquad (5)$$

where T_0 is the normal tissue temperature, κ is the thermal conductivity of the tissue and A is the heat transfer number at the surface. The magnitude of this heat transfer number can be increased by enhanced forced air or liquid flow at the surface

The thermal penetration depth δ_V is defined (3)

$$\delta_V = \sqrt{\frac{x}{Q}} \qquad (6)$$

where Q is the blood perfusion rate, i.e. the volume of perfused blood per unit volume of the tissue per unit time, and χ is the thermal diffusivity.

The total normalized dose for an interactive process between hyperthermic and photodynamic therapy may be written (4)

$$D = D_P + D_H + \beta D_P D_H \qquad (7)$$

where the coefficient β characterizes the interaction. Positive values for β correspond to synergistic interaction, $\beta = 0$ corresponds to the additive case, and negative values for β represent antagonistic interaction.

DISCUSSION

The possible enhancement of the depth of the selective necrosis of malignant tissue may be illustrated by some examples. The optimalization criterion has been to obtain a high probability for lethal damage to malignant cells, i.e. $S > 0.75$ simultaneously with a low probability, i.e. $S < 0.25$ for permanent damage to normal cells.

A typical example for the response probability is shown in Fig. 1a. The upper solid curve shows the response probability for malignant tissue versus distance from the irradiate surface. The lower solid curve shows the corresponding response for normal tissue. The absolute thermosensitivity for malignant tissue is taken to be $\xi_1 = 3.6.10^{-11}$ s^{-1} which corresponds to a response probability $S = 0.5$ for 30 min. exposure to 43.8^0 C. The value for normal tissue is put equal to $\xi_1 = 0.9.10^{-11}$ s^{-1} corresponding to $S = 0.5$ for 30 min. exposure to 45.8^0 C. The parameter ξ_2 which characterizes the temperature dependence of the hyperthermic lesion is taken $\xi_2 = 0.7(^0C)^{-1}$ which corresponds to a reduction in the required exposure time by a factor of two per degree temperature rise. The thermal conductivity $\kappa = 0.5$ W/mK and the thermal diffusivity $\chi = 1.2.10^{-7}$ m^2/s. The thermal penetration depth is put equal to $\delta_v = 8$ mm which corresponds to an intermediate value for the blood perfusion rate (3). The optical penetration depth for light at 514 nm wavelength is taken $\delta = 0.5$ mm which corresponds to the typical value found in several kinds of tumors, and the reflection coefficient is put $\gamma = 0.2$ (5). The parameter α which characterizes the photodynamic mechanism is taken $\alpha = 1600$ J/cm^2 for malignant cells, and the value for normal cells is taken $\alpha = 4800$ J/cm^2. These values correspond to a factor of 3 higher retention of photosensitizer in the malignant cell than in the normal cells. Both values for α correspond, however, to a very low administered drug dose; a dose 25 mg/kg DHE injected intraperiotonically 24 hours prior to exposure in a murine tumor model C3H/Tif in C3D2F1/Bom mice corresponds to $\alpha = 10$ J/cm^2 at 514 nm wavelength. The total exposure time to light is 30 min. The hyperthermic exposure time is reduced by an amount corresponding to the transient time of the temperature rise, i.e. about 5 min. The width of the probability distribution is taken $\sigma = 0.4$, and the heat transfer is $A = 50$ W/m^{2}K. This value for the heat transfer number corresponds to normal cooling conditions with very little cooling at the surface. The interactive parameter β is taken $\beta = 0$ corresponding to the additive case and the normal tissue temperature is $T_0 = 34^0$ C (4).

The results show that the response probability for malignant cells is higher than $S = 0.75$ to a depth of about 1.5 mm. The response probability for normal cells is less than $S = 0.25$ in the same region. The response is predominantly of hyperthermic nature. This is indicated by the upper broken curve which shows the probability for pure hyperthermic response of the tumor. The probability for pure photodynamic response of the tumor is correspondingly indicated by the lower broken curve.

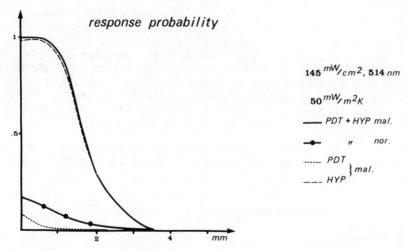

Fig. 1a. Response probability versus distance from irradiated surface.
Incident power density 145 mW/cm² at 514 nm wavelength.
Heat transfer number A = 50 W/m²K.

The depth of the region with selective necrosis may, as discussed
previously, be enhanced by utilizing forced cooling of the surface. This
modality is illustrated in Fig. 1b which shows the response probability in
the case of heavy forced air cooling of the surface, i.e. the heat
transfer number is A = 500 W/m²K. The other parameters are the same as
discussed in the example in Fig. 1a. The temperature rise is now
significantly reduced, and this reduction is most significant in the
surface layer. An increase of the optical irradiation from 145 mW/cm² to
535 mW/cm² initiates a hyperthermic response in the distal layers. This
increase in the optical fluence equally enhances the photodynamic response
in the surface region.

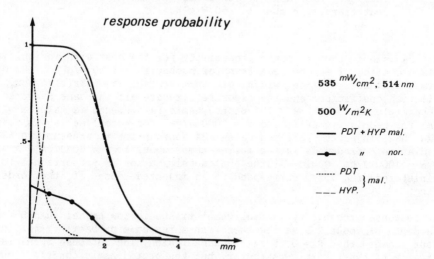

Fig. 1b. Response probability versus distance from irradiated surface.
Incident power density 535 mW/cm² at 514 nm wavelength.
Heat transfer number A = 500 W/m²K.

The upper solid line in this figure shows the response probability for malignant tissue, and the lower solid line shows the corresponding value for normal tissue. The probability for response of the tumor is higher than S = 0.75 to a depth of about 2 mm, and the response of the normal tissue is lower than S = 0.25 in the same region. The photodynamic contribution is significant to a depth of about 0.5 mm whereas the hyperthermic contribution is dominant in the distal layers. The example indicates some enhancement in the depth of selective necrosis by utilizing forced cooling, but the increase in depth is small and only about 0.5 mm.

This enhancement is, however, larger for irradiation with red light at 630 nm wavelength where the optical penetration depth is in tumors typically δ = 2 mm (5). This is illustrated in Fig. 2.

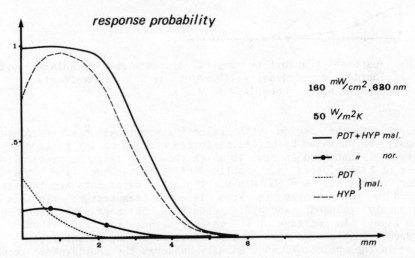

Fig. 2a. Response probability versus distance from irradiated surface. Incident power density 160 mW/cm² at 630 nm wavelength. Heat transfer number A = 50 W/m² K.

Fig. 2a shows the response probability for the case with insignificant surface cooling, i.e. the heat transfer number is A = 50 W/m² K. The other thermal parameters, the width of the probability distribution, the reflection coefficient and the exposure time are all the same as used in the example in Fig. 1. The photodynamic parameter α is taken α = 1000 J/cm² for neoplastic cells and α = 3000 J/cm² for normal cells. A dose 25 mg/kg DHE injected periotonically 24 hours prior to exposure to 630 nm irradiation corresponds in the murine tumor model to α = 30 J/cm². A value of α = 1000 J/cm² would, if the photosensitization is proportional to the administered drug dose, correspond to an injected dose of the order 1 mg/kg.

The response probability for malignant tissue is now higher than S = 0.75 to a depth of about 2.5 mm together with a response probability for normal tissue lower than S = 0.25 in the same region. The response mechanism is basically of hyperthermic character, but the photodynamic contribution is of importance in the surface layer.

The corresponding optimalized tumor response for heavy forced air cooling, i.e. the heat transfer number A = 500 W/mK is shown in Fig. 2b. An increase of the optical irradiation from 160 mW/cm² to 425 mW/cm² results in a selective tumor response to about 5 mm from the surface. The response is expected to be predominantly of photodynamic character to a depth of about 1.5 mm, and of hyperthermic character in the region deeper than about 2 mm.

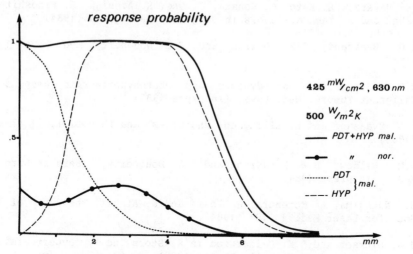

Fig. 2b. Response probability versus distance from irradiated surface. Incident power density 425 mW/cm^2 at 630 nm wavelength. Heat transfer number A = 500 W/m^{2}K.

CONCLUSIONS

These examples demonstrate that an optimalization of the hyperthermic and photodynamic mechanisms may enable an enhancement of the depth of selective necrosis. The possible increase in depth is small for 514 nm radiation, and typically less than 1 mm. The enhancement for 630 nm irradiation is, however, expected to be more significant. A conservative estimate indicates, as discussed in Fig. 2, that the depth might be increased from about 2.5 mm to about 5 mm.

A further increase in the depth may be obtained by simultaneous irradiation with two different wavelengths ex.g. 1060 nm and 630 nm. The optical penetration depth for 1060 nm irradiation is in the range 3-7 mm, i.e. typically 2-3 times larger than the corresponding value at 630 nm. The excitation of hematoporphyrin derivative from the 1060 nm irradiation is expected to be insignificant, and the photodynamic cell response will be initiated by the simultaneous 630 nm irradiation.

ACKNOWLEDGEMENT

The work has been supported by a NATO Research Grant no. 0342 under NATO Science Programmes and by a grant from Forretningsbanken, Trondheim. The author wishes to thank T. Dougherty, D. Doiron and A.E. Profio for valuable discussions, and B. Reitan for competent typing.

REFERENCES

(1) O.J. Balchum, D.R. Doiron, and G.C. Huth, Lasers in Surg. and Med. 4, 13 (1984).

(2) Y. Hayata, H. Kato, C. Konaka, J. Ono, R. Amenija, K. Kinoshita, H. Sahai and R. Yamada, Lasers in Surg. and Med. 4, 13 (1984).

(3) L.O. Svaasand, D.R. Doiron and T.J. Dougherty, Med. Phys. 10, 10 (1983).

(4) L.O. Svaasand, Photodynamic and Photohyperthermic Response of Malignant Tumors, Med. Phys. (In press 1985).

(5) L.O. Svaasand and R. Ellingsen, Photochem. and Photobiol. 41 (73-76), (1985).

(6) S.M. Waldow, B.W. Henderson and T.J. Dougherty, Lasers in Surg. and Med. (In press 1984).

(7) M. Nakajima, K. Motomura, A. Ihara and K. Atsumi, The Jour. of Japan Soc. for Laser Med. 4, 107 (1984).

(8) J.A. Dickson and S.K. Calderwood in K. Storm (ed.), "Hyperthermia" in Cancer Therapy" (G.K. Hall, Boston 1983), p. 63.

(9) W.M. Star, J.P.A. Marijnissen, A. van den Berg-Block and H.S. Reinhold in D.R. Doiron and C.J. Gomer (eds.), "Porphyrin Localization and Treatment of Tumors", (Alan R. Liss, New York, 1984), p. 637.

(10) M. Kreimer-Birnbaum, J.E. Klanning, R. Keck, P.J. Goldblatt, S.L. Britton and S.H. Selman in A. Andreoni and R. Cubeddu (ed.), "Porphyrins in Tumor Phototherapy" (Plenum, New York 1984), p. 235.

(11) T. Dougherty, T.J. Weishaupt and D.G. Boyle in S. Davis (ed.), Critical Reviews in Oncology/Hematology. Boca Raton, Fl.: CRC Press, (in press 1984).

DOSIMETRY METHODS IN PHOTORADIATION THERAPY

A. E. Profio,
L. R. Wudl and
J. Sarnaik

University of California, Santa Barbara

I.   INTRODUCTION

It is important to be able to specify the effective dose in photoradiation therapy (PRT) or photodynamic therapy (PDT) in order to compare results and to assure that the dose is sufficient to achieve the desired therapeutic effect while not damaging nonmalignant tissues excessively. The effective absorbed dose depends on the energy flux density or "space irradiance" at the dose point, the concentration of photosensitizing drug, and the concentration of molecular oxygen. Up to now, the clinician has not been able to specify the true, effective absorbed dose, although such terms as "light dose" (which is really an energy flux density, not a dose) have been used. The energy flux density at the dose point within the tumor is seldom known; instead the source power or irradiance is given. The dosage (mg drug per kg body weight) and the time delay after intravenous injection are specified, but the concentrations of the drug in tumor and normal tissue are not measured or even reproducible with the same dosage and time delay. The concentration of molecular oxygen is usually unknown, although a hypoxic region of a tumor would be protected from photodynamic cytotoxicity. Methods have to be developed to measure or calculate the contributions to the effective absorbed dose.

The dose formulation should apply to any photosensitizer, but the results and methods discussed here are intended specifically for hematoporphyrin derivative (HpD) and dihematoporphyrin ether (DHE), the photosensitizers now approved and used in treatment of tumors in human patients.

This paper has been updated from previous discussions of dosimetry in photodynamic or photoradiation therapy[1-5].

II.   DEFINITION OF DOSE

The effective absorbed dose is defined as

$$D^*(J/kg) = (a/\rho) \, C \, \phi \, K \, t \qquad (1)$$

35

where

    a = specific absorbance of photosensitizer at source wavelength, $m^{-1}(\mu g/g)^{-1}$

    ρ = tissue density, about 1030 kg m^{-3}

    C = concentration of photosensitizer in tissue, $\mu g/g$

    ϕ = energy flux density of light, W m^{-2}

    K = relative photodynamic effectiveness, ratio of absorbed dose required to achieve a specified biological endpoint under reference conditions, to the absorbed dose required to achieve the same endpoint under the prevailing conditions

    t = duration of irradiation, s

If the light is not monochromatic, the effective absorbed dose has to be integrated over the spectrum. (The relative photodynamic effectiveness, as well as the specific absorbance and energy flux density, are in general functions of wavelength.) If the energy flux density or concentration is not constant, the dose rate has to be integrated over the duration of the irradiation. Note that any thermal effects are excluded from the photochemical dose defined, but could be included by means of a time and temperature dependent factor if desired. Methods of determining the specific absorbance spectrum, concentration, energy flux density throughout the tumor and neighboring normal tissue, and the relative photodynamic effectiveness factor, are discussed in this paper.

III. SPECIFIC ABSORBANCE

Photocemistry is not possible unless light photons are absorbed. The specific absorbance of a substance (absorption coefficient per unit concentration) depends on wavelength and chemical form hence on bonding as influenced by solvent, pH, etc. For PRT, the specific absorbance is desired for physiological conditions. Apparently, for HpD or DHE, the photosensitizer is not active until bound in the cell, perhaps in mitochondrial membrane. Thus the absorbance should be measured in cells or a suspension of cells. This introduces a complication, because the scattering from cells interferes with the measurement of the absorption coefficient as a function of wavelength in a spectrophotometer. Using a like suspension of cells (but without photosensitizer) in a cuvette in the reference beam only partially alleviates the problem. Corrections have to be made, as in other turbid samples. It is also necessary to measure or control the concentration. Results obtained by our group for HpD are reproduced in Ref. 5. More measurements should be performed, especially for cells containing DHE. Our tentative value for the specific absorbance at the standard irradiation wavelength of 630 nm is $0.48m^{-1}$ per $\mu g/g$.

IV. CONCENTRATION

There is little information available on the concentration of HpD or DHE in human tumors and its dependence on dosage, time after injection, and physiological state or type of tumor. Some experiments have been performed in animal tumor models. The concentration is usually related to fluorescence, although a few experiments have been done with HpD labeled with carbon-14 or tritium (not applicable to humans because of the radiation dose). The information available indicates:

(1) The concentration is on the order of a few to several micrograms per gram tumor, and perhaps half to a tenth of that in most nonmalignant tissues (kidney, liver, and spleen being exceptions), at the usual dosages of 3.0 mg/kg (HpD) or 2.0 mg/kg (DHE), at 48-72 hours after injection.

(2) The concentration may peak at a few hours after injection and then remain essentially constant for days, or else decay over a few days, depending on the individual tumor, for reasons not clear at this time. Higher fluorescence but not necessarily higher photo-dynamically active concentration in cells may occur over a few hours after injection because of HpD or DHE remaining in the blood.

(3) Considerable variation occurs (on the order of a factor of two) even at the same dosage and time delay, between different tumors in one patient, and between different patients with the same kind of tumor.

(4) The concentration may decrease with irradiation, because of photodecomposition (photobleaching).

Obviously, it would be best to measure the concentration just before and possibly after or even during an irradiation, and adjust the duration t or the source power accordingly. Two methods are being developed to make such determinations: chemical extraction and fluorimetry on biopsy specimens, and in vivo measurements of the fluorescence. The biopsy specimen is lyophilized, ground, and the HpD or DHE extracted by a solution of perchloric acid and methanol as in the technique developed by El-Far and Pimstone[6]. A little detergent is added. The fluorescence is then measured on a spectrofluorometer, standardized against a solution of known concentration of HpD or DHE in the same solvent.

The in vivo fluorescence measurements are made with a fiberoptic probe and photomultiplier detector. The probe provides for illumination of a tumor with violet (400-410 nm) light to excert fluorescence. A lightguide transmits the red (600-720 nm) fluorescence of HpD/DHE to the detector. If contact or gaging distance is not feasible, then it is necessary to cancel the dependence of the signal on distance from probe to tumor. This can be accomplished by ratioing the red fluorescence signal to the green autofluorescence signal (as described in the paper by Profio and Balchum, these proceedings). Even better, a digital subtraction of autofluorescence background could be performed before the net red fluorescence is ratioed to the green fluorescence. It is difficult to make absolute determinations this way (the signal depends not only on concentration but thickness of the tumor, scattering and absorption in the tissue, and other factors), but the instrument can be used for relative but reproducible measurements of the concentration, e.g., as a function of dosage and time after injection.

V. ENERGY FLUX DENSITY

The energy flux density $\phi(W/m^2)$ at a point is also termed the space irradiance, or the energy fluence rate. It is equal to the power entering a differential sphere, divided by the cross sectional area of the sphere. It includes backscattered as well as forward scattered and unscattered light. It may also be defined as the integral of the angular energy flux density over a 4 pi solid angle. The angular energy flux density is the power crossing unit area perpendicular to the direction of the ray, per unit area and unit solid angle. Thus one has the option of measuring the angular flux density by a suitable collimated detector, in various directions, and integrating over solid angle[2]. The other approach is to use an isotropic detector imbedded in the tissue, e.g., a single fiberoptic fiber lightguide with a diffusing ball end. The detectors have to be calibrated againt a standard lamp.

Diffusion theory has been found to give a good prediction of the energy flux density as a function of position, if the scattering and absorption coefficients of the tissue can be measured, and if the tissue is reasonably homogeneous. More accurate solutions can be obtained using discrete ordinates or Monte Carlo transport methods and computer programs, originally developed for calculating the transport of neutrons and gamma rays. The real trick is to measure the absorption and scattering coefficients and angular distribution of scattering (or equivalently, the diffusion coefficient and diffusion length or its reciprocal, the attenuation coefficient). It does not seem fruitful to try to calculate scattering from electromagnetic theory as the relevant indexes and sizes and structures in tissue are not well defined.

The diffusion length L or attenuation coefficient $\alpha = 1/L$ can be determined by measuring the exponential distribution of the relative energy flux density in plane geometry, where

$$\phi(x) = \phi_o \exp(-\alpha x) \tag{2}$$

The measurement requires a thick slab of tissue (thickness and lateral dimensions several times the diffusion length, hence on the order of a 2 cm cube or more for red light). A collimated beam also effectively infinite in radius (in practice, at least 1 to 2 cm) is perpendicularly incident. The photodetector should be fitted with a filter or monochromator to reject tissue fluorescence. The relative energy flux density can be measured with an isotropic probe (or by integration over angle), imbedded at 3 or 4 x positions, along the centerline of the specimen and beam axis. A typical diffusion length for 630 nm light in tissue which is neither highly colored (absorbing) or scattering, is about 4 mm. It is considerably less in highly absorbing or scattering tissue, and at shorter wavelengths.

The diffusion coefficient, D (which includes the effect of anisotropic scattering, at least approximately) can be measured by adding a known amount of absorber, in steps, to the tissue, and measuring the diffusion length at each concentration of added absorber of known absorption coefficient. From the diffusion theory,

$$\alpha^2 = \mu_a/D \tag{3}$$

where μ_a is the absorption coefficient. Thus by plotting the square of the measured diffusion length against the added absorption coefficient, a linear graph is obtained with slope 1/D, and y intercept corresponding to the absorption in tissue only.

Another method of determining the diffusion coefficient, knowing the attenuation coefficient, is to measure the diffuse reflectance R in plane geometry (e.g. with a collimated beam and an integrating sphere at the surface of incidence). According to diffusion theory,

$$R = \frac{1 - 2D\alpha}{1 + 2D\alpha} \tag{4}$$

A typical value for D is 0.1 cm.

Then for plane geometry, the energy flux density can be calculated from the irradiance E (from the source) by

$$\phi(x) = \frac{4E}{1 + 2D\alpha} \exp(-\alpha x) \tag{5}$$

For infinite cylindrical geometry

$$\phi(r) = \phi_o K_o(\alpha r) \tag{6}$$

where

$$\phi = \frac{2S}{\pi r_o K_o(\alpha r_o) + 2D\alpha K_1(\alpha r_o)} \tag{7}$$

and

S = source power per unit length, W/m

K_o, K_1 = modified Bessel functions of zeroth and first order.

r_o = radius of line source.

For sphereical geometry with an isotropic source of radius a, emitting total power P, the energy flux density as a function of radius in an infinite medium is

$$\phi(r) = \phi_o(a/r) \exp -\alpha(r - a) \tag{8}$$

where

$$\phi_o = \frac{P}{4\pi \, Da(1 + \alpha a)} \tag{9}$$

Two and three dimensional geometries can be calculated with either diffusion theory or more exact theories, but the calculations require a computer.

It should be noted that the energy flux density may be up to four times the irradiance, because of backscattering and angular redistribution, and the definition of energy flux density compared to irradiance (power per unit area incident on a surface).

VI. RELATIVE PHOTODYNAMIC EFFECTIVENESS

The relative photodynamic effectiveness (RPE) factor, K, is intended to account for factors other than the absorbed energy per unit mass. It performs the same function as Relative Biological Effectiveness (RBE) in ionizing radiation dosimetry, and is defined relative to a standard or reference condition. RBE is defined with reference to x-rays, and is mostly dependent on the linear energy transfer (LET) and the state of oxygenation of the tissue. RPE can be defined with reference to a standard wavelength, standard photosensitizer, and standard state of oxygenation. The standard or reference conditions we propose are: 630 nm wavelength, DHE as standard photosensitizer, and normal oxygenation of tissue. Then under common conditions, K=1.

A dependence on wavelength may occur because the quantum yield for production of singlet oxygen (or other photoactivated chemical), depends on wavelength (this has not been established). Even if the quantum yield is independent of wavelength up to the wavelength corresponding to the minimum photon energy required for singlet oxygen production, the energy per photon varies inversely with wavelength. Hence K should include a wavelength ratio to convert to yield per unit energy absorbed.

A dependence of quantum or energy yield of singlet oxygen or other active chemical, on photosensitizer composition, is to be expected. For Photofrin (HpD), whose main active ingredient is DHE, a simple correction can be made for the difference in DHE concentration at the same overall concentration of the mixture of porphyrins.

Another factor, analogous to the Oxygen Enhancement Ratio (OER) of ionizing radiation, should be included in K to allow for less than normal oxygenation of the cells. The functional dependence will have to be determined by experiment.

The RPE and K will be determined from the results of experiments on tumor regression or "cure", as a function of wavelength, photosensitizer, and oxygenation. However, if an independent measurement of singlet oxygen concentration or photo-oxidation products can be made, then it may be possible to derive K from in vitro experiments. As most clinical treatments will be carried out with DHE at 630 nm and with normal oxygenation, one can use K=1.

VII. CONCLUSIONS

Methods have been developed to measure or calculate the photodynamic dose from specific absorbance vs. wavelength, concentration of photosensitizer, energy flux density as a function of position in irradiated tissue, and relative photodynamic effectiveness, in some situations. More research should be done on measuring the specific absorbance for DHE, the concentration in vivo as influenced by dosage and time after injection in human tumors and tissues, the diffusion coefficient and diffusion length in more tumors and tissues from which the energy flux density can be calculated, and the dependence of relative photodynamic effectiveness on oxygenation, wavelength, and photo-sensitizer.

ACKNOWLEDGMENT

This research was sponsored by the Public Health Service, National Cancer Institute, DHHS, under grant CA-31865.

REFERENCES

1. A. E. Profio and D. R. Doiron, "Dosimetry considerations in photo-therapy," Med. Phys. 8(2):190-196 (1981).
2. D. R. Doiron, L. O. Svaasand and A. E. Profio, "Light dosimetry in tissue: application to photoradiation therapy," in Porphyrin Photosensitization (D. Kessel and T. J. Dougherty, Eds.), pp. 63-76, Plenum, New York (1983).
3. A. E. Profio, M. J. Carvlin, J. Sarnaik and L. R. Wudl, "Fluorescence of hematoporphyrin derivative for detection and characterization of tumors," in Porphyrins in Tumor

   <u>Phototherapy</u> (A. Andreoni and R. Cubbedu, Eds.), pp. 321-337,
   Plenum, New York (1984).
4. A. E. Profio, "Dosimetry for Photoradiation Therapy," Proc. Inter-
    national Conf. on Lasers and Electro-Optics 1983, Vol. 37,
    <u>Medicine and Biology</u>, pp. 10-15, Laser Institute of America,
    Toledo, Ohio (1984).
5. A. E. Profio, "Physics of photodynamic dosimetry," Proc. Inter-
    national Conf. on Lasers and Electro-Optics 1984, Vol. 43,
    <u>Medicine and Biology</u>, pp. 45-51, Laser Institute of America,
    Toledo, Ohio (1985).
6. M. El-Far and N. Pimstone, "A Comparative Study of 28 Porphyrins
    and Their Abilities to Localize in Mammary Mouse Carcinoma:
    Uroporphyrin I Superior to Hematoporphyrin Derivative," in
    <u>Porphyrin Localization and Treatment of Tumors</u> (D. R. Doiron
    and C. J. Gomers, Eds.), pp. 661-672, Alan Liss, New York
    (1984).

FLUORESCENCE DIAGNOSIS OF CANCER

A. Edward Profio[a] and
Oscar J. Balchum[b]

[a]University of California, Santa Barbara
[b]University of Southern California, Los Angeles

I. INTRODUCTION

The fluorescence of hematoporphyrin-derivative (HpD) or dihematoporphyrin ether (DHE), together with the property of attaining a higher concentration in malignant tumors than in most normal tissues, form a basis for diagnosis of cancer. Malignant tumors exhibit greater fluorescence than the surrounding nonmalignant tissue when excited by light of appropriate wavelength, and the tumor can be detected by a suitable imaging or nonimaging system. This paper discusses the fluorescent agent and background fluorescence, excitation systems, imaging methods, and nonimaging methods for diagnosis of cancer.

The photodynamic property of HpD and DHE is a nuisance in diagnosis. A nontoxic drug which is more fluorescent than HpD or DHE and less photodynamically active would be preferable, but so far none has been developed and approved for human use. Detection is limited to tumors accessible to irradiation with the exciting light, and where the emitted fluorescence can be detected, but this in principle includes a large number of carcinomas of the respiratory system, gastrointestinal system, urogenital system, eye and skin. Most of the clinical experience has been with bronchogenic carcinoma[1,2], but there is no reason why similar methods and instrumentation could not be applied to other sites.

Both imaging and nonimaging techniques have been used for detection and localization of small as well as large tumors[3-7]. The imaging technique permits more precise localization, and contrast is independent of distance or angle. Nonimaging techniques suffer from field-averaging of the signal, but if compensated for variations in distance or angle, they can be made quantitative.

II. FLUORESCENT AGENTS

Hematoporphyrin-derivative, HpD, was marketed as Photofrin by Photofrin Medical, Inc. Because of a change in ownership, it is no longer certain that this agent will be available. Dihematoporphyrin ether, DHE, is still marketed as Photofrin II by the Photofrin Medical division of Johnson & Johnson. DHE is the photodynamically active main ingredient of HpD, which is a mixture of porphyrins. DHE is also fluorescent, but there are other fluorescent porphyrins in HpD. The concentration of DHE in saline in Photofrin II is about twice the concentration of DHE in Photofrin. It appears that fluorescence emission from tumors is less with Photofrin II than Photofrin, if the dosage has been adjusted for equal photodynamic effect. The fluorescence can be increased by increasing the dosage, and as an approximate but useful rule, the dosage of Photofrin II should be about two-thirds of the dosage of Photofrin, for at least equal fluorescence yield. The photodynamic effect will be somewhat increased but acceptable. A typical dosage is 2.0 mg Photofrin II per kg body weight, or 3.0 mg Photofrin per kg.

The actual dosage required for fluorescence diagnosis depends on the amount of HpD/DHE required to exceed background from autofluorescence. Side effects from the induced photosensitivity of the skin to sunlight (or the irradiated tissue to the exciting light) also have to be considered. Most tissues fluoresce when excited by short wavelength light, e.g. violet. If the fluorescence extends to the region of HpD/DHE emission (600-720 nm, red), it will be impossible to reject all of the autofluorescence by a red barrier filter. Image contrast will be degraded, and it may be impossible to distinguish a thin, low contrast tumor from the surrounding normal tissue. Likewise, the fluorescent power collected by a nonimaging probe at a suspected tumor may be too close to the power collected from a normal or control site, compared to noise and other fluctuations. (Contrast or tumor/control ratio depends on the thickness of the tumor because a tumor thin enough to transmit some of the exciting light will emit fluorescence corresponding to the lower concentration in the underlying normal tissue, as well as higher concentration in the tumor itself.)

Figure 1 plots the spectra measured in a lung cancer patient, using a fused quartz fiber to conduct light to a grating spectrograph and EG&G-PARC Optical Multichannel Analyzer (OMA) with intensified silicon CCD detector. The tumor was large (several mm) and it is easy to discern the HpD fluorescence emission above the autofluorescence, with a spectrum

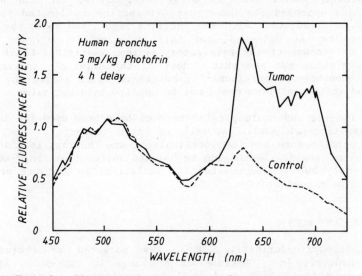

Fig. 1. Fluorescence spectra in lung cancer patient.

similar to the spectrum taken at a control site where the HpD concentration is much lower than in the tumor. The spectra have been normalized in the green region of the spectrum. As the tumor thickness decreases, the spectrum approaches that at the control site. For carcinoma in situ (CIS), the difference may be less than 20% in intensity near 630 nm.

Doubling the dosage (to 4.0 mg/kg, DHE) would improve contrast. This dosage has been used in a few lung cancer patients thus far, with no significant complications either from mucosa irradiated with the violet exciting light, or the skin from sunlight (although as with lower dosage, it is imperative to keep the skin covered with clothing or a sunscreen with a visible opaque ingredient such as titanium dioxide). A solar simulator has been built, based on a 200 W xenon arc lamp and "Air Mass 1" filter, for additional skin photosensitivity studies in patients who have been injected with HpD or DHE. Presently, a dosage of 2.0 mg DHE/kg is standard.

The concentrations of fluorescent agent in tumor and in surrounding tissue vary with time after injection. It was originally thought that both tumor and nonmalignant tissue took up HpD/DHE equally well, but that the agent was cleared from normal tissue more rapidly. Thus by waiting some time (days), the ratio of concentration in tumor (hence fluorescence yield) to concentration in tissue would be large, and yet there would still be a useful concentration in tumor. There is actually little data available for human beings. Measurements of fluorescence in fairly large (few mm) subcutaneous tumors in mice show a peak at about 4 hours after injection, followed by a valley at about 8 hours, then another increase to a broad maximum after about 12 hours[8]. The fluorescent agent may stay in the tumor for weeks. The fluorescence in muscle shows a smaller peak at 4 hours, then decays more or less monotonically to the background autofluorescence. The autofluorescence of normal tissue is important in fluorescence diagnosis but not in photodynamic therapy. Thus an optimum or usable delay for therapy may not correspond to the best delay for diagnosis.

Some results are now being obtained in human patients. Measurements of relative fluorescence intensity in subcutaneous metastatic breast tumors on the chest wall have been made by Dougherty's group at Roswell Park Memorial Institute. The tumors are always more fluorescent than skin. But there appear to be two classes of tumors: one in which the fluorescence reaches a maximum in about 4 hours and than stays fairly constant for at least 72 hours, and another in which the maximum is followed by a decrease. Similar results have been observed by the authors, using the ratio fluorometer, in lung cancer patients. For large tumors, an initial maximum at 3-4 hours may be followed by either a constant intensity out to at least 72 hours, or a decrease toward the autofluorescence background level (which is higher in large tumors than in normal bronchial mucosa). Repeat measurements suggest some HpD may remain in certain tumors for months. Only a few very small tumors have been measured thus far, but the data suggest that any initial maximum is small, and the fluorescence intensity either attains a maximum between 48 and 72 hours after injection, or remains fairly constant. We suspect that the peak at 3-4 hours may be due to HpD/DHE in the blood, as large tumors are well supplied with blood. The behavior after clearance from blood (half life 25 hours) must come from HpD/DHE bound in the cells, or possibly recycled. Additional measurements are in progress. Meanwhile, examination at 72 hours is probably acceptable.

III. EXCITATION SYSTEMS

Fluorescence may be excited in HpD/DHE throughout the visible spectrum. Maximum excitation is obtained using violet light (near 400 nm). However, as the maximum in the absorption of hemoglobin also occurs in the violet, the 1/e or 37% penetration depth in a typical tumor or tissue is only about 0.2 mm. It could be even less in highly absorbing (colored) tissue such as liver or kidney. They violet light is useful for excitation of fluorescence from only the surface layer. A great many carcinomas do reside on or in the surface. The small depth of penetration is an advantage for detection of tumors whose thickness is comparable to or less than the penetration depth (as for typical carcinoma in situ). Contrast is reduced if the exciting light is transmitted through the region of high HpD/DHE concentration, viz. the tumor.

The source of violet light may be a filtered mercury arc lamp (typically 200 watts), or a violet krypton ion laser set up to lase at 410 nm[9]. Even the laser emits red light, from the gas discharge, and has to be equipped with a monochromator or violet filter. In our experience, an irradiance of at least 15 milliwatts per square centimeter is needed at the tumor, hence the source power in the violet band should be on the order of 250 mW, allowing for losses in the filter or monochromator and the light delivery system. The laser is much to be preferred over the arc lamp for two reasons: red background can easily be made much lower (red:violet 1:100,000), and the coherent beam can be much more efficiently coupled to a small diameter fiberoptic lightguide for insertion in an endoscope. Contrast in tumor detection was greatly improved when the laser was substituted for the mercury arc lamp.

Laser light is transmitted through a 400 micrometre core, step index, fused quartz fiber, which has an overall diameter of 0.85 mm and fits in the biopsy channel of a standard fiberoptic bronchoscope. Similar arrangements can be made for other endoscopes. Fused quartz is preferred over glass to minimize excitation of fluorescence within the lightguide, and because violet transmission is better. A drawback is the small numerical aperture of fused quartz fibers, which means the illuminated spot is small, and could be mistaken for a region of localized fluorescence. At present the authors use a negative (diverging) microlens cast in a tube from polyester plastic, attached to the end of the fiber. This is not ideal, as the lens attenuates the violet power by 50%, some fluorescence is excited in the plastic, and the concave surface tends to clog with mucus and debris. Other light delivery systems with good power transmission, low fluorescence, and constant irradiance over a spot corresponding to the field of view of the imaging instrument, should be investigated.

IV. IMAGING SYSTEMS

The essentials of an imaging system are an instrument (such as an endoscope) to form the image, and a red barrier filter to reject reflected violet as well as most of the autofluorescence background. In applications of the bronchoscope, and most likely with other endoscopes, an image intensifier is also required, as the fluorescence is weak and light losses in endoscopes are not negligible. The gain required from the image intensifier depends on the endoscope and violet irradiance, hence on the application. Probably a violet irradiance of 30 mW per square centimeter, for a typical exposure time of 30 seconds or so, would do no harm. Even so, a gain of some 10,000 to 20,000 is desirable for cystoscopes and other rigid endoscopes, while 30,000 to 50,000 is desirable for a flexible fiberoptic bronchoscope.

Electrostatic image intensifiers are available commercially for night vision, and can be adapted to visualization of a fluorescent image. The earlier, "first generation" type uses an electrostatic lens between photocathode and output phosphor to accelerate and focus the photoelectrons. The later "second" and "third generation" types obtain gain by secondary electron emission in a microchannel plate between photocathode and phosphor. These devices are small and light, but the gain available in commercial instruments is only about 15,000. A three-stage first generation device intensifies brightness some 40,000 times. The output phosphor is green, because the eye is most sensitive to green. The third generation differs from the second generation in using a gallium arsenide photocathode which is more sensitive to red light, but these devices are still developmental and expensive. The fluorescence bronchoscope devised originally[3] uses a first generation image intensifier, while the alternate fluorescnece and whitelight viewing instrument (flipflop) uses a third generation intensifier modified for higher gain[5].

Considerable study has gone into specifying the optimum barrier filter characteristics. From the measured HpD/DHE fluorescence spectrum, and the autofluorescence spectrum, maximum contrast is achieved for a 40 nm bandwidth filter centered at 690 nm. Contrast was somewhat lower for a center wavelength of 630 nm, and much lower if the entire red spectrum (600-720 nm) was transmitted. Unfortunately, the 690 nm filter gives the dimmest image. Interference filters were used, which have a transmission at the center wavelength of only about 50%. A nonfluorescent bandpass or possibly a cuton filter with close to 100% transmission would reduce the gain required from the intensifier. If autofluorescence background is subtracted, the entire red emission spectrum of HpD and DHE could be used.

A digital background image subtraction system has been devised. Figure 2 is a block diagram of the system. The image from the endoscope is transmitted through either a red or green filter, and then focused on the photocathode of the image intensifier. The filters are mounted in a solenoid operated slide so they can be changed in less than 0.2 second. The output of the intensifier is coupled to a Silicon Intensified Target (SIT) video camera by means of a beamsplitter and fiberoptic imageguide (not shown). The video signal is fed to a distribution amplifer. One output of the distribution amplifier is connected through a coaxial switch in the controller to the digital frame store (Quantex DS 20), when the red filter is in place. The other output of the distribution amplifier is fed to a video amplifier with adjustable gain, and then to the digital frame store, when the green filter is in place. The Quantex DS 20 can store an image and then subtract it from the next incoming image after a command is given. In operation, the green-filtered image is stored as background, and subtracted from the red-filtered image, in a very short time. (One would like to subtract the red background from the red HpD/DHE emission plus background, but this cannot be done at the same location. A fraction or multiple of the green image is used as a measure of the red background, assuming the spectrum shape is reasonably constant.) The procedure is to set the difference image to black at a control or nontumor site, by adjusting the video gain on the green channel, leave this gain fixed, and then search for a tumor by noting a difference image above the black level.

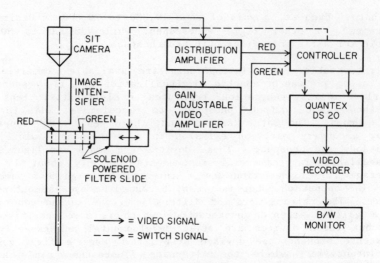

Fig. 2.   Diagram of digital image subtraction system.

Figure 3 illustrates early results in human lung cancer patients with the digital image subtraction system.   The photographs were taken from the video monitor.   The top row shows the control site red image on

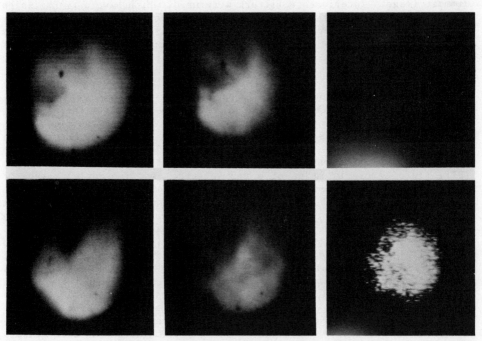

Fig. 3.   Photographs of control site (upper row) and tumor site (bottom) row) fluorescence taken (left to right) with red filter, green filter, and the difference (red-green).

the left, the green image in the middle, and the (red-green) image on the right, which has been adjusted to be black. The bottom row shows, left to right, the tumor site red, green, and (red-green) images. The difference image now is brighter than background. This system is in clinical use, and the real test will be detection of carcinoma <u>in situ</u> and other small tumors difficult to distinguish from background.

V.   NONIMAGING SYSTEMS

Nonimaging systems are probes in which the fluorescent light is collected, detected, and the relative power displayed in some way, but which do not provide an image of the fluorescence distribution. One system[7] provides only a reflectance, whitelight illuminated image, and detects the fluorescence power in a photomultiplier tube. The endoscopist is guided by an audible signal whose pitch is related to the signal intensity. The signal intensity depends on the HpD/DHE fluorescence and autofluorescence, but is also extremely sensitive to the distance between the distal end of the fiberoptic lightguide, and the suspected tumor site.

Figure 4 is a diagram of a ratioing fluorometer which was developed to cancel the dependence on distance, angular orientation, and source power[6]. Best results are obtained when the red fluorescence is ratioed to the green autofluorescence, rather than reflected violet. The light is collected by fiberoptic lightguide, and the red separated from the remainder of the spectrum by a dichroic mirror (beamsplitter). After narrowband filtering, the red is detected by a photomultiplier tube. The green is isolated by a filter and detected in another photomultiplier tube. After amplification and autoranging (controlled by the green signal), the red signal is divided by the green signal. The ratio is displayed on a meter and may also be applied to a voltage controlled oscillator for an audible indication. The instrument is calibrated against a red and green fluorescing standard, and if necessary the gain adjusted by means of the high voltage on the photomultiplier tubes. A typical control site will have ratio of 0.3 units, a small tumor after injection of HpD will read 0.5 units, and a large tumor over 1.0 unit. The units are arbitrary but reproducible, and the ratio of the red:green ratios (tumor/control) is significant. Thus a small tumor may be 1.5 times the control, and a large tumor 3 or more times the control. Additional measurements are in progress.

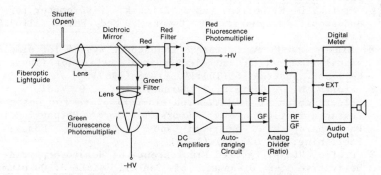

Fig. 4.   Diagram of ratio fluorometer.

VI. CONCLUSION

HpD or DHE (Photofrin II) may be used to detect small tumors by the characteristic red fluorescence, excited by violet light from a krypton ion laser. The tumor fluorescence may be visualized with the aid of a red barrier filter and image intensifier. Contrast can be improved with digital background image subtraction. A nonimaging probe may also be applied, and quantitative measurements of fluorescence intensity and contrast are possible if dependence on distance and angle are cancelled by ratioing.

ACKNOWLEDGMENTS

The digital image subtraction system was assembled and tested by Felix Carstens as part of the requirements for an M.S. degree at the University of California, Santa Barbara. We are grateful to Mr. Ed Drollinger for loan of the EG&G Princeton Applied Research Corporation optical multichannel analyzer used in the fluorescence spectrum measurements. The research was sponsored by the National Cancer Institute, DHHS grant CA-25582, and by the Clinical Research Group at the University of Southern California School of Medicine.

REFERENCES

1. O. J. Balchum, D. R. Doiron, A. E. Profio, and G. C. Huth, "Fluorescence Bronchoscopy for Localizing Early Bronchial Cancer and Carcinoma in Situ," in Recent Results in Cancer Research, Vol. 82, 97-120, Springer-Verlag, Berlin (1982).
2. E. G. King, G. Man, J. LeRiche, R. Amy, A. E. Profio, and D. R. Doiron, "Fluorescence Bronchoscopy in the Localization of Bronchogenic Carcinoma," Cancer 49:777-782 (1982).
3. D. R. Doiron, A. E. Profio, R. G. Vincent, and T. J. Dougherty, "Fluorescence Bronchoscopy for Detection of Lung Cancer," Chest 76:27-32 (1979).
4. A. E. Profio, D. R. Doiron, and E. G. King, "Laser fluorescence bronchoscope for localization of occult lung tumors," Med. Phys. 6(6):523-525 (1979).
5. A. E. Profio, D. R. Doiron, O. J. Balchum, and G. C. Huth, "Fluorescence bronchoscopy for localization of carcinoma in situ," Med. Phys. 10(1):35-39 (1983).
6. A. E. Profio, D. R. Doiron, and J. Sarnaik, "Fluorometer for endoscopic diagnosis of tumors," Med. Phys. 11(4):516-520 (1984).
7. J. H. Kinsey and D. A. Cortese, "Endoscopic system for simultaneous visual examination and electronic detection of fluorescence," Rev. Sci. Instrum. 51(10):1403-1406 (1980).
8. A. E. Profio, M. J. Carvlin, J. Sarnaik, and L. R. Wudl, "Fluorescence of Hematroporphyrin-Derivative for Detection and Characterization of Tumors," in Porphyrins in Tumor Phototherapy (A. Andreoni and R. Cubbedu, Eds.), Plenum Publishing, pp. 321-337 (1984).
9. A. E. Profio, "Laser Excited Fluorescence of Hematoporphyrin Derivative for Diagnosis of Cancer," IEEE J. Quantum Elec., QE-20(12):1502-1507 (1984).

STUDIES OF HPD AND RADIOLABELLED HPD IN-VIVO AND IN-VITRO

W. Patrick Jeeves*+, Brian C. Wilson*+,
Gunter Firnau+, and Kay Brown

* Ontario Cancer Treatment & Research Foundation
+ McMaster University, Hamilton, Ontario, Canada

INTRODUCTION

Successful tumor sterilization by Photodynamic Therapy (PDT) depends on delivery of a light dose appropriate for the concentration of photoactive HPD in the tumor at the time of irradiation. Since the concentration of HPD in target cells or tissues is an important determinant of response to PDT, we have radioactively labelled HPD with the positron emitter Copper-64 to allow non-invasive in-vivo quantitation of HPD concentration by external nuclear detectors. Imaging of HPD distribution is also made possible by Positron Emission Tomography (PET). Labelling with the pure gamma-emitting Copper-67 would permit standard nuclear scintigraphy or Single Photon Emmission Computed Tomography (SPECT) in patients, the longer half-life of ^{67}Cu (58 hours compared to 13 hours for ^{64}Cu) being advantageous in this application.

Incorporation of radiocopper into the tetrapyrrol ring of HPD has been achieved by disaggregation of Photofrin I^{TM} (Photofrin Medical Inc.) followed by subsequent re-aggregation, through manipulation of pH. Using our protocol, ^{64}Cu-HPD of radiopharmaceutical quality has been produced with specific activity typically of 250 μCi/mg. Furthermore, we have recently manufactured ^{14}C-HPD for use as a more reliable control than ^{3}H-HPD for in-vivo studies of the biodistribution characteristics of Cu-HPD.

MATERIALS AND METHODS

Copper-64 Labelling of HPD

Natural (^{63}Cu) copper wire was dissolved in nitric acid in quartz vials. The dried copper nitrate ($^{63}Cu(NO_3)_2$) was then converted to the desired radioactive species ($^{64}Cu(NO_3)_2$) by neutron bombardment in the McMaster University Nuclear Reactor. After irradiation, the material was solubilized in dilute 0.1N HCl, transferred to a Teflon dish, and evaporated to a solid $^{64}CuCl_2$ precipitate.

Chlorine was removed by repeated addition and evaporation of distilled water. 1.0mg ^{64}Cu was dissolved in 0.2ml distilled H_2O (i.e. to 15.6 µmol) and subsequently added to 28mg HPD as isotonic Photofrin I solution (i.e. 46.4 µmol), in a quartz vial containing a stirring bar. The pH of the mixture was immediately raised to 11.5-12.0 with 1.0N NaOH, stirred for 2 minutes, then adjusted to pH7.0 with 1.0N HCl. The pH was never allowed to fall below 6.5. The solution was forced through a 0.22 micron filter into sterile vials. At all stages of this procedure, the HPD was protected from light. The final concentration of the copper-labelled HPD preparation (^{64}Cu:HPD 1:3) was assessed by comparing its absorbance at 535nm to that of a standard solution of unlabelled Photofrin I in 10mM cetyltrimethylammonium bromide (CTAB). The specific activity of a standard preparation of ^{64}Cu-HPD was typically 250µCi/mg, as determined using a sodium iodide well-counter.

<u>Carbon-14 Labelling of HPD</u>

The method used to manufacture ^{14}C-HPD is a modified version of the procedure of Gomer and Dougherty (1979), starting with whole blood taken from rabbits previously injected with ^{14}C-glycine.

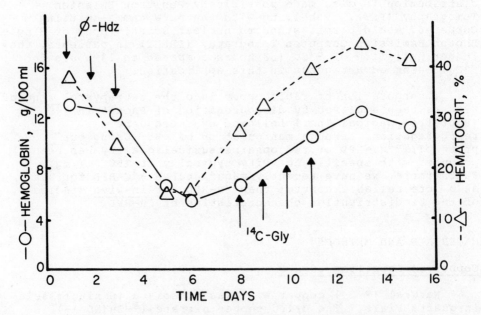

Fig. 1. Time course of rabbit hemoglobin and hematocrit after subcutaneous administration of phenylhydrazine hydrochloride. Phenylhydrazine (ϕ-Hdz) and subsequently carbon-14 labelled glycine (^{14}C-Gly) were given at the times indicated.

A 3.5kg New Zealand White rabbit was rendered anemic in order to stimulate hemoglobin synthesis, by 3 daily subcutaneous injections (0.35ml/kg body weight) of a 2.5% saline solution of neutral phenylhydrazine hydrochloride (Altman et al, 1948). The levels of hematocrit and hemoglobin were closely monitored (Borsook et al, 1954), and when the hemoglobin level began to rise (see Figure 1 for schedule), the rabbit was injected intravenously with neutral ^{14}C-glycine (56 mCi/mmol, Amersham Corp.) on each of 4 successive days (total amount, 3.25 mmol, total activity 4mCi). When the level of hemoglobin had reached a plateau, the blood was harvested by carotid vein cannulation after the animal was anaesthetized with sodium pentobarbital (30 mg/kg body weight).

A 1:3:7.5 (v/v) mixture of blood :2.0% Sr $Cl_2.6H_2O$ in glacial acetic acid : acetone (Labbe and Nishida, 1957; Falk,1964) was stirred for 30 minutes in a 2 litre round bottomed flask, then brought to a rolling boil for 5 min. The resulting suspension was cooled, then filtered twice by aspiration on Whatman No.2 paper. The residue was washed twice with a total of 1 volume of a 1:3 mixture of 2% $SrCl_2.6H_2O$ in glacial acetic acid: acetone (freshly mixed just prior to use). The acetone was removed by distillation and rotary evaporation, the temperature never exceeding 100°C. The remaining solution was centrifuged to precipitate the hemin, which was washed twice with 50% aqueous acetic acid, once with 95% ethanol, and once with diethyl ether.

The hemin was washed into a 200 ml thick-walled Erlenmeyer flask, with a total of 100 ml hydrobromic acid/acetic acid (% $HB_r \leq 38\%$, Fisher Scientific), and the flask covered with a loose-fitting glass lid. The solution was stirred at room temperature for 24 hr, then poured into 400ml distilled H_2O. This was stirred for 20 minutes, then transferred to a large separatory funnel and overlaid with 2 volumes of diethyl ether. To this was added 125g solid sodium acetate, and the funnel was immediately vigorously agitated. The ether layer was collected, and the acid layer was re-extracted with 1 volume of fresh ether. The combined ether extracts were washed with 10-15 ml 0.1 mM aqueous $NaCO_3$. The aqueous phase was re-extracted with fresh ether, the ether extract was dried on a rotary evaporator with the aid of a cold trap, and the resulting hematoporphyrin (HP) crystals washed twice with water.

The equivalent of 100mg HP was dissolved in a mixture of 1.9ml glacial acetic acid and 0.1ml sulfuric acid (Lipson et al, 1961; Kessel, 1985), stirred for 1 hour at room temperature, and then poured into 12ml 3% aqueous sodium acetate solution. The pH was adjusted to 4.5, the HP-acetate mixture was collected by centrifugation, then washed 3 times with distilled H_2O. The crystals were either processed immediately, or were lyophilized and stored at -20°C in a dessicator.

Alkaline hydrolysis of the HP-acetate was conducted under conditons designed to maximize the yield of the active HPD component dihematoporphyrin ether (DHE) (Kessel and Cheng, 1985). The HP-acetate mixture was suspended in 0.1N NaOH at a

concentration of 40mg/ml, the mixture was stirred for 15 min.
at room temperature, and the same amount of 0.1N NaOH was
added. Stirring was continued for 15 min., then the pH was
carefully adjusted to pH 7.0 with HCl. The solution was
adjusted to a final concentration of 10mg/ml with distilled
H_2O, and stored at $-20^{\circ}C$. At all times following hemin
isolation, the preparation was protected from light. The
specific activity of the ^{14}C-HPD preparation used in this
study was $0.05\mu Ci/mg$, as determined by liquid scintillation
counting of a known weight of the compound in a Beckman
LS-3133T liquid scintillation counter.

Chromatographic and Spectral Analyses

The aggregation status of Cu-HPD was compared to that of
unlabelled Photofrin I by passing 3 mg samples of each (in
aqueous solution) through a 20x1cm P-10 gel chromatographic
column (Bio-Rad Laboratories, Richmond, California), eluting
with saline, and collecting 0.5 ml fractions. Samples of each
fraction, using a solvent composed of methanol: H_2O (9:1),
were monitored for absorbance at 535 nm using a Hitachi Model
100-60 spectrophotometer. This instrument was also used to
measure absorption spectra.

The efficiency of ^{64}Cu incorporation into HPD
components, and the effect of labelling on HPD composition,
was assessed by reverse-phase thin-layer chromatography using
Whatman $KC_{18}RP$ plates and a solvent composed of methanol:
tetrahydrofuran: water (60:6:34) containing 1.0mM
tetrabutylammonium phosphate (TBAP). The resultant
chromatogram was monitored for radioactivity using a Model
7201 Radiochromatogram Scanner (Packard Instrument Co.).

In-Vivo Biodistribution of R*-HPD

Tumours were induced in the flanks of 1-3 week old Syrian
golden hamsters by subcutaneous injection of $5x10^6$
adenovirus Type-5 transformed hamster embryo fibroblasts
(Williams, 1973), and were allowed to develop and grow for 4-6
weeks prior to use.

Each radiolabelled HPD preparation, R*-HPD, was
administered intraperitoneally at doses of D=2-10mg/kg body
weight. At appropriate times after HPD injection, groups of
animals were sacrificed and selected tissues dissected for
counting immediately after sacrifice. Three samples were
normally taken for each tissue.

For ^{64}Cu activity measurements, samples up to 500mg
each were used. These were counted in plastic tubes
containing 1ml H_2O, using a gamma counter (Nuclear Chicago).

For beta counting, tissue samples no greater than 100mg
were placed into glass liquid scintillation vials and
dissolved in 1ml Soluene-350[TM] (Packard Instrument Co.) for
2-5 days at room temperature, or for 3-5 hours at $50^{\circ}C$ in
a shaking water bath. For blood, 0.1ml samples were
solubilized in 1ml of a 1:1 (v/v) Soluene-350/isopropanol
mixture. After solubilization, 0.5ml 30% hydrogen peroxide
solution was added to each sample and the sample incubated at
room temperature for 15 minutes with intermittent swirling.

When the reaction had gone to completion, 15ml of a 1:9 (v/v) mixture of 0.5N HCl/Instagel™ (Packard Instrument Co.) was added to each sample and the sample thoroughly mixed to homogeneity. Previous experiments indicated that this was the optimum liquid scintillation cocktail for beta counting in our system, in order to eliminate artefact resulting from chemiluminescence. Samples were dark and temperature equilibrated for 1 hour in a Beckman liquid scintillation counter, prior to counting. Internal standardisation was used to correct for quenching and counting efficiencies. Two weeks (>20 half-lives) delay was used before beta counting of ^{64}Cu-containing samples, in order to allow for complete decay of the ^{64}Cu. Standard aliquots of the injectate were also counted to convert the tissue counts to equivalent HPD concentration:

$$[\underline{C}] = \mu g \ HPD/g \ tissue.$$

The biodistribution results are then expressed in terms of:

$$specific \ uptake \ rati\acute{o} \ (\underline{SUR}) = [\underline{C}]/\underline{D}$$

In-vitro Uptake and Photocytotoxicity of HPD and R*-HPD:

The cell line used for all in-vitro experiments was the V79 Chinese hamster lung line. Monolayer cultures were maintained and propagated in α-minimal essential medium (α-MEM, GIBCO, New York), supplemented with 10% fetal bovine serum, 50 U/ml penicillin G and 50 μg/ml streptomycin. Cells were grown in 75 cm^2 plastic screwcap flasks, incubated at 37°C in a humid incubator (90-100% humidity) in 5% CO$_2$ and 95% air.

Spheroid cultures were grown according to published protocols (Sutherland and Durand, 1976). Cells growing in monolayer were detached using 0.05% trypsin-EDTA (GIBCO), and 5x10^6 cells were plated into a 100mm non-tissue culture plastic petri dish with 10ml α-MEM + 15% FBS. After 24 hours, these cells were collected using 0.25% trypsin-EDTA, and 7.5x10^5 cells were seeded into a 250 ml glass spinner culture bottle (Johns Scientific, Toronto) containing 75 ml α-MEM + 5% FBS. The spinner culture was maintained on a non-heating magnetic stirring platform at a spin speed of 180-190 r.p.m., in an air-convection incubator at 37°C. On each of the 4th and 6th days of spinner culture incubation, a further 75 ml of α-MEM + 5% FBS was added to the flask. The culture medium was subsequently replaced daily with 200 ml fresh growth medium, after first carefully aspirating the spent medium from the flask in which the spheroids had been allowed to settle to the bottom under gravity.

For studies of HPD uptake, monolayer or spheroid cultures were incubated for various times with α-MEM + 1% FBS and either 2.0, 5.0, or 10.0 μg HPD/ml. For fluorescence measurements on HPD-treated cells, monolayers or spheroids were dispersed with 0.25% trypsin to form single cell suspensions. These suspensions were then analysed for cellular HPD fluorescence yield using an Ortho Spectrum III™ laser flow cytometer (Ortho Diagnostic Systems Inc.), comprising an argon-ion laser (488nm, 50mW operating power) for fluorescence excitation and a red high pass filter

(50% transmission at 630nm) for HPD fluorescence detection.
For radioactivity measurements, the cellular concentration of
HPD was determined using gamma or beta counting, as for the
animal tissue samples.

For the initial survival experiments, spheroids grown in
spinner culture were plated into 60mm plastic tissue culture
dishes. After incubation with labelled or unlabelled HPD,
cultures were thoroughly washed prior to light irradiation.
The light source used was a 1000W tungsten-filament lamp,
filtered to remove wavelengths below 590 nm. After
irradiation, spheroids were then re-incubated for 4 days in
α-MEM + 10% FBS to permit colony development. Cultures were
subsequently fixed, stained with crystal violet, and the
colonies counted to determine survival fractions.

We have recently designed and constructed an apparatus
that allows uniform irradiation of spheroids in spinner
culture under more strictly controlled conditions (Figure 2).
This irradiation chamber maintains the normal spheroid growth
conditions while irradiation is being carried out. The light
source consists of a cylindrical array of 20W fluorescent
lamps, with acetate colour filters appropriate for providing
visible light of the desired wavelength spectrum. The lamp
array is mounted with a rigid, highly-reflecting cylinder to
maximize the light power received by the culture flask. The

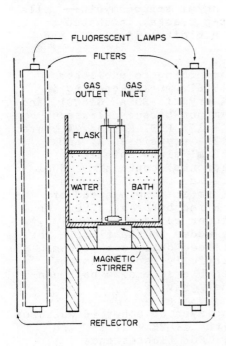

Fig. 2. Apparatus for irradiation of spheroid spinner
cultures with visible light under controlled
conditions. The optical filters can be changed to
select different wavelength bands, while the
intensity can be altered by the number of fluorescent
tubes energised.

sample flask is centrally mounted within a transparent perspex water bath whose temperature can be constantly maintained. During irradiation, the spheroid suspension can be constantly stirred, and maintained in equilibrium with gas of specific composition/concentration (ie. % O_2).

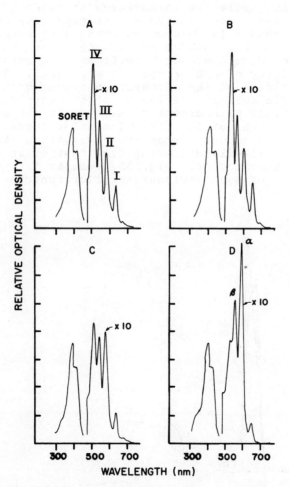

Fig. 3.   Effects of copper incorporation on the absorption spectrum of Photofrin I.   The figure shows absorption spectra in 10mM CTAB of (A) Photofrin I, with bands labelled according to Falk (1964); (B) Photofrin I subjected to pH manipulation of labelling procedure, but without addition of copper; (C) Cu-HPD (Cu:HPD≈ 1:3) as produced in the standard labelling procedure; (D)   Photofrin I labelled with excess copper, showing α and β metalloporphyrin bands.

RESULTS AND DISCUSSION

Analysis of Copper-64 Labelled HPD

Figure 3 shows the effects of copper incorporation on the absorption spectrum of HPD. The absorption spectum is not altered (Fig. 3B) by the pH manipulation to which the parent HPD solution (Fig. 3A) is subjected, ie. the Soret band, as well as bands I-IV, remain unchanged in their relative positions and intensities. An identical absorption spectrum is also obtained for ^{14}C-HPD manufactured according to our protocol (data not shown). Upon addition of copper in the molar ratio of Cu: HPD = 1:3 used in our standard labelling procedure (Fig. 3C), the intensity of band III (and that of the Soret band) remains identical to that observed for unlabelled HPD, while the intensities of bands II and IV are markedly altered. Whereas the slight reduction in the intensity of band I (at 630nm) suggests a reduced photodynamic potential of Cu-HPD, we have previously found that our labelling procedure does not significantly alter the in vitro photocytotoxicity of HPD (Firnau et al, 1984). Since the shape and position of the porphyrin Soret absorption band is a sensitive indicator of relative porphyrin aggregation (Brown and Shillcock, 1976; Andreoni et al, 1982), the observation that the Soret band of Photofrin I remains unchanged by the copper labelling suggests that the aggregation state of Cu-HPD should not be markedly different from that of unlabelled HPD. Addition of excess copper (Fig. 3D) results in a spectrum more characteristic of square-planar metalloporphyrins, in which

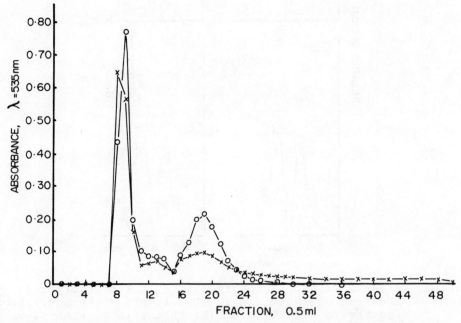

Fig. 4. Gel exclusion chromatographs of HPD and Cu-HPD. Samples of aqueous solutions of Photofrin I (O 3.00mg) or Cu-HPD (X 2.93mg, Cu:HPD≈1:3) were loaded onto gel columns, eluted with saline, and 0.5 fractions collected for measurement of absorbance at 535nm.

the α-band derives from bands I and III of the unchelated
porphyrin ring, and the β-band derives from bands II and IV
(Falk, 1964). It is seen that chelation in this case is not
complete, since bands I and IV remain detectable. For the
Cu-HPD routinely produced and used by us (Fig. 3C), the final
concentration of a given preparation is assessed by
measurement of the absorbance at 535 nm (band III) against a
standard curve of unlabelled HPD in 10 mM CTAB.

Figure 4 shows an analysis of Cu-HPD by gel-exclusion
chromatography, with the eluate fractions monitored for
absorbance at 535 nm. The biphasic exclusion profile of
Cu-HPD is very similar to that for Photofrin I in the absence
of added copper, although the measurable tail at higher
fraction numbers suggests that total re-aggregation has not
been achieved. Similar elution curves for ^{64}Cu-HPD, showing
correlation of radioactivity and optical density (data not
shown), have demonstrated that the ^{64}Cu is associated with
the main, highly aggregated photoactive components of HPD.

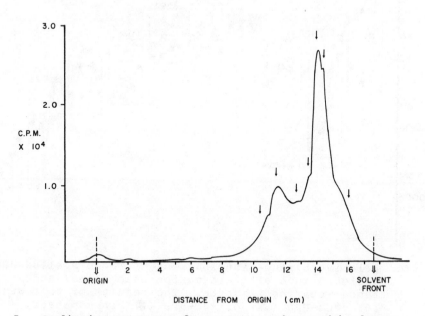

Fig. 5. Radiochromatogram for reverse-phase thin-layer
chromatography of ^{64}Cu-HPD. The figure shows the
co-migration of radioactive and fluorescent HPD
porphyrin species. The 2 non-fluorescent ^{64}Cu-HPD
components are believed to be labelled monomers.

In order to assay the efficiency of incorporation of the ^{64}Cu into component HPD molecules, the labelled HPD solution was analysed by measurement of the radioactivity profile on reverse-phase thin-layer chromatographs of the ^{64}Cu-HPD mixture. A typical radiochromatogram is shown in Figure 5, indicating that essentially no free copper remained in solution after labelling. The two additional non-fluorescent radioactive components observed for ^{64}Cu-HPD are believed to be ^{64}Cu-HPD monomers.

<u>In-vivo Tissue Distribution Studies</u>

Figure 6 shows the variation of the ^{64}Cu-HPD specific uptake ratio (<u>SUR</u>) with time after HPD injection into hamsters, and also presents preliminary results for our ^{14}C-HPD. These data are compared with results of similar studies using [3]H- and ^{14}C-labelled HPD in tumour-bearing mice (Gomer and Dougherty, 1979; Gomer et al, 1982). Although

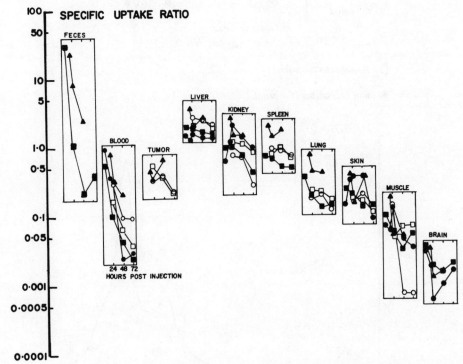

Fig. 6.   Relative tissue distributions of radiolabelled HPD at various times after injection into normal and tumour-bearing hamsters. The results of this study for ^{64}Cu-HPD (▲) or ^{14}C-HPD (●) in hamsters, are compared with results obtained by other investigators for ^{14}C-HPD (○,Gomer and Doughtery, 1979) and ^{3}H-HPD (□,Gomer and Dougherty, 1979; ■, Gomer et al, 1982) in mice.

only part of the study has been completed at this time, it can be seen that the tissue distribution of Cu-HPD is, overall, comparable to that of carbon-14 labelled HPD in a normal and tumour-bearing hamster model.

Current studies with ^{14}C-HPD in this and other animal models will be used to establish definitively whether the Cu labelling significantly alters the biodistribution characteristics of HPD.

In-Vitro Uptake and Photocytotoxicity Studies

Monolayer and multicell spheroid cultures of the V79 Chinese hamster cell line are being used to study the uptake kinetics and distribution of HPD and R*-HPD, and to correlate with cytotoxicity resulting from treatment with HPD + light. This is being done as a function of spheroid size, HPD concentration, and HPD incubation time. Multicell spheroid tumour models have been used extensively in studies of radiation and chemo cytotoxicity (Sutherland and Durand, 1976), and preliminary investigations of PDT have also been reported by others using this approach (Christensen et al, 1984). Since there are oxygen and metabolite gradients from the outside proliferating cell layers to the inner hypoxic or necrotic regions, spheroids represent a good intermediate model between monolayer cell cultures and poorly-vascularized, solid tumours in-vivo.

Figure 7A shows HPD fluorescence flow cytometry profiles for V-79 cells in monolayer exposed for 16 hours to different HPD concentrations. The profiles consist of relatively well-defined single peaks, whose shapes are consistent with the size distribution of the cells (data not shown). The mean fluorescence intensity of the distributions is directly proportional to the concentration of HPD in the incubation medium (Figure 8), although there is some distribution broadening with increasing HPD concentration, suggesting a wider variation in cellular fluorescence intensity (and therefore HPD content) at higher HPD concentrations. The dashed curve of Figure 8 shows the percentage of HPD-incubated cells in which the fluorescence intensity is greater than that in untreated cells (i.e. autofluorescence).

Figure 7B shows similar HPD fluorescence profiles for cells from spheroids which have been disaggregated following incubation in medium containing 10 μg/ml HPD for different times. These profiles differ somewhat from those for monolayer cells in two respects. Firstly, the distributions for spheroid cells are more skewed, exhibiting "tails" which extend to higher fluorescence intensity and which become more pronounced with increasing time of HPD incubation. This feature suggests a greater variation of HPD content amongst spheroid cells compared to that seen for monolayer cells. Secondly, while the mean distribution intensity increases with incubation time up to 72 hours, it nonetheless remains at all times less than that observed for monolayer cells incubated for only 16 hours. These observations suggest that the uptake of HPD into spheroids may be diffusion-limited, and we are currently carrying out similar analyses on HPD-incubated spheroids from which cell layers have been sequentially

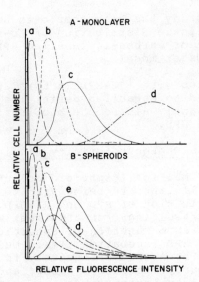

Fig. 7. Fluorescence flow cytometry profiles of HPD uptake
by V-79 cells. (A) monolayer culture, 16 hour
incubation in HPD concentrations of (a)2, (b)5,
(c)10, (d)20 µg/ml; (B) spheroids incubated in
10 µg/ml HPD for (a)8, (b)24, (c)48, (d)72hr; curve
(e) is for monolayer, 10 µg/ml for 10 hr. The
narrow peak in each case is a non-HPD control.

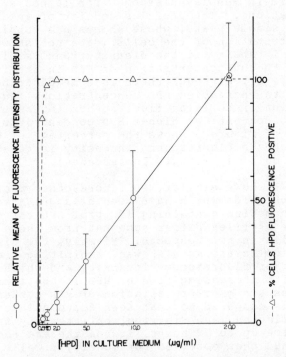

Fig. 8. HPD uptake in V-79 cells in monolayer, corresponding
to the profiles of Fig. 7A: ◯ relative mean of the
fluorescence intensity distribution, △ % of cells
for which the fluorescence is significantly above
background.

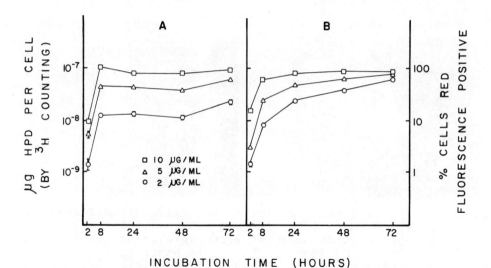

Fig. 9.   HPD uptake kinetics into V-79 spheroid cells as a
           function of time of incubation in various
           concentrations of HPD. HPD uptake was measured by:
           (A) radioactivity counting with ^{3}H-HPD; (B)
           fluorescence flow cytometry.

trypsinized prior to flow cytometry to test the possibility
that a gradient of HPD may exist throughout the volume of a
spheroid. Figure 9 compares the uptake kinetics of different
concentrations of ^{3}H-HPD into spheroids, as measured by beta
counting and flow cytometry of cells derived from completely
dissociated, HPD-incubated spheroids. A more detailed study
of the HPD uptake kinetics in spheroids is currently under
way, using ^{14}C- and ^{64}Cu-HPD.

       Spheroids of V-79 Chinese hamster cells have also been
used to study cell killing with both HPD and ^{64}Cu-HPD in the
presence of light (Firnau et al, 1984). Figure 10 shows
stained sections of spheroids before and after treatment, and
typical spheroid survival curves are shown in Figure 11. The
irradiation chamber (Figure 2, described in Materials and
Methods) which allows the spheroids to be exposed to light
under strictly controlled conditions while in spinner culture,
will be used to study spheroid cell killing as a function of
HPD concentration and light dose/dose rate, as well as factors
such as temperature and oxygen tension before, during, and
after irradiation.

SUMMARY

       This paper presents results of preliminary studies of the
in-vivo uptake and biodistribution of copper-labelled HPD, and
of the in-vitro uptake and photocytotoxicity of HPD and
radiolabelled HPD. The results indicate that:

1) Cu-HPD is chromatographically very similar to HPD except
    for two minor additional non-fluorescent (monomeric)
    components.

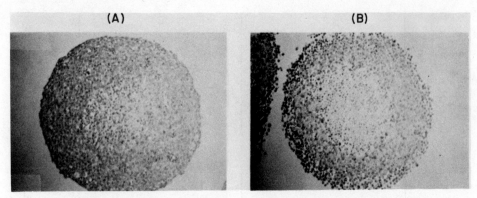

Fig. 10. Histological sections of spheroids (A) before and
(B) 4 hours after HPD treatment with HPD + light.
Spheroids of diameter 250μm were incubated in medium
containing 20μg/ml HPD for 16 hours, and
subsequently irradiated with red light
($1000J/m^2, \lambda > 590nm$). 5μm histological sections of
spheroids were prepared and stained with hematoxylin
and eosin.

2) The tissue distributions of Cu-HPD and ^{14}C-HPD are,
overall, comparable to each other in a normal and
tumour-bearing hamster model. Current studies with
[14]C-HPD in this and other animal models will be used to
establish definitively whether the copper labelling
significantly alters the detailed biodistribution
characteristics of HPD

3) The multicell spheroid model will allow detailed
investigation of the uptake kinetics of HPD, by radioassay
and/or fluorescence flow cytometry. Furthermore, the
photocytotoxicity of HPD + light may be measured under
variable physiological conditions, and correlated with HPD
uptake distributions, as a means of investigating the
mechanism(s) of action of PDT.

Clinical Applications

These results point to the potential usefulness of
copper-labelled HPD for planning and assessing clinical PDT,
and a preliminary study has been carried out in which PET head
scans were obtained in 6 patients with brain tumours. Tracer
doses of ^{64}Cu-HDP (8 mCi) were injected intravenously, and
scans made 24 hours later. In the scans, 2 out of 2
meningiomas and 1 out of 4 gliomas showed increased [64]Cu
activity in the area of the tumour. The observation that not
all gliomas could be visualized is consistent with other
reports that HPD may not accumulate in all such lesions
(Wharen et al, 1983), possibly due to failure to cross the
blood-brain barrier (Wise and Taxdal, 1967). We have likewise
observed very low uptake in the normal hamster brain (see
Figure 6). In assessing the uptake of HPD into one of the
meningiomas, the blood marker ^{125}I-albumin was also used in
order to determine if the concentration of ^{64}Cu in the
tumour was the result of active HPD uptake or simply due to

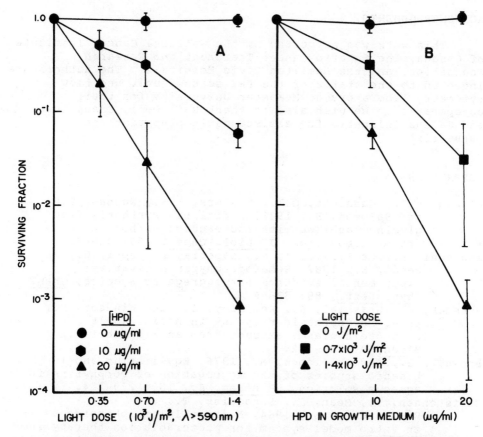

Fig. 11. Effect of HPD + red light on colony-forming ability
of plated multicell spheroids. Spheroids
(diameter ≃ 250 μm) plated into tissue culture dishes
were incubated for 16 hours in medium containing
different concentrations of HPD, and subsequently
irradiated with various light doses as shown.
Survival is expressed as the percentage of treated
spheroids capable of forming a macroscopic colony at
4 days after treatment, as a function of either (A)
light dose or (B) HPD concentration.

increased blood flow to the region. After surgery (1 day
after the PET scan), samples of tumour and of blood were
counted for ^{64}Cu and ^{125}I activity. The ratio of
^{64}Cu:^{125}I was 2-3 times higher in the periphery of the
tumour than in blood, but the same as blood in the centre of
the tumour.

While anecdotal, these findings illustrate how
quantitative PET (or SPECT with ^{67}Cu-HPD) may enable
in-vivo, non-invasive quantitation of HPD concentration in
tumour and normal tissues.

ACKNOWLEDGEMENTS

    This work was supported by the National Cancer Institute
of Canada, the Ontario Cancer Treatment and Research
Foundation, and the Hamilton Civic Hospitals. The authors are
indebted to the staffs of the PET scanner unit and flow
cytometry laboratory at McMaster University for their
co-operation. We wish also to thank Gail Oddi, Diane Lowe,
and Galena Szlapetis for assistance in preparing the
manuscript.

REFERENCES

Altman, K.I., Casarett, G.W., Masters, R.E., Noonan, T.R.,
        and Salomon, K., 1948, Hemoglobin synthesis from
        glycine labelled with radioactive carbon in
        its α-carbon atom. J. Biol. Chem., 176: 319-325.
Andreoni, A., De Silvestri, S., Laporta, P., Jori, G., and
        Reddi, E., 1982, Hematoporphyrin derivative:
        experimental evidence for aggregated species, Chem.
        Phys. Lett., 88: 33-36.
Borsook, H., Graybiel, A., Keighley, G., and Windsor, E.,
        1954, Polycythemic response in normal adult rats to a
        nonprotein plasma extract from anemic rabbits,
        Blood, 9: 734-742.
Brown, S.B., and Shillcock, M., 1976, Equilibrium and
        kinetic studies of the aggregation of porphyrins in
        aqueous solution, Biochem. J., 153: 279-284.
Christensen, T., Moan, J., Sandquist, T., and
        Smedshammer, L., 1984, Multicellular spheroids as an
        in vitro model system for photoradiation therapy in
        the presence of HPD, in: "Porphyrin Localization and
        Treatment of Tumors", D. Doiron, and C. Gomer, eds.,
        Alan R. Liss, Inc., New York: 381-390.
Falk, J.E., 1964, "Porphyrins and Metalloporphyrins.
        Elsevier Publishing", Amsterdam.
Firnau, G., Maass, G., Wilson, B.C., and Jeeves, W.P.,
        1984, ^{64}Cu labelling of hemataporphyrin derivative
        for non-invasive in-vivo measurements of tumour
        uptake, in: "Porphyrin Localization and Treatment of
        Tumors", loc cit: 629-636.
Gomer, C.J., and Dougherty, T.J., 1979, Determination of
        ^{3}H- and ^{14}C- hemataporphyrin derivative
        distribution in malignant and normal tissue, Cancer
        Res. 39: 146-151.
Gomer, C.J., Rucker, N., Mark, C., Benedict, W.F., and
        Murphree, A.L., 1982, Tissue distribution of
        ^{3}H-hematoporphyrin derivative in athymic "nude"
        mice heterotransplanted with human retinoblastoma,
        Invest. Opthalmol. Vis. Sci., 22: 118-120.
Labbe, R.F., and Nishida, G., 1957, A new method of hemin
        isolation, Biochim. Biophys. Acta, 26: 437.
Lipson, R.L., Baldes, E.J., and Olsen, A.M., 1961, The use
        of a derivative of hematoporphyrin in tumor
        detection, J. Natl. Cancer Inst., 26: 1-8.
Kessel, D., 1985. Personal communication.
Kessel, D., and Cheng, M.L., 1985, Studies on the
        biological and biophysical properties of
        dihematoporhyrin ether, the tumour-localizing
        component of HPD, Cancer Res., in Press.

Sutherland, R.M., and Durand, R.E., 1976, Radiation
        response of multicell spheroids - an in-vitro tumour
        model, Current Topics in Radiation Res., 11: 87-139.
Wharen, R.E., Anderson, R.E., and Laws, E.R., 1983,
        Quantitation of hematoporphyrin derivative in human
        gliomas, experimental central nervous system tumors,
        and normal tissues, Neurosurgery, 12: 446-450.
Williams, J.F., 1973, Oncogenic transformation of hamster
        embryo cells in-vitro by adenovirus type 5, Nature,
        243: 162-163.
Wise, B., and Taxdal, D.R., 1967, Studies of the blood-brain
        barrier utilizing hematoporphyrin, Brain Research, 4:
        387-389.

PHOTODYNAMIC EFFECTS AND HYPERTHERMIA IN VITRO

Terje Christensen*, Lars Smedshammer, Anne Wahl
and Johan Moan

Norsk Hydro's Institute for Cancer Research
The Norwegian Radium Hospital
Montebello, 0310 Oslo 3, Norway

ABSTRACT

Cells from the established human line NHIK** 3025 were labelled
with hematoporphyrin derivative in vitro. Subsequently, the cells were
treated with light and hyperthermia. The cells could be irradiated
either before, during or after the incubation at a hyperthermic temperature. It was shown that hyperthermia given shortly after the light
exposure gave a synergistic killing effect. In spite of some loss of
porphyrins from the cells, the light sensitivity increased 20 min after
a light irradiation. At later times, the cells apparently repaired some
of the photodynamic damage at 37°C. At higher temperatures, the
repair was inhibited.

INTRODUCTION

There are two reasons for investigating the interactions between
the photodynamic effect of porphyrins and hyperthermia: Some local
heating of the tissue due to absorption of light energy is inevitable.
Furthermore, local hyperthermia can be used in combinations with other
forms of cancer therapy to increase their efficiency. It has been shown
that heating of the tumor to above 40°C can take place during a
standard regimen of photodymanic therapy (PDT) (Svaasand et al. 1983,
Kinsey et al. 1983).

Preliminary reports indicate that hyperthermia combined with PDT
can increase the effect above that of PDT alone on mouse tumors (Waldow
and Dougherty, 1984, Reports in this book). Our own studies have
indicated that there is a synergistic interaction between PDT and hyperthermia in vitro when the cells are heated shortly after the light

---

* Present address: Department of Toxicology, The National Institute
  of Public Health, Geitmyrsveien 75, N-0462 Oslo 4.

** Abbreviations used: PDT: photodynamic therapy, Hpd: hematoporphyrin
   derivative, NHIK: Norsk Hydro's Institute for Cancer Research, MEM:
   minimal essential medium.

exposure (Christensen et al. 1984). In this chapter we will show how experiments can be performed where cells are exposed to Hpd plus light in different combinations with hyperthermia. Results of such experiments will be presented and discussed with reference to the possibility of using the combinations in clinical trials.

CELL CULTIVATION

Cells from the established line NHIK 3025 were used. They were originally derived from a human carcinoma in situ of the uterine cervix (Nordby and Oftebro 1969, Oftebro and Nordby 1969). Until 1978 the cells were cultivated in Puck's medium E 2a (Puck et al. 1957) containing 30% serum (human and horse). The cells used in this study were transferred to MEM with 10% newborn calf serum in 1978 and recultivated in this medium twice a week. The medium was supplemented with 100 u/ml penicillin, 100 μg/ml streptomycin and L- glutamin. All medium components were supplied from Gibco, Paisley, Scotland. Frozen samples of the stock culture were kept in case of contamination or altered properties of the cells. The stock culture has been replaced twice with frozen samples since 1978. No infections with mycoplasma have been observed by extranuclear fluorescence after staining with Hoechst 33285.

The cells showed an epitheloid morphology, had a mean population doubling time of 24 h (Christensen et al. 1983), did not form tumors in nude mice after injection of $3 \cdot 10^6$ cells (Rofstad, personal communication), and were aneuploid.

Normally, the cells were subcultivated twice a week in the experimental period. Therefore they were in exponential growth most of the time. Routinely, the cells were detached and dispersed in 0.25% trypsin (Difco 1:250), whereafter about $0.5 \cdot 10^6$ cells/75 cm^2 tissue culture flask (Falcon) were inoculated. The medium was equilibrated with 5% CO_2 and the cells were cultured at 37°C in a thermostated room or a CO_2-incubator (National).

For experimental purposes, the cells were trypsinized as described above, and carefully dispersed into a single cell suspension. After centrifugation (5 min, 1000 r.p.m.), the number of cells was determined by counting in a haemocytometer. Proper dilutions were made in MEM and the cells were seeded in tissue culture flasks (Falcon, 25 cm^2) or tubes (Nunc, no. 156758, 5.5 cm^2 flat area). The number of cells were regulated to give between 1 and 20 colony forming cells per cm^2 after treatment. This was achieved by inoculating between 20 and 2000 cells/cm^2, depending on the treatment.

The survival of the cells was scored by counting the number of colonies formed after 10-12 days of incubation. Only colonies of 40 cells or more were taken into account in the measurements. The fraction of untreated cells forming colonies (plating efficiency) was determined and used to calculate the absolute value of the surviving fraction. The plating efficiency was normally 60-90% when the cells were seeded in tissue culture flasks or dishes. Some cells tended to attach to and form colonies on the curved walls of the tubes. Therefore, the number of cells on the flat part of the tubes was somewhat lower than the total number of colony-forming cells.

At the end of the incubation period, the cells were rinsed in 0.9% NaCl, fixed in absolute ethanol and stained with methylene blue.

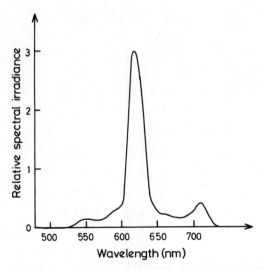

Fig. 1.   The corrected spectrum of the lamps with filter
(Christensen et al. 1984).

INCUBATION WITH HPD

The cells were incubated for 22 h in growth medium (including 10%
serum) with 25 μg/ml Hpd. Photofrin I, the crude alkali-treated
sulphuric- and acetic acid derivative of hematoporphyrin, was obtained
from Oncology Research and Development, U.S.A. After incubation the
cells were transferred to fresh MEM. This has been shown to remove
about 40% of the cellular porphyrins during a 30 min incubation and
leave a constant amount bound to the cells for at least 24 h (Christensen et al. 1983). The ratio between bound porphyrin and cellular
protein was, however, slowly reduced due to cell multiplication. The
photosensitivity was reduced to about the same extent as the porphyrin
contents after removal of the porphyrin. For all practical purposes the
photosensitivity can be regarded as constant in a 4 h period after the
removal of Hpd (Christensen et al. 1983).

LIGHT IRRADIATION

The cells were irradiated with light from a bank of 4 fluorescent
tubes (Philips TDL/83). When combined with a red filter (Cinemoid 35,
Rand Stand Electric, U.K.), the lamps produced the emission spectrum
shown in Fig. 1.

The fluence rate was measured with a calibrated UDT model 11A
photometer with detector 1223 or 1222. The sensitivity of the detectors
and the transmittance of the monochromators used (Bausch & Lomb) varied
with the wavelength. This has been corrected for in Fig. 1.

During the treatment the tubes containing the cells were immersed
in a waterbath as described below. The irradiance at the posistion of
the cells in the tubes was 12 Wm^{-2}.

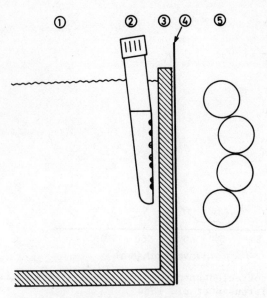

Fig. 2. The experimental irradiation setup (Christensen et al.
1984).
1) Waterbath, 2) Culture tube, 3) Perspex wall,
4) Red filter, 5) Fluorescent lamps.

HYPERTHERMIA

The cells were heated by vertical immersion in waterbaths (Heto,
Denmark) made of transparent Perspex (Fig. 2). A rack holding serveral
tubes in position was placed in each waterbath. Two waterbaths could be
placed in front of the lamp simultanously. Therefore, the cells could
be irradiated at selected temperatures, and it was also possible to
shift quickly between two temperatures. It took less than 6 min before
the medium above the cells reached the new temperature after a tube had
been transferred from one waterbath to another.

CYTOTOXIC EFFECTS OF COMBINATIONS BETWEEN THE PHOTODYNAMIC EFFECT OF HPD
AND HYPERTHERMIA

When a constant dose of light and Hpd is combined with a constant
dose of hyperthermia, the cellular survival varies with the sequence of
the two treatments. A typical example is shown in Fig. 3, where
$1.1 \cdot 10^4$ J/m^2 (15 min) was combined with heating at 42.5°C for
2 h. The effect was maximal when the hyperthermia followed immediately
after the light exposure. In the experiments shown in Fig. 3 both

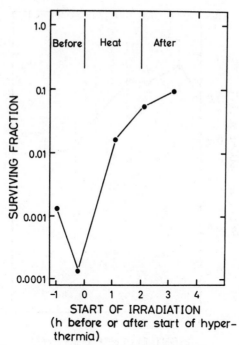

Fig. 3. Effect of the sequence of exposure to 2 h heat at $42,5^{\circ}C$ and $1.1 \cdot 10^4$ Jm^{-2} red light.

agents were applied in doses separately inactivating roughly 50% of the cells. It has been shown that the same type of interaction takes place when the hyperthermia (42.5 or $45^{\circ}C$) is delivered in sub-lethal doses after light irradiation. The opposite, a sub-lethal dose of Hpd plus light combined with a near-lethal dose of hyperthermia, does not lead to any potentiation of the effect of hyperthermia.

By comparing experiments where different doses of both agents were used, it was concluded that the combined effect was synergistic when the cells were heated shortly after the irradiation (Christensen et al. 1984). To determine the type of interaction, Berenbaum's method was used (Berenbaum 1980).

Several observations indicated that hyperthermia may inhibit the repair of photodynamic damage: the high efficiency of post-irradiation hyperthermia, the shape of the dose response curves with a reduction of the shoulders as well as the lack of synergistic interaction when the light dose was low. To elucidate this further, split-dose experiments were performed. Methodologically, these are difficult to carry out and the results may be difficult to interpret. The Hpd uptake is particularly important to control. It was attempted to do so by making homogenates of the cells at different times during the experiments and assaying the fluorescence. When the cells were kept at $37^{\circ}C$ in the dark, the porphyrin contents seemed to be relatively constant for the

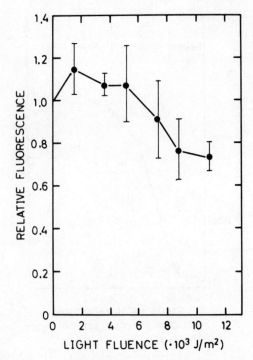

Fig. 4. Fluorescence, expressed relative to the fluorescence from
cells kept in the dark, from cell homogenates exposed to
the indicated light fluence. The fluorescence was
measured 2 h after irradiation (400 nm ex., 635 nm em.).

first hours after removal of Hpd (Christensen et al. 1983). The fluorescence from cell homogenates is reduced after light irradiation of the
whole cells (Fig. 4). This indicates that light can promote leakage of
porphyrins from the cells. Alternative explanations are photobleaching
of the porphyrins or light induced changes in the fluorescence quantum
efficiency. The former effect was tested, and no indications of photobleaching at the doses used were found. The latter explanation also
seems to be improbable, since binding of the Hpd to a detergent (Cetyl
trimethyl ammonium bromide) added to the cell homogenates did not change
the differences in fluorescence between samples stored in the dark and
irradiated samples.

As indicated by Fig. 5, splitting the light dose in two affects the
survival in a complicated manner. After a dark incubation of 20 min to
1 h at 37°C or 41°C between the exposures, the survival is lower
than for the whole dose given in one exposure. The sensitivity is
higher during that period, even though the porphyrin contents is reduced
(Fig. 4). If the dark period is extended to 3 h, the survival increases
for cells incubated at 37°C. At 41°C the survival seems to be
constant for dark periods between 20 min and 3 h (Fig. 5). These
results were confirmed by construction of full dose-response curves
after different dark incubation periods at different temperatures

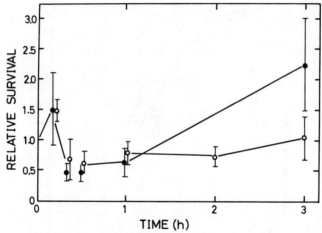

Fig. 5. Survival of cells after exposure to first
$1.1 \cdot 10^4$ Jm^{-2} and then $0.7 \cdot 10^4$ Jm^{-2} light spaced
by incubation in the dark for the indicated time. Solid
symbols refer to incubation at $37^{\circ}C$ and circles to
incubation at $41^{\circ}C$. The survival is expressed relative
to the survival after exposure to a single fluence of
$1.8 \cdot 10^4$ Jm^{-2}.

(Fig. 6). The shoulder on the curve reappeared after incubation at
$37^{\circ}C$, while the increased sensitivity after incubation at $41^{\circ}C$ is
related to the lower curve.

CONCLUSIONS

1. There is an apparent synergistic interaction between the
   photodynamic effect of Hpd and hyperthermia when the cells are
   heated shortly after the exposure to light.

2. The cells loose more porphyrins after a brief light irradiation
   than when kept in the dark.

3. Hyperthermia seems to inhibit the repair of photodynamic damage.

4. If the present results are relevant under in vivo conditions such
   as in the therapeutic situation, they suggest that one should first
   expose the tumor to light and then to hyperthermia.

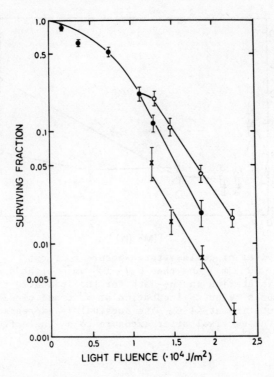

Fig. 6. Dose response curves for cells exposed to either a single
light irradiation (solid symbols) or two exposures spaced
by 3 h incubation in the dark. The incubation
temperatures were 37°C (circles) or 41°C (crosses).

REFERENCES

1. M. C. Berenbaum, Adv. Cancer Res. 35:269 (1980).
2. T. Christensen, T. Sandquist, K. Feren, H. Waksvik, and
    J. Moan, Br. J. Cancer 48:35 (1983).
3. T. Christensen, A. Wahl, and L. Smedshammer, Br. J. Cancer 50:
    85 (1984).
4. J. H. Kinsey, D. A. Cortese, and H. B. Neal, Cancer Res. 43:
    1567 (1983).
5. K. Nordbye and R. Oftebro, Exptl. Cell Res. 48:458 (1969).
6. R. Oftebro and K. Nordbye, Exptl. Cell Res. 58:459 (1969).
7. T. T. Puck, S. J. Ciecicura, and H. W. Fisher, J. Exptl. Med.
    106:145 (1957).
8. L. O. Svaasand, D. R. Doiron, and T. J. Dougherty, Med. Phys.
    10:10 (1983).
9. S. M. Waldow and T. J. Dougherty, Rad. Res. 97:380 (1984).

PRELIMINARY STUDIES WITH IMPLANTED POLYVINYL ALCOHOL SPONGES AS A MODEL
FOR STUDYING THE ROLE OF NEOINTERSTITIAL AND NEOVASCULAR COMPARTMENTS OF
TUMORS IN THE LOCALIZATION, RETENTION AND PHOTODYNAMIC EFFECTS OF
PHOTOSENSITIZERS

Richard C. Straight and John D. Spikes

Research Service, Veterans Administration Medical Center
and Department of Biology, University of Utah, Salt Lake
City, Utah

INTRODUCTION

Although several thousand cancer patients have been treated by
porphyrin-sensitized photodynamic therapy (PDT) to date (Dougherty,
1984), the fundamental mechanisms involved in this unique and interesting
application of photochemotherapy to the palliation and eradication of
tumors are poorly understood. For example, little is known about the
mechanism(s) of the localization and retention of porphyrins in tumors
and in certain non-malignant tissues (Tsutsui et al., 1975; Evensen et
al., 1984) or of those structural aspects of the porphyrin molecule
essential for efficient retention. Related to this, much more
information is needed on the microscopic sites of porphyrin retention in
tumors. Finally, much remains to be learned about the processes by which
a tumor that retains porphyrin is destroyed on illumination during PDT.

With respect to porphyrin retention in tumors, some investigators
feel that the malignant cells of the tumor selectively retain the
hematoporphyrin derivative (HPD) porphyrins. However, there is no
convincing evidence that various kinds of tumor or transformed cells in
culture consistently retain substantially higher levels of HPD porphyrins
than non-tumorous cells (Bugelski et al., 1981; Henderson et al., 1984;
Henderson et al., 1985). Also, in studies with several animal solid
tumor models, the majority of the retained porphyrin appears to be in the
extracellular matrix of the tumor (including the capsule) (Bugelski et
al., 1981; Spikes and Straight, 1985a), with smaller amounts in the
macrophages and necrotic areas, and bound to interstitial fibers and to
endothelial cells or perivascular fibers of the microvasculature (Spikes
and Straight, 1985a; Straight et al., 1985; Musser and Datta-Gupta, 1984;
Waner, 1985; Bonnett et al., 1984; Straight et al., 1984). In this
connection it should be recalled that the rapidly proliferating
capillaries of the tumor vasculature are "leakier" than normal (Denekamp,
1984); as a result, there is an increased rate of extravasation of plasma
components, including proteins and other colloids, into the tumor. Also,
as a result of this property, injected, negatively charged, "colloidal
dyes" (such as Evans blue, Pontamine sky blue, Vital new red, Congo red,
etc.) leak out of tumor capillaries and effectively stain (retention) the
tumor tissue (Weil, 1916; Ludford, 1929; Duran-Reynals, 1939; Goldacre
and Sylvén, 1962). Unfortunately, these very efficient tumor localizers

are not photodynamically active in vitro or in vivo (Straight and Spikes, 1985). As in the case of efficiently retained porphyrins, little if any of the negatively charged, colloidal dye is found in the malignant cells of the tumor; almost all of it is retained in the extracellular matrix (phagocytic cells, interstitial and vascular compartments) of the tumor. It should be noted that many negatively charged porphyrins are highly water soluble but may tend to aggregate or bind to colloids in plasma and like negatively charged, colloidal dyes may leak into the tumor tissue and be retained by binding to constituents of the extracellular matrix. However, many of the two carboxy porphyrins found in HPD are relatively hydrophobic. As the lipid solubility of a dye increases, the possibility increases that it will penetrate living cells, bind efficiently at some subcellular site (plasma membrane, mitochondrial membrane, nucleic acid, etc.) and exert a direct photodynamic action (PDA) on the normal or malignant cell (Kessel, 1984). Finally, it should be mentioned that the large amounts of newly synthesized collagen, and perhaps other proteins (fibrin, fibronectin), in the interstitial compartment of tumors, bind porphyrins more strongly than the same proteins in normal tissues (Musser et al., 1982).

In summary, much more work needs to be done on the mechanisms of the in vivo localization and retention of photosensitizing molecules in solid tumors. It is becoming clear, however, that these phenomena will probably have to be explained in terms of the anatomy, physiology and biochemistry of the whole tumor rather than just on the basis of some special array of properties of the malignant cells of the tumor (Spikes and Straight, 1985b). The extracellular vascular and interstitial compartments of tumors may play a predominant role, not only in the localization and retention of photosensitizers, but also in their phototherapeutic effect (Spikes and Straight, 1985a).

Some progress has been made recently in our understanding of the mechanisms of tumor destruction in PDT. In one study, clonogenic assays of cells from EMT6 tumors in mice receiving HPD-sensitized PDT show that light dosages giving tumor destruction do not kill the malignant tumor cells if the tumor is removed shortly after illumination; however, cells in tumors left in place after illumination start dying progressively with time (Henderson et al., 1984). Illumination of L1210 solid tumors in mice sensitized with HPD and certain other porphyrins leads to their destruction. However, if tumors from porphyrin-sensitized animals are removed and the cells illuminated with the same light doses in vitro, no cell killing occurs (Musser and Datta-Gupta, 1984). These results probably reflect the rather low porphyrin levels in the malignant cells of at least some kinds of tumors in animals receiving PDT. What mechanisms then are involved in the damage and destruction of tumors in PDT? Several workers have now shown that, within a few minutes after the illumination period, there is a significant decrease in the rate of blood flow through tumors in animal models followed by a complete cessation (Henderson et al., 1984; Selman, et al., 1984; Star et al., 1984; Straight and Spikes, 1985). This suggests that ischemia may ·be a major factor in tumor killing as a result of PDT and that the vascular compartment may be an important target of PDT. Actually, it was observed very early that photodynamic treatment had marked effects on capillaries in normal tissues. For example, Hausmann (1911) and Levy (1929) found that illumination of the ears of mice sensitized with hematoporphyrin rapidly resulted in edema and circulatory stasis. Similar results with hematoporphyrin were found in studies on circulation in the frog tongue and the mesentery of the rat (Castellani et al., 1963) as well as with the mesentery of rats injected intravenously with uroporphyrin (Allison et al., 1966). Illumination of porphyrin-sensitized animals also results in an increased permeability of the capillary walls (Bonnett et al.,

1984; Straight et al., 1985). Also, a selective photosensitized destruction of the endothelial cells and basal lamina of the skin microvasculature is observed in a griseofulvin induced protoporphyric mouse model (Konrad et al., 1975); the surrounding skin is not directly damaged by protoporphyrin photosensitization.

The information presented above suggests that more detailed studies should be made of the role of the neovascular and neointerstitial components of tumors in the localization, retention and photodynamic effect of porphyrins and other photosensitizers and the pattern of tumor damage resulting from PDT as well as the damage to the normal skin microvasculature observed as a phototoxic side effect of PDT. We have examined the use of implanted polyvinyl alcohol sponges in mice as a model for studying the role of these host derived compartments of the extracellular matrix of tumors in photosensitizer localization, retention and photodynamic action independent of the malignant cells that are present in actual tumors. This paper describes the methodology used as well as some of the preliminary results obtained.

Sterile polyvinyl alcohol sponge (PVAS) implants are known to initiate a series of destructive and reconstructive tissue processes in the host similar to the initial phases of solid tumor growth or the initial stages of wound repair (Grindlay and Waugh, 1951; Boucek and Noble, 1955; Edwards et al., 1957; Davidson et al., 1985). At first, there is a period of destructive processes followed by repair and reorganization processes. The "extracellular" space of the sponge rapidly fills with plasma components (fluid, plasma colloids, crosslinked fibrin, fibronectin); inflammatory cells; newly formed interstitial stroma (collagen); and cells and components of newly formed microvasculature, in an orderly, sequential fashion (Edwards et al., 1957). These components of the extracellular matrix produced by the host in response to the inert, sterile sponge are similar to those produced in response to solid tumor growth and are potentially important targets for PDT.

MATERIALS AND METHODS

Polyvinyl Alcohol Sponge Implants in Swiss-Webster Mice

IVALON (polyvinyl alcohol-formal) sponge blocks measuring 6 in x 4 in x 3 in were obtained from Unipoint Industries, 416 South Elm Street, High Point, North Carolina 27260. The biologically inert sponge consists of polymerized polyvinyl alcohol crosslinked with formaldehyde. The sponge blocks were washed in running water for 4-6 h with frequent squeezing out of the excess water in order to remove the formaldehyde and other soluble materials. The washed sponge blocks were squeezed dry and frozen (-80°C) and cut into 0.5 cm sheets on an autopsy saw. When these sheets are washed and dried they yield discs about 0.2 cm thick. The cut slabs were dried with a hair dryer and, by using a cork borer, implant discs measuring 0.8 cm in diameter were cut from the sheets, washed, dried, grouped by weight within ± 0.5 mg and autoclaved or gas sterilized. The method of PVAS preparation was adapted from that used by Dr. Stephen Woodward (1985). Whole or half discs were implanted subcutaneously on the right and/or left flanks of albino Swiss-Webster mice. The mice (25 ± 1 g) were anesthetized with sodium pentobarbital (~ 30-50 µg/g). A small slit was cut in the skin about 3 cm distal to the site of implant and the sterile sponge was inserted subcutaneously in the fascia between the skin and body wall with forceps. Organization of repair and granulation tissue into the interstices of the sterile sponge has been well documented by histology (Edwards et al., 1957; Bole and Robinson,

1962; Herrmann and Woodward, 1966 and Davidson et al., 1985). The sponge discs were readily dissected from the subcutaneous implant site leaving any pericapsular tissue intact or the capsule could be removed. The discs were cut in half perpendicularly, exposing a vertical surface for histological section. Half of the sponge was fixed in phosphate buffered formalin and processed (hematoxylin–eosin stain) for light microscopy. The other half was frozen and processed for photosensitizer and dye analysis or for frozen sections and fluorescence microscopy.

Photosensitizers and Colloidal Dyes

Hematoporphyrin derivative (HPD) was obtained from Photofrin Medical Inc. (Photofrin I), Cheektowaga, New York as a frozen solution (5 mg/cc, 30 cc per bottle). A high molecular fraction (DHE) of HPD was obtained as Photofrin II (Photofrin Medical), a frozen solution (2.5 mg/cc, 30 cc per bottle) or obtained by sephadex LH20 column chromatography (DHE/LH20) eluted with tetrahydrofuran (THF), methanol and water (2:1:1) by volume using as starting material HPD obtained from Porphyrin Products, Inc., P.O. Box 31, Logan, Utah. HPD and DHE dissolved in base (0.1 N NaOH and neutralized with 0.1 N HCl were characterized by high performance liquid chromatography (HPLC) as to their relative percentages of monomers (hematoporphyrin, HP; isomeric hydroxyethylvinyldeuteroporphyrin, HVP; and protoporphyrin, PP) and a high molecular weight fraction (HMF) as shown in Table 1.

HPLC was carried out using a silanized (trimethylchlorosilane) column (0.5 cm x 30 cm) containing reverse phase, C1 packing (hand packed, μ bondapak) and a Waters 6000A and M45 solvent delivery system and rheodyne injector valve. A Schoeffel Model SF770 absorbance detector operating at 400 nm and a Shoeffel Model 970 fluorescence detector exciting at 400 nm and detecting fluorescence at wavelengths > 550 nm were used. Data was recorded using a Waters Data Module. All HPLC solvents and buffers were degassed and filtered through a 0.45 μm Millipore filter before use. The starting solvent system was 70% methanol and 30% buffer, 0.05 M acetic acid adjusted to pH 4.0 with triethylamine, TEA. The system was run isocratically for 10 min then changed to the solvent system THF: methanol (MEOH); water (2:1:1 by volume) and run isocratically. Protoporphyrin content, as percent of total porphyrin, was determined separately on a Waters C18 reverse phase column.

Table 1. Relative Amounts (%) of Porphyrins in HPD and DHE Preparations by Analytical Reverse Phase HPLC

Preparation	HP[a]	HVP[b]	PP[c]	HMF[d]
[e]HPD (Porphyrin Products)	28%	38%	9%	24%
[f]HPD (Photofrin I)	22%	30%	12%	36%
[e]DHE (LH20 Column)	4%	8%	6%	82%
[f]DHE (Photofrin II)	6%	2%	8%	84%

[a]HP = hematoporphyrin. [b]HVP = hydroxyethylvinyl deuteroporphyrin (isomeric). [c]PP = protoporphyrin. [d]HMF = high molecular weight fraction. [e]The HPD from Porphyrin Products was the starting material for the preparation of DHE using a Sephadex LH20 column. [f]The Photofrin I and Photofrin II were used as obtained from Photofrin Medical, Inc.

The porphyrin constitutents, as determined by HPLC, reverse phase analyses of the HPD and DHE preparation used are shown in Table 1. Each preparation is a mixture mainly of four porphyrin fractions present in different relative amounts.

Tetraphenylporphine sulfonate (TPPS) was obtained from Midcentury, P. O. Box 217, Posen, Illinois and from Porphyrin Products, Inc. Solutions (5 mg/cc x 30 cc) were prepared in sterile saline and stored frozen. Reverse phase HPLC (C18, Waters μ bondapak eluted with 10%–100% methanol: water adjusted to pH 4 with acetic acid) analysis of freshly prepared solutions showed that they contained about 58% $TPPS_4$ and 42% ($TPPS_3$, $TPPS_2$, TPPS). The material was used as obtained from the suppliers.

Zinc phthalocyaninetetrasulfonate (ZPS) was obtained as a gift from Dr. Dean F. Martin, Chemical and Environmental Management Services Center, Department of Chemistry, University of South Florida, Tampa, Florida. The material is reported to contain 56% ZPS_4. Impurities included inorganic salts (31%, 2:1 $NaCl/Na_2SO_4$); di, tri and mono sulfonates; and 5% moisture (Barltrop et al., 1983). A solution (5 mg/cc) was prepared in sterile saline and used without purification. The solution had a strong absorption peak at 715 nm and fluoresced (excitation, 632 nm) strongly at 760 nm.

Evans blue (C.I. 23860), 80% dye content, was obtained from Sigma Chemical Co. Evans blue (EB) is a tetrasulfonate, acidic disazo dye, classified as a direct dye because of its affinity for cellulose fibers. Such dyes are, by definition, anionic dyes, mostly of sulfonated disazo, trisazo or polyazo types. Some anionic monoazos, stilbenes, oxazins, thiazols and phthalocyanines also act as direct dyes, binding strongly to hydroxyl containing fibers such as cellulose and collagen. Evans blue solutions were prepared in sterile saline and used without further treatment. This dye has been used extensively as a vital stain for tumors (Duran-Reynals, 1939) and in general as a "colloidal" dye for detecting disruption of the microvascular system in live animals.

Fluorometric and Spectrophotometric Assays for Dyes and Photosensitizers

Polyvinyl alcohol sponge and tumor or tissue samples were harvested from mice; they were frozen, lyophilized and then extracted to recover the porphyrin, phthalocyanine, or EB for fluorometric or spectrophotometric quantitation per gram dry weight of sample. In each case, samples from five mice were pooled and assays done on the pool. For porphyrin analyses, purified water was added to lyophilized tissue samples to give 50 mg tissue per ml water and the sample was homogenized using a Brinkman polytron. An aliquot (200 μl) of homogenate was added to a microfuge tube containing 500 μl of 10% TCA. The sample was mixed and centrifuged to pellet the precipitate. It was important not to centrifuge too long as the firm pellet was difficult to resuspend. The supernatant was examined (B-100A Blak-Ray Lamp, 395 nm, 7000 $\mu W/cm^2$; Ultra-Violet Products, Inc.) for porphyrin fluorescence and then discarded as it was always free of fluorescent porphyrin. The pellet was resuspended in 200 μl water and vortexed until homogeneously dispersed. The pH of the resuspension was about 2. The suspension was extracted with reagent grade (J. T. Baker) methyl ethyl ketone (MEK) by adding 250 μl MEK and vortexing the tube. The MEK (top layer) was removed and transferred to a glass test tube. The suspension was extracted again with MEK (250 μl) and the second MEK layer pooled with the first. The extracted tissue pellet was saved and further extracted as described below to recover bound porphyrin. The aqueous phase was discarded as it contained non-porphyrin, interfering, fluorescent

compounds. Note that a different procedure, described below, was used
for the more water soluble TPPS, ZPS and EB. The MEK extracts were then
acidified with aqueous HCl (15%, 1.0 ml). After mixing, the aqueous
layer containing the acidified, extracted porphyrins, settled to the
bottom of the tube. This layer was collected and analyzed
fluorometrically using an Aminco-Bowman spectrophotofluorometer. For
HPD/DHE (two-carboxy porphyrins), the excitation wavelength was 412 nm
and the emission wavelength was 610 nm. The assay was standardized with
standard solutions of HPD or DHE. An initial reading was taken and then
1 μg (10 μl) of standard (HPD or DHE) was added and a second reading
was made to determine if fluorescence quenching was occurring. The
sample reading was corrected accordingly. The MEK extracted tissue
pellet was then solubilized with 0.5 ml hyamine hydroxide, 10-X (Packard
Instrument Company). Methanol (1.0 ml) was added to the hyamine solution
and the solution was measured for porphyrin fluorescence and quenching
(excitation, 402 nm, emission, 628 nm) as described. The corrected
values for the MEK/HCl extract and for the hyamine solution were added to
give the total porphyrin content per gram dry weight of the sample.

    Tumor tissue and polyvinyl alcohol sponge samples containing water
soluble TPPS, ZPS and EB were not extracted into MEK, but were
solubilized in either hyamine or 88% formic acid (reagent grade,
Mallinckrodt) and immediately assayed fluorometrically (TPPS) or
spectrophotometrically (ZPS, EB). These more highly charged compounds
did not bind to the sponge matrix in vivo but they did bind in vitro. EB
was bound to PVAS matrix in vitro most strongly. Polyvinyl alcohol
sponge was complete solubilized to an optically clear solution by 88%
formic acid. Neutralization of the acid to pH 3-4 resulted in
reformation of solid sponge.

Method Used for Measuring Dye and Photosensitizer Retention in Implanted PVAS

    The PVAS were removed at different times (6 h – 15 days) after
implantation for histologic examination and for analysis of
photosensitizer and/or dye content. The photosensitizer or dye was
injected I.P. with a dose of 40 mg/kg 24 hrs before the PVAS were
harvested except in the case of PVAS harvested at 6 h and 12 h where
photosensitizer or dye was injected I.P. immediately after implant. Care
must be taken when harvesting the PVAS not to squeeze out the fluid and
tissue components; when this occurs, substantial errors are introduced
for analyses expressed on a dry weight basis. After 4-6 days, a
connective tissue capsule forms and becomes continuous with the fibrous
connective tissue of the sponge interspaces. All analyses were done with
the capsule intact as part of the sponge tissue. Each data point was
from a pool of 5 individual mouse samples. A sample of subcutaneous
connective tissue was taken from the side of the mouse opposite the PVAS
implant to provide control values for the retention of photosensitizer
and dye in normal tissue. In another series of experiments, PVAS were
implanted as described and after 3, 6 and 12 days, the mice were injected
with different doses (5-50 mg/kg) of photosensitizer or dye 24 h prior to
sponge removal. The sponges were analyzed as described previously.

Methods for Examining Photodynamic Effects on PVAS In Vivo

    Mice were implanted with PVAS as described, but for these experiments
two sponges were implanted (bilaterally) in each mouse, one sponge
serving as a non-irradiated control. Three days, 6 days or 12 days after
implantation, the mice were injected I.P. with photosensitizer at a dose
of 40 mg/kg. Twenty-four h later, the mice were anesthetized and a
midline incision was made from pelvis to sternum in order to expose the

encapsulated sponge attached to the skin (occasionally the sponge would
be attached to the body cavity). Care was taken to avoid disrupting the
rich neovascularization that always surrounded the encapsulated sponge.
One (right side) of the exposed sponges was irradiated with an argon-dye
laser (Aurora-Lexel, Cooper Laser Sonics) at 632 nm using an intensity of
50 mW/cm^2 intensity. The exposure times (1-30 min) determined the
light dose (3-90 J/cm^2). In order to determine the effect of
irradiation on the normally leaky PVAS vasculature, the mice were
injected (0.1 ml) with EB (5 mg/ml solution in sterile saline) in the
right exposed iliac vein (26 gauge needle) with the aid of a surgical
microscope (10X). The irradiated and nonirradiated PVAS were observed
for 30 min after dye injection and then were collected and bisected with
sharp scissors. One-half of each PVAS was fixed in phosphate buffered
formalin for histology and one-half was frozen for dye and
photosensitizer analyses. Five irradiated and nonirradiated PVAS from
five individual mice were pooled for each data point.

RESULTS AND DISCUSSION

Host Response of Mice to Implanted PVAS

     In general, the host response to implanted PVAS in mice followed the
same pattern as that observed in rats by others (Davidson et al., 1985;
Boucek and Noble, 1955). The implanted PVAS rapidly fills with fluid and
swells to about 1 cm x 0.3 cm. The dry weight of photosensitizer or dye
retained in the PVAS appears to be a function of the dry weight of fluid
colloids and tissue deposited in the PVAS at any interval after
implantation. Since the dry weight of the newly deposited tissue is a
function of the dry weight of the implanted PVAS, care must be taken to
use PVAS of uniform size and weight for quantitative studies. Single
PVAS of the size used contained enough material for most analyses but the
reproducibility of the data was improved by analyzing a pool of five PVAS
harvested from five individual mice.

     Gross observation and routine histology (hematoxylin-eosin staining)
of PVAS harvested at 6 h to 12 days showed a characteristic host
response. The PVAS implant was filled with plasma, cellular and
extracellular matrix components generated by the host in a sequential
manner as described by Edwards et al., 1957. At 6 h, the periphery of
the PVAS was swollen with fluid; at 2 days, the entire PVAS was filled;
and at 4-6 days a fibrous, connective tissue capsule began to form around
the PVAS. Also, the number and size of host blood vessels supplying the
PVAS and connective tissue capsule were remarkably increased and could be
readily observed grossly by comparing the blood vessels in the
subcutaneous tissue on the side of the mouse containing the PVAS with the
opposite side without PVAS. The neovascularization was remarkable and
was similar to that seen with transplanted mouse tumors (e.g., S180 tumor
at 4-6 days).

     Microscopic examination showed a moderate inflammatory response in
the surrounding host connective tissue. After 24-48 h, the interstices
of the PVAS were filled with fluid and an amorphous, semisolid substance
containing stringy (probably crosslinked fibrin-fibronectin) and granular
material and numerous polymorphonuclear neutrophils. Inflammation and
changes in the size and number of blood vessels was a consistent and
prominent early finding in the connective tissue surrounding the PVAS.
The host response was rapid and by 2-4 days, numerous fibroblasts
appeared scattered throughout the space with a higher density in the
periphery of the PVAS; inflammatory cells disappeared. By 4-6 days new
collagenic fibers appeared and numerous microcapillaries were seen in the

new connective tissue in the PVAS. By 12 days, much of the material deposited earlier was replaced by loosely organized connective tissue and collagen fibers. At 12-15 days, the increase in new microvasculature appeared to have reached a plateau.

It should be pointed out that the tissue infiltrated sponges lend themselves to a number of analytical procedures for quantitatively determining the constituents (fibrin, fibronectin, collagen, elastin, fatty acids, phospholipid, cholesterol, DNA, RNA, hexosamine, amino acids, etc.) of the newly formed granulation tissue (Boucek and Noble, 1955; Bole and Robinson, 1962) as well as added materials such as photosensitizers, dyes and drugs and, perhaps, the products of photodynamic actions.

Preliminary Observations on the Localization and Retention of HPD, DHE, TPPS, ZPS, and EB in Newly Formed Granulation Tissue in Implanted Polyvinyl Alcohol Sponges

HPD, DHE, TPPS, ZPS and EB localization and retention in newly formed granulation tissue as a function of time after implantation of PVAS. In this experiment, PVAS were harvested from host mice at different time intervals after implantation as shown in Table 2. In the mice injected with HPD, DHE or TPPS, observation using a Blak-Ray (395 nm) light at the time of harvest showed a remarkable retention of brightly fluorescent porphyrin in the PVAS as compared to surrounding tissue. PVAS in mice injected with ZPS or EB were selectively stained dark blue (EB) or blue-green (ZPS). A similar selective fluorescence and staining was also seen with S180 tumors in mice coloring the connective tissue capsule of the tumor blue. Fluorescence microscopy showed that TPPS was evenly distributed throughout newly formed PVAS-tissue (6 h - 3 day implants); although more brightly fluorescent fibers could be seen in a less fluorescent background. Light microscopy showed that ZPS and EB were similarly distributed. In sharp contrast, HPD and DHE showed a spotty distribution of fluorescence similar to the pattern found in S180 tumors in mice. The fluorescence apeared to be distributed along fibers in the longer interval (6-12 days) PVAS where histological examination indicated that collagenic fibers were deposited. In shorter interval (4-6 days) PVAS the HPD spotty fluorescence was also seen around what appeared to be remnants of capillaries.

The quantitative data shown in Table 2 correlates with the direct observation of the PVAS and shows that the amount of photosensitizer or dye retained in PVAS-tissue samples 24 h after injection varied depending on the time interval after PVAS implantation. Both the photosensitizers and the dye were more highly concentrated in the short interval (6-12 h) PVAS and again in the 2-6 day PVAS. Thus, the amount of material retained in µg per gram dry weight of newly formed tissue in the PVAS showed two peaks with the first occurring about 12 h after implant when the PVAS was filling with plasma fluid, colloids (fibrin, fibronectin) and inflammatory cells; the second peak occurred at about 6 days when new vasculature was being formed most rapidly and collagen fibers were being deposited. The amount of photosensitizer and dye in PVAS tissue begins to decline in 6-8 day sponges and is progressively lower in 12 day and 15 day sponges. The levels found in transplanted, S180 mouse tumors also vary with time, but the patterns have not been studied systematically.

Preliminary HPLC analysis of porphyrins extracted from PVAS (3 day implants) that were harvested from mice injected (I.P.) 24 h previously with HPD at 40 mg/kg dose showed the following distribution: HP - 48%; HVP - 16%; PP - 36%; HMF - <1%. This distribution of porphyrins is different from that found in the HPD that was injected (Table 1).

Table 2. Amount of HPD, DHE, TPPS, ZPS, and EB Retained by Newly
Formed Granulation Tissue in Implanted PVAS as a Function
of Time After Implantation.[a]

Time Interval	[b] Normal Tissue HPD	[c] HPD	DHE	DHE (LH20)	TPPS	ZPS	EB
6 h	10	86	68	60	121	58	99
12 h	8	88	60	56	109	89	107
1 day	6	42	78	63	91	60	93
2 days	4	87	106	99	117	77	130
3 days	5	94	129	121	132	98	141
6 days	3	110	125	128	140	114	156
12 days	<1	75	82	80	111	31	121
15 days	<1	30	35	36	98	14	89

[a]PVAS of uniform size (0.8 cm x 0.2 cm) and weight (9.5 ± 0.5
mg) were implanted subcutaneously in Swiss-Webster mice (25 ± 1
g; five mice per group). Photosensitizers and dye were injected
I.P. at 40 mg/kg, 24 h prior to PVAS harvest except for the 6 h
and 12 h harvest where they were injected immediately after
implant. Five pooled sponges were assayed for each value
reported as μg/g dry weight of tissue. [b]Normal mouse
subcutaneous tissue from the side opposite the PVAS implant.
Values are μg per gram dry weight of tissue and represent the
values for mice injected with HPD. Levels of the other compounds
(DHE, TPPS, ZPS and EB) in normal tissue were similarly low.
Similar low values were found for the other photosensitizer and
Evans blue. [c]HPD, DHE, DHE (LH20), TPPS, ZPS and EB are
Photofrin I, Photofrin II, high molecular fraction of HPD from
LH20 column, tetraphenylporphine- sulfonate, zinc phthalocyanine
sulfonate and Evans blue respectively. The values are μg per g
dry weight of newly formed granulation tissue.

HPD, DHE, TPPS, ZPS, and EB retention in newly formed granulation
tissue in 3 day, 6 day and 12 day PVAS implants as a function of the dose
injected. The amount of dye and photosensitizer retained in newly formed
PVAS granulation tissue increased in proportion to injected dose for each
interval as shown in Table 3. The 6 day PVAS implants contained dye or
photosensitizer in amounts directly proportional to the injected dose
from 5-50 mg/kg. In contrast, the 3 day and 12 day PVAS implants show a
dose saturation phenomenon suggesting that the mechanism(s) and
component(s) responsible for retention of these compounds vary as a
function of host response in stages of tissue repair and reconstruction
similar to the responses of the host to solid tumor growth. Further
studies with this model may be used to determine the identity of the
newly formed tissue components with the highest affinity for negatively
charged "colloidal" dyes and photosensitizers.

Photodynamic Studies with Implanted PVAS

The effect of illumination at different light doses (3-90 J/cm^2)
(632 nm dye laser, intensity 50 mW/cm) on capillary blood flow in the
microvasculature of PVAS implanted for 6 days was examined. For these
experiments, the mice were injected with 25 mg/kg of HPD 24 h prior to
illumination. In situ visual observations were made with a 10-25X
surgical microscope using anesthetized mice with two implanted sponges as

Table 3. Amounts HPD, DHE, TPPS, ZPS and EB Retained by Newly Formed Granulation Tissue in Three Day, Six Day and 12 Day Implanted PVAS as a Function of Dose Injected

[a]Dose Injected	[b]HPD	TPPS	ZPS	EB
Three Day PVAS				
5	60	75	32	80
10	77	90	44	102
25	82	115	81	120
50	100	130	104	145
Six Day PVAS				
5	18	22	15	31
10	40	56	30	66
25	68	92	60	140
50	130	189	120	198
12 Day PVAS				
5	20	29	18	28
10	36	42	21	50
25	71	81	39	101
50	82	116	51	142

[a]Dose in mg/kg of photosensitizer or dye injected I.P., 24 h prior to PVAS harvest. [b]The amount of photosensitizer and dye retained in PVAS tissue in μg per gram dry weight of newly formed tissue.

described in the Materials and Methods section. Immediately following irradiation with a dose of greater than 45 J/cm^2, there was decrease in blood flow rate and a clumping and stasis of the red blood cells; in some cases, reverse flow and a blanching of the blood vessels occurred, especially at higher light doses (90 J/cm^2). With higher light intensities (> 100 mW/cm^2), these latter effects occurred at even lower total light doses. Light doses above 45 J/cm^2 tended to cause the fluid in the PVAS to gel; this might result from a photodynamic-induced crosslinking of proteins. Histologic examination of hematoxylin-eosin stained sections of illuminated PVAS showed that light doses of even less than 30 J/cm^2 increased the clumping of red blood cells in the capillaries. Illumination of blood vessels in PVAS in the absence of photosensitizer had no effect.

The amount of EB leaking into 6 day PVAS implants in irradiated and non-irradiated control PVAS is shown in Table 4. The data show that for similar PVAS-tissue levels of HPD, the amount of injected EB able to leak into and be retained in irradiated 6 day PVAS decreases with light dose compared to the unirradiated control PVAS that contains about the same level of EB after each dark period. The slightly lower value after the longer (30 min) dark period may be due to general stress on the animal affecting blood flow. Irradiation of PVAS in the absence of HPD had no apparent effect on EB levels.

EB has been used extensively in nonphotodynamic animal experiments as a tool for assessing the integrity of the animal's vascular system. We have used EB to measure the photodynamic (sensitized by HPD, DHE, TPPS, ZPS) disruption of normal capillaries in mouse brain, ear and tail as a

Table 4. Effect of Photodynamic Action on the Vascular System
of PVAS Tissue Containing HPD[a]

[b] Light Dose (J/cm^2)	[c] HPD $(\mu g/g)$	[d] EB-T $(\mu g/g)$	[e] EB-C $(\mu g/g)$
3	58	73	88
30	64	40	76
60	56	32	80
90	48	18	71
45	0	82	—

[a]The photosensitizer, HPD (Photofrin I) was injected (I.P.) at
25 mg/kg in Swiss- Webster mice 24 h prior to irradiation of PVAS
(6 day implant) which were surgically exposed in situ in
anesthetized mice. [b]Cooper Laser Sonics, Aurora Lexel
argon-dye laser operated at 632 nm, 50 mW/cm^2, was used as the
light source. The light doses were obtained by exposing the PVAS
tissue (0.8-1.2 cm spot) for 1 min - 30 min using a 400 μ
straight cut fiber about 4 cm from the surface of the sponge.
[c]HPD porphyrins were assayed fluorimetrically in the irradiated
sponges and the non-irradiated controls. The values are μg
porphyrin per g dry weight of PVAS tissue. The data shown are
from irradiated sponges (pool from five mice). At the higher
light doses destruction of the porphyrin may be responsible for
the lower measured levels of fluorescent porphyrin. This was not
seen in non-irradiated controls. The zero value at 45 J/cm^2 is
the case where no HPD was injected. [d]Evans blue was assayed
spectrophotometrically in irradiated (EB-T) and non-irradiated
(EB-C) controls. The Evans blue (20 mg/kg) was injected I.V.
(exposed right iliac vein) of the anesthetized mouse immediately
after irradiation. The sponge tissue was harvested 30 min later
and assayed for EB levels ($\mu g/g$) as a measure of blood flow
blockage in irradiated compared to non-irradiated PVAS.

measure of the phototoxicity of different photosensitizers (Straight et
al., 1985). EB can also be used as shown in this experiment with
implanted PVAS to demonstrate the occlusion of leaky capillaries.

These results using implanted PVAS are consistent with earlier
studies on the effects of porphyrin sensitized photodynamic action on
normal capillaries (Allison et al., 1966; Levy, 1929; Hausman, 1911;
Castellani et al., 1963) and on tumor blood flow (Henderson et al., 1984;
Selman et al., 1984; Star et al., 1984).

CONCLUSION

These preliminary studies indicate that implanted polyvinyl alcohol
sponges are a useful in vivo model for studying the effect of PVAS
granulation tissue (neovascular and interstitial compartments of loose
connective tissue) on the localization, retention and photodynamic action
of negatively charged HPD, DHE, TPPS, ZPS and EB in the absence of the
malignant cells present in solid tumors. The PVAS provides a tool for
quantitatively determining the levels of photosensitizers retained in
granulation tissue at different stages of formation. The neointerstitial
and neovascular components of granulation tissue are similar to those

found in the extracellular matrix of solid tumors. All of the compounds studied are also efficiently retained in transplanted mouse tumors suggesting that similar mechanisms may be involved in localization and retention of these compounds in both solid tumor transplants and PVAS implants.

The observation that irradiation of implanted PVAS containing HPD interfered with the blood flow to the PVAS compared to non-irradiated controls suggests that the PVAS model may be useful for studying the effect of PDT on the neovascular compartment of tumors. Also, the gelation of PVAS fluid following irradiation suggest that there are other targets in the interstitial compartment that may be important in PDT.

ACKNOWLEDGMENTS

This work was supported in part by the Veterans Administration Medical Research Program, the University of Utah Laser Institute, Cooper Laser Sonics, Inc. and Grant # PDT-259 from the American Cancer Society. The authors appreciate the collaboration of John A. Dixon and the excellent technical assistance of Arthur Claerhout, Martha Wolfe and Tony Wayne. We also appreciate the assistance of Dr. Davidson, Dr. Woodward and Mr. Kamerath with the polyvinyl alcohol sponges.

REFERENCES

Allison, A. C., I. A. Magnus and M. R. Young (1966) Nature 209, 874-878.

Barltrop, J., B. B. Martin and D. F. Martin (1983) Microbios 37, 95-103.

Bole, G. G. and W. D. Robinson (1962) J. Lab. Clin. Med. 59, 713-729.

Bonnett, R., M. C. Berenbaum and H. Kaur (1984) In Porphyrins in Tumor Phototherapy (Edited by A. Andreoni and R. Cubeddu), pp 67-80, Plenum Press, New York.

Boucek, R. J. and N. L. Noble (1955) Arch. Path. 59, 553-558.

Bugelski, P. J., C. W. Porter and T. J. Dougherty (1981) Cancer Res. 41, 4606-4612.

Castellani, A., G. P. Pace and M. Concioli (1963) J. Path. Bacteriol. 86, 99-102.

Davidson, J. M., M. Klagsbrun, K. E. Hill, A. Buckley, R. Sullivan, P. S. Brewer and S. C. Woodward (1985) J. Cell Biol. 100, 1219-1227.

Denekamp J. (1984) Prog. Appl. Microcirc. 4, 23-38.

Dougherty, T.J. (1984) In Porphyrin Localization and Treatment of Tumors (Edited by D. P. Doiron and C. J. Gomer), pp 75-87. Alan R. Liss, Inc., New York.

Duran-Reynals, F. (1939) Am. J. Cancer 35, 98-107.

Edwards, L. C., L. N. Pernokas and J. E. Dunphy (1957) Surg. Gynec. Obst. 105, 303-309.

Evensen, J. F., J. Moan, A. Hindar and S. Sommer (1984) Porphyrin Localizational and Treatment of Tumors (Edited by D. P. Doiron and C. J. Gomer), pp 541-562. Alan R. Liss, In., New York.

Goldacre, R. J. and B. Sylvén (1962) Br. J. Cancer 16, 306-322.

Grindlay, J. H. and J. M. Waugh (1951) Arch. Surg. 45, 288-297.

Hausmann, W. (1911) Biochem. Z. 30, 276-316.

Henderson, B. W., S. M. Waldow, T. S. Mang, W. R. Potter, P. B. Malone and T. J. Dougherty (1985) Cancer Res. 45, 572-576.

Henderson, B. W., T. J. Dougherty and P. B. Malone (1984) In Porphyrin Localization and Treatment of Tumors (Edited by D. P. Doiron and C. J. Gomer), pp 601-612, Alan R. Liss, Inc., New York.

Herrmann, J. B. and S. C. Woodward (1966) Surgery 59, 559-565.

Kessel, D. (1984) Biochem. Pharmacol. 9,1389-1393.

Konrad, K., H. Hönigsmann, F. Gschnait and K. Wolff (1975) J. Invest. Derm. 65, 300-310.

Levy, A. G. (1929) J. Path. Bact. 32, 387-399.

Ludford, R. J. (1929) Proc. Royal Soc. London Ser. B 104, 493-511.

Musser, D. A. and N. Datta-Gupta (1984) J. Natl. Cancer Inst. 72, 427-433.

Musser, D. A., J. M. Wagner and N. Datta-Gupta (1982) Res. Commun. Chem. Pathol. Pharmacol. 36, 251-259.

Selman, S. H., M. Kreimer-Birnbaum, J. E. Klaunig, P. J. Goldblatt, R. W. Keck and S. L. Britton (1984) Cancer Res. 44, 1924-1927.

Spikes, J. D. and R. C. Straight (1985a) Med. Biol. Environ., in press.

Spikes, J. D. and R. C. Straight (1985b) In Photodynamic Therapy of Tumors and Other Diseases (Edited by G. Jori and C. Perria), Libreria Prodetto, Padova, Italy. In press.

Star, W. M., J. P. A. Marijnissen, A. E. Van Den Berg-Blok and H. S. Reinhold (1984) In Porphyrin Localization and Treatment of Tumors (Edited by D. R. Doiron and C. J. Gomer), pp 637-645, Alan R. Liss, Inc., New York.

Straight, R. C. and J. D. Spikes (1985) Unpublished results.

Straight, R. C., J. A. Dixon and J. D. Spikes (1985) Lasers in Surg. Med. 5, 139.

Straight, R. C., J. A. Dixon and J. D. Spikes (1984) Photochem. Photobiol. 39, 67S.

Tsutsui, M., C. Carrano and E. A. Tsutsui (1975) Ann. N.Y. Acad. Sci. 244, 674-684.

Waner, M. (1985) Univ. of Cincinnati Med. Center (personal communication).

Weil, R. (1916) Cancer Res. 1, 95-106.

Woodward, S. C. (1985) Personal communication.

QUANTATIVE STUDIES OF TETRAPHENYLPORPHINESULFONATE AND HEMATO—

PORPHYRIN DERIVATIVE DISTRIBUTION IN ANIMAL TUMOR SYSTEMS

James W. Winkelman

Department of Pathology
State University of New York-Upstate Medical Center
750 East Adams Street
Syracuse, New York 13210

This review will accumulate and compare the very few reports in the literature which quantify the concentration of tetraphenylporphinesulfonate (TPPS) and hematoporphyrin derivative (HPD) in tumors and other tissues after parenteral administration to tumor bearing animals.

The actual data upon which is based the notion that porphyrins selectively or preferentially accumulate, localize or concentrate in tumors is extremely skimpy. Yet upon it an enormous and rapidly growing body of work aims to exploit that property. Reports in great abundance have appeared that study the organic and photochemistry of porphyrins, their solution and aggregation properties, interactions with serum constituents and other biochemicals of interest, behavior and effects with cultured normal, transformed, and malignant cell lines of human and animal origin, and utility in photodynamic therapy (PDT) of tumors in animals and man.

Prior to 1960 none of the references to accumulation of porphyrins in tumor included any quantitative data to support the contention that more porphyrin was actually present in tumors than in other tissues.[1,2,3,4,5,6,7,8,9] In those reports, the intensity of visible red fluorescence characteristic of porphyrins was taken as indicative of the presence of porphyrins presumably in proportion to their concentration. It has subsequently been proven[10,11,12] that visible porphyrin fluorescence is very dependent upon the background color of the tissue and actually bears no reliable relationship to actual concentration.

Winkelman and Rasmussen-Taxdal[13] were the first to prove by quantitative chemical measurements that a porphyrin, hematoporphyrin (HP), actually attained higher concentration in an experimental animal tumor than in the surrounding muscle. Twenty-four hours after injection, a 40 mg dose (160 mg/kg) produced levels of 6-10 µg HP/gm tumor. Twenty mg (80 mg/kg) doses produced 1.4-3 µg HP/gm tumor. The tumor was the Walker 256 carcinosarcoma, grown in the anterior chamber of the eye or subcutaneously for 11 days.

In 1962 Winkelman published the first studies that employed TPPS in the same animal-tumor system[14] that had been employed in the prior studies with HP. A summary of the quantities of TPPS and HPD that localized in the tumor employed by Winkelman in that and his previous study is shown in Table I. To this day, these data provide virtually the only direct comparison of

Table I. Comparative Tumor Tissue Concentrations of HP vs TPPS

	Dose[a]		Tumor tissue level
	mg	mg/kg	µg/gm wet weight
HP[13]	20	80	2.5
	40	160	8.4
	80	320	7.5
TPPS[14]	1.0	4	12
	3.3	13.2	35
	10	40	150
	33	132	370
	75	300	355

[a]Following i.p. injection into 250 gm Sprague-Dawley rats bearing 11-day old transplanted Walker 256 carcinosarcoma.

tumor localization by TPPS with HP. It shows an approximately 50 fold superiority of TPPS as a tumor localizer. In that study TPPS attained a higher concentration in tumor than in any other tissue, including liver, spleen, lung and kidney.

Over the ensuing years, sporadic reports have essentially substantiated the finding that TPPS is a better tumor localizing porphyrin than any others studied. Studies showing the positive and favorable relative tumor tissue uptake of TPPS comparable to other porphyrins were summarized by Tsutsui[15].

Carrano et al.[16] also studied TPPS at doses of 100, 200 and 250 mg/kg, administer i.p. or i.v. to BALB/c Tex mice with subcutaneous rhabdomyosakrcomas. Tumor to other tissue concentrations measured 48 hours later showed ratios greater than 1 for all tissues other than kidney. The relative concentrations in kidney declined as dose was lowered, but the lowest dose, 100 mg/kg was still much higher than that used by others. The authors commented that tumor to other tissue concentration ratios were highest at 96 hours, but no data for that time were presented.

Musser, et. al.[17] studied TPPS uptake in particular transplantable tumors tissue. Distribution varied as a fraction of injected dose, interval between injection and killing, animal species, and tumor. At doses of 10 and 5 mg/kg of TPPS to L1210 tumor bearing mice, the kidney and lung had higher and the liver only very slightly lower concentrations after 48 hours. However, at a dose of 1 mg/kg the tumor had 19.3 µg/gm, kidney 7.8 µg/gm, liver 0.8 µg/gm, and lung no measurable TPPS. At 48 hours following injection 10 mg/kg, i.v., sarcoma 180 had higher concentration of TPPS than any other tissue than kidney. Data were not presented to show whether injection of lower doses might also have elevated the tumor to all other tissue concentration ratios to greater than 1 for that animal-tumor system as it had for the other studied by those authors.

Zanelli et al. in 1981[18] studied numerous C^{14} and S^{35} labelled non-naturally occurring porphyrins in tumor bearing mice. Their data showed that the tumor localization was most favorable with TPPS and exceeded that of the other porphyrins tested and of Ga^{67}.

Fairchild, et al. in 1983[19] studied TPPS in seven different animal-tumor models, including mice and rats, and found abundant but variable tumor concentration in each tumor. Using doses of ~40 mg TPPS/kg, and studying the tissue concentration of tumor and nine other tissues, they found TPPS concentrations ranging from 45-184 µg/gm in the various tumors. No other tissue was as high as tumor in the cases of the Walker 256 carcinosarcoma in rat and an adenocarcinoma in the C57 mouse. Kidney concentrations were higher than tumor in 4 other animal-tumor systems; amelanotic melanoma in BALB mice, melanotic Harding-Passey melanoma in BALB mice, T-9 glioma (intracranial) in CDF rats, and EMT sarcoma in BALB mice. Concentration in kidney were reported from 81-309 µg/gm. The subcutaneous T-9 glioma in CDF rats showed lower concentration in tumor than in kidney, spleen and lung.

Determination of the distribution of S^{35}, H^3 and C^{14} radioisotopes incorporated directly into the ring structure of TPPS by radioisotopic counting technics[20], aimed at confirming the earlier distribution studies, in which quantification had been by chemical methods. S^{35} and H^3 extensively dissociated from TPPS and distributed independently. But C^{14} TPPS distributed as had cold TPPS. A similar conclusion was also reached by Gomer and Dougherty for C^{14} HPD although they found that the H^3 derivative of HPD was also stable.[12] These data are summarized in Table II, along with the appropriate, and only quantitative results for the corresponding cold or non-radioactively labelled porphyrins.

The results with TPPS and C^{14} TPPS are more or less internally consistent. TPPS is generally higher in tumors than all other tissues. Results with HP, HPD, C^{14} HPD and H^3 HPD show some striking inconsistencies. Distribution studies among tumor and other tissues based on C^{14} or H^3 show that the levels obtained in tumor are low both absolutely and relative to several other major tissues. Liver, kidney and spleen showed 9-3 times greater concentration of HPD than does tumor.

In view of these data, the very terminology and syntax of the study of porphyrins in tissues needs reconsideration. It is clear that neither TPPS or HPD "selectively accumulates" or "selectively localizes" in tumor, if by that is meant, to the exclusion or virtual exclusion of other tissues. "Preferential accumulation" in tumors, compared to all or most other tissues may occur for TPPS. But studies with various animal tumor systems have already shown considerable differences among tumors and other tissues, particularly lung and kidney. HPD does not, apparently, "preferentially accumulate", if the earliest study of Dougherty et al[21] is omitted. Even the term "uptake", devoid of modifiers such as selective or preferential, may be an inappropriate term to employ. It could be thought to imply a specific active mechanism, which is as yet not known to exist, or a specific process particular to the tissue of interest, which is also not known. "Distribution" is certainly not an incorrect term as a description for the findings to date. Although it is pallid, "distribution" may be the most utilitarian and prudent term to employ until more is known about the fate of parenterally administered porphyrins to tumor bearing animals or man.

Does it make any difference at all whether porphyrins are preferentially or selectively localized in tumors? Perhaps not, with respect to its exploitation in PDT. If PDT depends or relies upon the presence of porphyrin enhancing the photo induced killing of cells, that purpose would be served by porphyrins in, on, or near the tumor cells regardless of its concentration in other tissues. If laser or other light sources delivered radiant energy to fields limited neatly to the tumor, then only the lack of field definition would introduce a negative feature to porphyrin sensitized PDT to the extent that porphyrin was in adjacent tissue. However, the potential for exploitation of actual selective accumulation of porphyrin by tumors may well extend to other diagnostic or therapeutic applications than PDT. If preferential or selective uptake of porphyrins actually did occur in tumors and if it was conferred by a biological mechanism unique to tumors

Table II. Comparison of Cold and C¹⁴ TPPS and Cold, H³ HPD and C¹⁴ HPD
Distribution in Animal Tumor Systems

UNITS	TPPSa μg/gm	C¹⁴ TPPSb dpm/gm	C¹⁴ TPPSc % inject dose/gm	HPd μg/gm	HPDe μg/gm	H³ HPDf μg H³ HPD/gm	C¹⁴ HPDg μg C¹⁴ HPD/gm
TISSUE							
Tumor	375	3,785	9.5	8.4	42	2.47	3.4
Liver	65	1,090	5.5	-	12	41.16	29.1
Kidney	328	2,450	12	-	-	21.58	8.4
Spleen	42	780	3	-	-	14.60	9.9
Lung	-	-	7.5	-	-	4.30	2.0
Muscle	11.7	510	<3	0	-	1.51	1.9
			(undetectable)				
Skin	-	-	<3	-	3-4	4.87	1.9
Brain	9.3	397	-	-	-	0.44	-
Blood	-	-	8	-	-	1.95	2.8
Control eye	-	-	-	-	-	0.61	-

a. Dose: 50.0 mg/kg, i.v.
   Animal-tumor: 250 gm Sprague-Dawley rats with Walker 256-carcinosarcoma[20]

b. Dose: 4.6 x 10⁻² μc C¹⁴ in 10 mg TPPS
   Animal-tumor: 250 gm Sprague-Dawley rats with Walker 256-carcinosarcoma[20]

c. Dose: 100 μg C¹⁴-TPPS, i.v. Results at 24 hours.
   Animal-tumor: CBA mice - CBA carcinoma NT.[18]

d. Dose: 160 mg/kg, i.p.
   Animal-tumor: 250 gm Sprague-Dawley rat with Walker 256 carcinosarcoma[13]

e. Dose: 10 mg/kg body weight, intraperitoneal
   Animal-tumor: mouse, SMTF-transplantable mammary carcinoma[21,22]

f. Dose: 20 mg/kg, i.p., 10-12 μCi/mouse. Results at 24 hours.
   Animal-tumor: Athymic "nude" mice - human retinoblastoma.[23]

g. Dose: 10 mg/kg. Measurements made 24 hrs. after injection.
   Animal-tumor: mouse-mammary sarcoma.[12]

or a mechanism that is either exaggerated or suppressed in tumors compared to non-tumor tissues, then an understanding of the phenomenon could be of potential significance far beyond the particular manifestation of porphyrin accumulation that revealed it.

REFERENCES

1. A. Policard, Etudes sur les aspects offerts par des tumeurs experimentales a la lumiere de Woods, Compt. Rend Soc. Biol., 91:1423 (1924).
2. J. Korbler, Untersuchung von Krebsgewebe im fluoreszerregenden Licht. Strahlentherapie, 41:510 (1931).
3. H. Auler, and G. Banzer, Untersuchungen uber die Rolle der Porphyrine bei Geschwulstkranken Menschen und Tieren, Ztschr. Krebsforsch., 53:65 (1942).
4. F.H.J. Figge, G.S. Weiland, and L.O.J. Manganiello, Cancer detection and therapy. Affinity of neoplastic, embryonic and traumatized regenerating tissues for porphyrins and metallo-porphyrins, Proc. Soc. Exper. Biol. & Med., 68:640 (1948).
5. K. Bingold, W. Stich, and H. Cramer, Poephyrine und Krebs; I. Untersuchungen uber den Porphyringehalt maligner Tumoren beim Menschen, Ztschr. Krebsforsch., 57:653 (1951).
6. L.O.J. Manganiello, and F.H. J. Figge, Cancer detection and therapy. II. Methods of preparation and biological effects of metallo--porphyrins, Bull. School of Med. Univ. Maryland, 36:1 (1951).
7. G.C. Peck, H.P. Mack, and F.H.J. Figge, Cancer detection and therapy. III. Affinity of lymphatic tissue for hematoporphyrin, Bull. School Med. Univ. Maryland, 38:124 (1953).
8. D.S. Rasmussen-Taxdal, G.E. Ward, and F.H.J. Figge, Fluorescence of human lymphatic and cancer tissue following high doses of intravenous hematoporphyrin, Cancer, 8:78 (1955).
9. S. Schwartz, K. Absolon, and H. Vermund, Some relationships of porphyrins, x-rays and tumors, Univ. Minnesota M. Bull., 27:5 (1955).
10. S. Schwartz, M.H. Berg, I. Bossenmaier, and H. Dinsmore, Determination of porphyrins in biological materials, Methods of Biochem. Anal., 8:221 (1960).
11. J. Winkelman, and J.E. Hayes, Jr., Distribution of endogenous and parenterally administered porphyrin in viable and necrotic portions of a transplantable tumour, Nature, 200(4909):903 (1963).
12. C.J. Gomer, and T.J. Dougherty, Determination of [^{3}H]- and [^{14}C] Hematoporphyrin derivative distribution in malignant and normal tissue, Can. Res., 39:146 (1979).
13. J. Winkelman, and D.S. Rasmussen-Taxdal, Quantitative determination of porphyrin uptake by tumor tissue following parenteral administration, Bull. John Hopkins Hospital, 107:228 (1960).
14. J. Winkelman, The distribution of tetraphenylporphinesulfonate in the tumor-bearing rat, Can. Res., 22:589 (1962).
15. M. Tsutsui, C. Carrano, and E.A. Tsutsui, Tumor localizers: Porphyrins and related compounds (Unusual metalloporphyrins XXIII), Ann. N.Y. Acad. Sci., 244:674 (1975).
16. C.J. Carrano, M. Tsutsui, and S. McConnell, Localization of meso-tetra (p-sulfophenyl) porphine in murine sarcoma virus-induced tumor-bearing mice, Cancer Treat. Rep., 61:1297 (1977).
17. D.A. Musser, J.M. Wagner, and N. Datta-Gupta, Distribution of tetraphenyloporphine sulfonate and tetracarboxyphenylporphine in tumor bearing mice, J. Natl. Cancer Inst., 61:1397 (1978).
18. G. D. Zanelli, and A.C. Kaelin, Synthetic porphyrins as tumor-localizing agents, Brit. J. Radiol., 54:403 (1981).

19. R.G. Fairchild, D. Gabel, M. Hillman, and K. Watts, The distribution of exogenous porphyrins in vivo: Implications for neutron capture therapy, in Biology and Medicine, R.G. Fairchild, and G.L Brownell, eds., Massachusetts Institue of Technology, Cambridge, Mass., 1983, 266.

20. J. Winkelman, G. Slater, and J. Grossman, The concentration in tumor and other tissues of parenterally administered tritium- and 14C-labeled tetraphenylporphinesulfonate, Can. Res., 27:2060 (1967).

21. T. J. Dougherty, G.B. Grindley, R. Fiel, K.R. Weishaupt, and D.G. Boyle, Photoradiation therapy. II. Cure of animal tumors with hematoporphyrin and light, J. Nat. Cancer Inst., 55:115 (1975).

22. T.J. Dougherty, G. Lawrence, J.H. Kaufman, D. Boyle, K.R. Weishaupt, and A. Goldfarb, Photoradiation in the treatment of recurrent breast carcinoma, J. Natl. Cancer Inst., 62:231 (1979).

23. C. J. Gomer, N. Rucker, C. Mark, W.P. Benedict, and A.L. Murphree, Tissue distribution of ^{3}H-hematoporphyrin derivative in athymic "nude" mice heterotransplanted with human retinoblastoma, Invest. Ophthalmol. Vis. Sci., 22:118 (1982).

CORRELATION OF TUMOR BLOOD FLOW TO TUMOR REGRESSION AFTER HEMATOPORPHYRIN DERIVATIVE (HPD) PHOTODYNAMIC THERAPY TO TRANSPLANTABLE BLADDER TUMORS[1]

Stephen H. Selman[2], Martha Kreimer-Birnbaum, Rick W. Keck, Andrew J. Milligan, Peter J. Goldblatt, and Stephen Britton

Departments of Surgery (Urology), Biochemistry, Pathology, Physiology and Radiology, Medical College of Ohio and the Research Department, St. Vincent Medical Center, Toledo OH

INTRODUCTION

Hematoporphyrin derivative (HPD) photodynamic therapy (PDT) has proved effective in the treatment of selected neoplasms (1). The effectiveness of this form of therapy rests on the retention of the systemically administered HPD in neoplastic tissue and its photoactivation with visible light resulting in 'photodynamic' tumor destruction. Although it is generally agreed that singlet oxygen liberated during HPD-photodynamic therapy is responsible for the biologic damage created by PDT, the mechanisms of cell death have not been clearly defined (2). Previous studies in our laboratory have demonstrated a rapid and sustained decrease in tumor blood flow after HPD-photodynamic therapy (3,4). The present study was undertaken to correlate changes in tumor blood flow with tumor regression after HPD-PDT.

MATERIALS AND METHODS

Animals

Four week-old Fischer 344 rats (Charles River Breeding Laboratories, Boston MA) were housed 3 to a cage and provided with laboratory chow and water ad libitum.

Hematoporphyrin Derivative

One gram of HPD (Porphyrin Products, Logan UT) was dissolved in 50

[1]Supported by grants from the American Cancer Society, B.R.S.SO-7-RR05700-12-13, the F.M. Douglas Foundation, the Geiger Foundation for Cancer Research and the NIH R23CA38754-01. Permission for reproduction of portions of this manuscript has been granted by the Journal of Urology, Williams & Wilkins Co., Baltimore MD.

[2]Requests for reprints: Medical College of Ohio, Division of Urology, C. S. 10008, Toledo OH 43669.

ml of 0.1 N NaOH. The solutions was stirred for 1 hr at room temperature, neutralized to pH 7.1 with 0.1 N HCl and adjusted to a final volume
of 200 ml with 0.9% NaCl. The solution was then made isotonic with solid
NaCl, and sterilized by passage through a Swinnex 25 filter unit, 0.45 μm
(Millipore Corp., Bedford MA).

Tumor

The FANFT {N[4-(5-nitro-2-furyl)-2-thiazolyl]-formamide} induced
urothelial tumor line (AY-27) was used for this investigation. This tumor
line was kindly provided by Dr. Samuel M. Cohen, University of Nebraska,
Omaha. The tumor line has been maintained in vivo since 1981 by periodi-
cally propagating small pieces into syngeneic animals.

For implantation, tumors were harvested from donor animals, gently
minced in 5 ml of Hanks balanced salts solution (GIBCO, Grand Island NY),
washed, placed in 5 ml of RMPI 1640 medium (GIBCO) to which 0.01 ml of
antimycotic antibiotic solution (penicillin, 10,000 units/ml; Fungizone,
25 μg/ml; Streptomycin, 10,000 μg/ml) had been added. This solution was
then passed through a series of #21 gauge needles, and 0.2 ml injected
subcutaneously. Tumors were palpable within 1 week of implantation and
had grown to approx. 1 cm diameter by 3 weeks. At this size, such tumors
are not necrotic.

In the present experiments, 2 tumors were grown in each animal, 1
below the xiphoid, the other above the pubis. After injection of HPD, 1
tumor was treated with light while the other was shielded; the latter
served as an internal control. Prior to phototherapy, all animals were
anesthetized with sodium pentothal administered intraperitoneally (65 mg/
kg). Anesthesia was continued during the subsequent blood flow measurements.

Phototherapy Unit

A 500 watt Quartzline lamp (GE-CBA) in a Kodak slide projector
fitted with a red filter (Corning 2418) which excludes light <590 nm was
employed. The light beam was deflected 90 degrees with a mirror and
focused on the intact skin over the tumor with a convex lens. The light
intensity at the surface of the phototreated tumor was measured with a
radiometer (UDT #351). The tumor temperature was monitored with a #24
gauge hypodermic transistor prove (YSI #524X) placed percutaneously
beneath the surface of the tumor. The body-core temperature of the rat
was measured with an intrarectal thermistor probe (YSI #401). Tumor
temperature was maintained within 2°C of body-core temperature by direct-
ing a jet of cool air over the tumor during PDT.

Regional Blood Flow Determination

Regional blood flow determinations were made using a modification of
the reference sample method described by Malik (5). A papered PE 50
catheter (Clay-Adams) filled with heparinized saline (10 U/ml) was advanced into the left ventricle via the right common carotid artery. Continuous monitoring of the arterial pulse pressure via a Statham
transducer connected to a polygraph (Beckman Dynograph R511A) confirmed
proper placement of the left ventricular catheter as determined by a
change in pulse width. This was subsequently verified at necropsy. Both
femoral arteries were exposed and cannulated with PE 10 catheters. The
right was connected to a Statham transducer for continuous blood pressure
monitoring while the left catheter was connected to a Harvard syringe-
driven withdrawal pump. A total of 3.6 x 10^5 microspheres (15 μm
diameter) labeled with either [103]Ru or [141]Ce (New England Nuclear, Boston

MA) were injected into the left ventricle over a 10 sec. period. the microspheres were suspended in 10% dextran containing 0.01% Tween 80. Prior to injection, the microspheres were sonicated and mechanically agitated. Starting 10 sec. before and continuing for 60 sec. after completion of each injection of the microspheres, a continuous sample of blood was withdrawn from the left femoral artery at a rate of 0.3 ml/min. This sample of blood is referred to as the 'reference organ'. Animals were killed by intravenous injection of a saturated solution of KCl. The tumors were immediately excised, cleared of surrounding skin and subcutaneous tissue, weighed and placed in scintillation vials for radioactive counting. Kidneys, liver and spleen were harvested at the same time, for blood flow determinations. The radioactivity of the 'reference organ', tumor and visceral organs were measured with a Beckman Biogamma Counter (#5310). Blood flow to the tumor (F_T) was calculated from the equation

$$F_T = (F_r/R_r) \times R_T$$

where F_r is the flow to the reference organ, R_r the radioactivity of the reference organ, and T_T the radioactivity of the tumor. Visceral blood flow was determined by substituting radioactivity of the visceral organ for R_T in the formula.

Statistical Analysis

All results are presented as the mean ± standard error of the mean. The statistical significance of differences between blood flows or tumor weights of treated vs. untreated tumors was evaluated with Student's t test for paired values.

EXPERIMENTAL DESIGN

I. Tumor blood flow: Correlation with HPD dose

Forty-two animals were randomly divided into 7 groups of 6. Each group was injected via the dorsal tail vein with graded doses of HPD: 0, 1, 3, 5, 7, 10 or 20 μg/g body weight (BW) 24 hr prior to phototherapy. Light dose was kept constant at 360 J/cm² (200 mw/cm² x 30 min) of red light. Blood flow was determined 24 hr after completion of phototherapy for both treated and control tumors.

II. Tumor regression: Correlation with HPD dose

In this experiment, 35 animals were divided into 7 groups of 5. In each group, HPD was injected via the dorsal tail vein in doses of 0, 1, 3, 4, 5, 10 or 20 μg/ml BW. Phototherapy (360 joules/cm$_2$) was given 24 hr later to 1 of the 2 implanted tumors as described above. The animals were then housed 2 per cage and given water and chow ad libitum for 3 weeks, then anesthetized and killed with a saturated KCl injection. The phototreated and control tumors were then excised, cleared of surrounding skin and subcutaneous tissue, dessicated under vacuum for 96 hr and weighed.

RESULTS

Our previous study showed no significant difference in mean blood flow between the 2 tumors in animals receiving neither HPD nor light treatment (4). This was confined in the present series of experiments since mean blood flow to the 2 tumors in the animals receiving no HPD, in

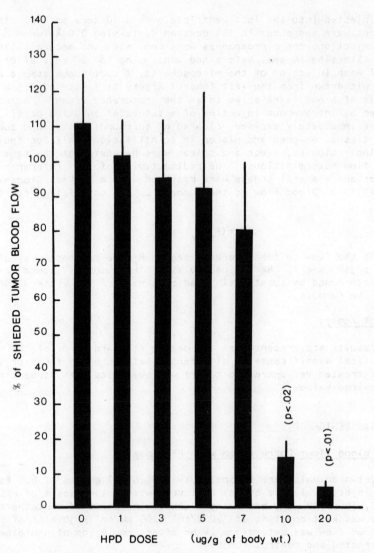

Figure 1. Experiment I. Blood flow as a function of HPD dose. All animals were treated with 360 joules/cm² of light.

experiment I, was 0.546 ± 0.129 ml/min/gm and 0.565 ± 0.160 ml/min/gm. To test for an even distribution of microspheres, blood flow to the right and left kidneys in each animal was measured and compared. Mean blood flow to the right kidney was 4.26 ± 0.025 ml/min/gm, while the mean flow to the left was 3.94 ± 0.025 ml/min/gm. These results are in good agreement with our previously reported values (4).

As seen in Fig. 1, there were no significant differences in mean blood flow to the 2 tumors in animals treated with HPD at doses of 0, 1, 3, 5 and 7 µg/ml BW and light (experiment I). At an HPD dose of 10 µg/gm BW, mean blood flow to the phototreated tumor (0.075 ± 0.033 ml/min/gm)

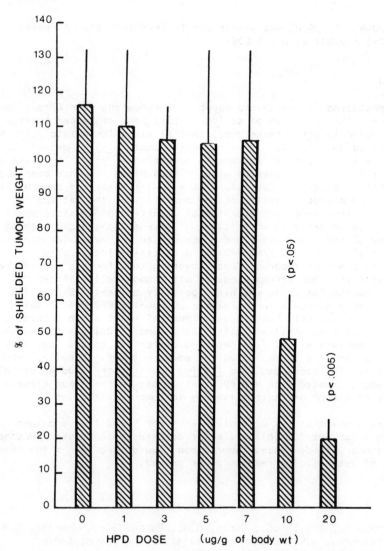

Figure 2. Experiment II. Tumor weight as a function of HPD dose. Dry tumor weight was determined 3 weeks after phototherapy (350 houles/cm²).

was significantly less than to the shielded control tumor (0.711 ± 0.189 ml/min/gm; $p < 0.02$). Similarly, at 20 μg/gm BW, mean blood flow to the phototreated tumor, (0.031 ± 0.012 ml/min/gm) was significantly less than to the shielded control tumor (0.525 ± 0.116 ml/min/gm; $p < 0.01$).

In experiment II (Fig. 2), doses of HPD of 7 μg/ml or less produced no significant difference in the mean dry weights of the phototreated vs. control tumors. At an HPD dose of 10 μg/ml and 360 joules/cm² of light, the mean weight of the phototreated tumor (0.099 ± 0.019 g) was significantly less than the untreated control (0.261 ± 0.047 g; $p < 0.05$). Similarly, at an HPD dose of 20 μg/ml, the mean weight of the treated tumors (0.045 ± 0.009 g) was significantly less than the untreated controls (0.253 ± 0.027 g; p < 0.005).

DISCUSSION

HPD-photodynamic therapy is capable of rendering significant tumor destruction. The mechanism of action of this form of cancer therapy is unknown. Using in vitro techniques, several sites for cytotoxicity have been suggested (6-10). But it is not clear that in vitro mechanisms of cytotoxicity accurately reflect events occurring in vivo. That the combination of light and porphyrins might effect blood flow has been suggested by other studies. Castellani et al. (11) described sequential changes in the microcirculation of the mesentary of the rat and the tongue of the frog after intraperitoneal injection of hematoporphyrin and exposure of these areas to white light. These changes began with agglutination of red blood cells and proceeded to intravascular stasis. Bugelski et al. (12) used [^{3}H]-HPD for drug tissue localization in transplantable murine tumors. These investigators found a 5-fold grain/density ratio in tumor stroma compared with tumor cells. After PDT, damage to endothelial cells was histologically evident. Bicker (13) measured a rapid decrease in intratumor pO_2 during HPD-PDT, leading this investigator to suggest that the cells most affected by this treatment were the vascular endothelial cells of the tumor. Using a 'sandwich' technique, Chambers et al. (14) reported disruption of the capillary microcirculation in a transplanted rat mammary tumor. Finally, Henderson and Dougherty (15) concluded from in vivo and in vitro studies of EMT-6 and RIF tumors treated with HPD-PDT that damage to the tumor circulation may be one of the major factors leading to tumor destruction.

In the present study, disruption of tumor blood flow and tumor regression after HPD-PDT follow similar dose-response curves. Whether anoxia secondary to decreased tumor blood flow during PDT is the sole mechanism of cytotoxicity remains to be proved.

REFERENCES

1. Dahlman, A., Wile, A.G., Burns, R.G., Mason, G.R., Johnson, F.M. and Berns, M.W., Laser Photoradiation Therapy of Cancer, Cancer Res., 43:430-484, (1983).
2. Weishaupt, K.R., Gomer, C.J., and Dougherty, T.J., Identification of Singlet Oxygen as the Cytotoxic Agent in Photo-inactivation of a Murine Tumor, Cancer Res. 36:2326-2329, (1976).
3. Selman, S.H., Keck, R.W., Klaunig, J.E., Kreimer-Birnbaum, M., Goldblatt, P.J. and Britton, S.L., Acute Blood Flow Changes in Transplantable FANFT-Induced Urothelial Tumors Treated with Hematoporphyrin Derivative and Light, Surg. Forum, 34:676-678, (1983).

4. Selman, S.H., Kreimer-Birnbaum, M., Klaunig, J.E., Goldblatt, P.J., Keck, R.W., and Britton, S.L., Blood Flow in Transplantable Bladder Tumors Treated with Hematoporphyrin Derivative and Light, Cancer Res., 44:1924-1927, (1984).

5. Malik, A.B., Kaplan, J.E., and Saba, T.M., Reference Sample Method for Cardiac Output and Regional Blood Flow Determination in the Rat, J. Appl. Physiol., 40:472-475, (1976).

6. Berns, M.W., Dahlman, A., Johnson, F.M., Burns, R., Sperling, D., Guiltinan, M., Siemens, A., Walter, R., Wright, W., Hammer-Wilson, M., and Wile, A., In Vitro Cellular Effects of Hemato-Porphyrin Derivative, Cancer Res. 42:2325-2329, (1982).

7. Moan, J., Johannsen, J.V., Christensen, T., Espevik, T., and McGhie, J.B., Porphyrin Sensitized Photoinactivation of Human Cells in Vitro., Am. J. Pathol., 109:184-192, (1982).

8. Kessel, D., Effects of Photoactivated Porphyrins at the Cell Surface of Leukemia L1210 Cells, Biochemistry, 16:3433-3449, (1977).

9. Moan, J., and Christensen, T., Photodynamic Effects on Human Cells Exposed to Light in the Presence of Hematoporphyrin. Localization of the Action Dye, Cancer Letters, 11:209-214, (1981).

10. Blazek, E.R., and Hariharan, P.V., Alkaline Elution Studies of Hematoporphyrin Derivative Photosensitized DNA Damage and Repair in Chinese Hamster Ovary Cells, Photochemistry and Photobiology, 40:5-13, (1984).

11. Castellani, A., Pace, G.P., and Concioli, M., Photodynamic Effect of Hematoporphyrin on Blood Microcirculation, J. Pathol. Bacteriol., 86:99-102, (1963).

12. Bugelski, P.J., Porter, C.W., and Dougherty, T.J., Autoradiographic Distribution of Hematoporphyrin Derivative in Normal and Tumor Tissue of the Mouse, Cancer Res., 41:4606-4612, (1981).

13. Bicker, H.I., Impact of Microcirculation and Physiologic Considerations on Clinical Hyperthermia, in: "13th International Cancer Congress, Part D," Research and Treatment, Alan R. Liss, Inc., (1983), 235-245.

14. Star, W.M., Marijinissen, J.P.A., Vanden Berg-Blok, A., and Reinhold, H.S., Destructive Effect of Photoradiation on the Microcirculation of a Rat Mammary Tumor Growing in "Sandwich" Observation Chambers, in: "Porphyrin Localization and Treatment of Tumors," D. Doiron and C.J. Gomer eds., Alan R. Liss, Inc., New York, (1984), 637-645.

15. Henderson, B.W., Dougherty, T.J., and Malone, P.B., Studies on the Mechanism of Tumor Distruction by Photoradiation Therapy, in: "Porphyrin Localization and Treatment of Tumors," D. Dorion and C.J. Gomer eds., Alan R. Liss, Inc., New York, (1982), 601-612.

THE BIOLOGICAL EFFECTS OF PHOTODYNAMIC THERAPY ON NORMAL SKIN IN MICE —

I. A LIGHT MICROSCOPIC STUDY

Chuannong Zhou, Weizhi Yang, Zhixia Ding, Yunxia Wang
Hong Shen, Xianjun Fan and Xianwen Ha

Cancer Institute, Chinese Academy of Medical Sciences
Beijing, People's Republic of China

INTRODUCTION

Photodynamic therapy (PDT) of cancer using hematoporphyrin derivative (HpD) and visible light irradiation has been applied in clinical
practice for several years and data on photobiology and photomedicine
have been accumulated (1-3). Even so, much remains to be learned about
the destructive mechanism of PDT. In order to better understand whether
PDT may cause damaging effects to normal tissue in vivo, and the possible mechanism of its action, we have made a light and electron microscopic study on the effects of photodynamic treatment to normal mouse
ears. In this paper we describe the results of our light microscopic
observation and make a brief discussion on the significance of our findings.

MATERIALS AND METHODS

The hematoporphyrin derivative was prepared by the Beijing Phar-
maceutic Industry Institute and was administered intraperitoneally (50
mg/kg) two days prior to laser beam exposure (630 nm, 200 mW/cm², 5 minutes) to the left ears of mice (total number = 117) divided into four
groups: HpD plus laser, laser alone, HpD alone, and controls. Specimens
were taken for examination at 0, 5, 10, 20, 40, 60, 90 minutes, 2, 3, 4,
6, 12, 24 hr, and 2, 7, 14 days after PDT treatment, 2-3 mice/group.
Both ears of each mouse were fixed with 4% formaldehyde and 1% glutaraldehyde, dehydrated with series of ethanolic solutions, and embedded in
paraffin. Histological sections (1 micron) were stained with
hematoxylin and eosin and examined under the microscope.

RESULTS

In the HpD + laser group, both the left and right ears of 48 mice
were examined. Grossly, the irradiated ears became slightly red 1 hr
after photodynamic treatment, markedly red and slightly thickened after
3 hr, dark purplish red and considerably swollen after 12 hr, and dark
brown and dry after 24 hr. Most of the dry, shrunken, crust-like irradiated ears completely sloughed off within two days after photodynamic
treatment.

Microscopically, dilation and congestion of capillaries and small blood vessels were found as early as 40 minutes after PDT (Fig. 1). In some capillaries, segregation of erythrocytes occurred. At 1-4 hr after irradiation, most of the capillaries were markedly dilated and filled with crowded erythrocytes in their lumen (Fig. 2). Edema of the loose connective tissue and inflammatory cell infiltration, mainly with neutrophils and sometimes a few eosinophils, occurred 1 hr after PDT treatment (Fig. 3). Remarkable degeneration of nerve fibers could be noticed 6 hr after irradiation (Fig. 4). These changes increased progressively and reached the highest degree about 24 hr after light exposure. Swelling or degeneration of striated muscles appeared 1 to 4 hr after irradiation in some mice. Severe degeneration and necrosis of striated muscles could be seen in a few ears 6 hr after light exposure. The crust-like damaged ears taken 24 hr after PDT showed that the skin was widely necrotic, but sometimes quite well preserved epidermis and sebaceous glands could still be observed (Figs. 4,5). One to two weeks after PDT, the injured areas of the damaged ears were still edematous and markedly infiltrated by a great number of neutrophils (Fig. 6).

In the group treated with laser alone, the ears of 31 mice were examined. Most of the irradiated ears were slightly reddened within 1 to 2 hr and they returned to normal during a few hr. Microscopically, slight to moderate dilation and congestion of microvasculature could be found within 1 hr which disappeared gradually within 4 - 6 hr after irradiation. No irreversible alterations of the skin were found in the ears in this group.

In the group treated with HpD alone, ears of 35 mice were examined and in comparison with the ears of the control group, neither macroscopic nor microscopic changes were found.

DISCUSSION

1. In this experiment, the HpD was given at a dose of 50 mg/kg, much higher than that used clinically in patients (1,2,4). Yet, no histological changes were found in the ear skin of mice treated with HpD alone.

2. Irradiation with the laser alone caused only slight or moderate dilation and congestion of microvasculature, which disappeared completely within several hours after irradiation. No irreversible damages of the skin were caused by light irradiation alone.

3. It was demonstrated that severe skin damage and necrosis can be caused by photodynamic treatment. This result is consistent with the observations by Dougherty et al (4) on the skin response in patients after photodynamic therapy: If light treatment was given too soon after HpD administration or if the light dose was too high, skin necrosis occurred. To obtain the maximum damage to neoplastic tissues, while protecting the normal surrounding skin from serious damages, it is of great importance to choose the proper HpD and light doses and interval between HpD administration and light exposure.

4. Different tissues in the skin responded to the photodynamic treatment differently: The vasculature lesions occurred early and significantly, and the fibroblasts and nerve fibers behave similarly. but the epidermis cells and chondrocytes responded much less and slowly. This fact is interesting and certainly could be explained in different ways. Differences of sensitivity may exist between various types of cells. This possibility may be related to the membrane damage of

106

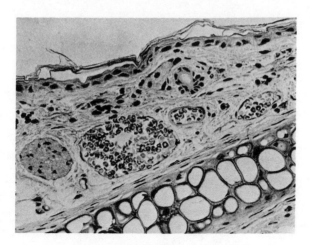

Figure 1

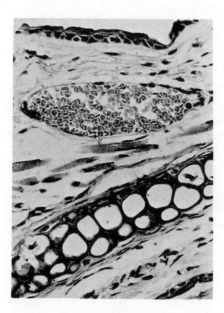

Figure 2

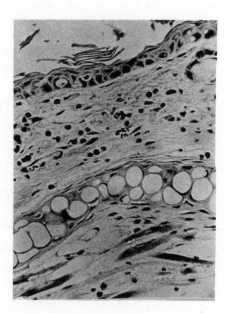

Figure 3

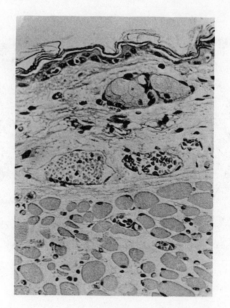

Figure 4

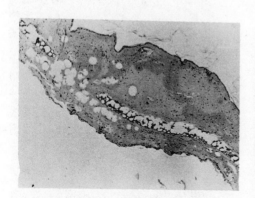

Figure 5

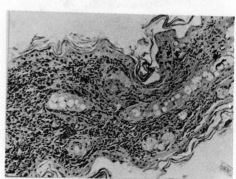

Figure 6

various cell types (5,6). Alternatively, the oxygen concentration in different types of cells may be an important factor. This possibility is supported by the observation that the removal of oxygen from the atmosphere over cell cultures prevents destruction due to HpD plus light and clamping off blood flow in the legs of mice with leg tumors before and during illumination prevents tumor response due to HpD plus light (4). Much remains to be explored before the exact mechanism of PDT action in vivo becomes clarified.

SUMMARY

A light microscopic investigation of the biological effects of photodynamic therapy using hematoporphyrin derivative and red light irradiation has been carried out. HpD alone, even used at a dose of 50 mg/kg, did not cause histologically discernible changes. The irradiation alone used in this experiment caused only slight or moderate dilatation and congestion of blood vessels, which disappeared completely within several hours after light exposure. In the PDT treated mouse ears severe degeneration and necrosis of skin tissue occurred within 1-2 days. The importance of adequate drug and light doses and proper interval chosen for obtaining best therapeutic effects and protection of surrounding normal tissues was pointed out. The possible mechanism of the PDT action in vivo was briefly discussed.

REFERENCES

1. Dougherty, TJ, JE Kaufman, A Goldfarb, KR Weishaupt, D Boyle and A Mittleman (1978) Cancer Res. 38:2628-2635.
2. Ha, XW, XM Sun, JG Xie, XJ Fan and H Shen (1983) Chin. Med. J. 96:754-757.
3. Spikes, JD (1983) In Photoimmunology, Ed. ML Kripke and WL Morison, pp. 23-49. Plenum Press, New York.
4. Dougherty, TJ, KR Weishaupt and DG Boyle (1982) In Cancer, Principles and Practice of Oncology Ed. VT DeVita, S Hellman and SA Rosenberg, pp. 1836-1844. Lipincott, Philadelphia.
5. Pooler, JP and DP Valenceno (1981) Med. Phys. 8:614-628.
6. Spikes, JD (1982) In The Science of Photomedicine, Ed. JD Regan and JA Parrish, pp, 113-144. Plenum Press, New York.

THE BIOLOGICAL EFFECTS OF PHOTODYNAMIC THERAPY ON NORMAL SKIN IN MICE —

II. AN ELECTRON MICROSCOPIC STUDY

Chuannong Zhou, Weizhi Yang, Zhixia Ding, Yunxia Wang,
Hong Shen, Xianjun Fan, and Xianwen Ha

Cancer Institute, Chinese Academy of Medical Sciences
Beijing, People's Republic of China

INTRODUCTION

Photodynamic therapy (PDT) using hematoporphyrin derivative (HpD) and light irradiation has been applied in the clinical treatment of tumors for several years (1,2). Several investigations have been carried out on the destructive effects of PDT to tumor cells (1-5). It has also been reported that under certain circumstances damage to normal tissues can also be caused by photodynamic treatment (3,5). But there has been little reported on the morphological or ultrastructural effects of PDT. We have previously made a light microscopic study on the in vivo effects of PDT to normal ear skin in mice (6). In this paper, we report the ultrastructural changes of the ear observed after PDT treatment.

MATERIALS AND METHODS

All the specimens were taken from the same animals of the experiment described in our previous paper (6). The mouse ears were taken and immediately immersed in the fixative solution containing 1% glutaraldehyde and 4% formaldehyde in 0.1 M phosphate buffer (pH 7.4). Then they were immediately cut into small pieces about 1 mm² and fixed for at least two hr in the same fixative. After washing in the buffer, the specimens were postfixed with 1% osmium tetroxide for two hr. The tissues were dehydrated with a series of acetone solutions and embedded in Epon 812. Ultrathin sections were cut with glass knives, double stained with uranium acetate and lead citrate, and examined with a JEM-100B electron microscope.

RESULTS

The ultrastructure of skin in the normal mouse ears observed in this experiment was similar to those described before (7). In the skin of ears of the experimental group, swelling and deformation of some mitochondria in endothelial cells, axons, Schwann cells and fibroblasts; swelling of endothelial cells and decreased density of their cytoplasmic matrix; and dilation of fibroblast rough endoplasmic reticulum could be sometimes found as early as 10 min after PDT treatment. These changes became gradually more serious during the first day after light exposure.

At four to six hr after PDT, more swollen endothelial cells appeared, and aggregated heterochromatin often occurred in their nuclei. Most of the capillaries were dilated and crowded erythrocytes often filled their lumen. Neutrophil infiltration often occurred near the capillaries. At 12 hr after PDT treatment, a number of capillaries became severely degenerated or necrotic. In some capillaries the endothelial cells were ruptured and sometimes an erythrocyte penetrating through the gap of the rupture could be observed (Fig. 1).

At 12 hr after photodynamic treatment, necrosis of capillaries was more often observed; sometimes only some debris was left (Fig. 2). The fibroblasts in the skin were also often damaged significantly or even necrotic (Fig. 3). Their mitochondria were often injured early, the amount of ribosomes was decreased and sometimes large vacuoles occurred in their nuclei. In the necrotic fibroblasts, the cytoplasmic organelles were seriously damaged and the plasma membrane was often ruptured.

The loose connective tissue became considerably edematous with some cell debris floating in the electron transparent matrix (Figs. 2,4). The structure of collagen fibers seemed unchanged. Inflammatory cells, mainly neutrophils and sometimes lymphocytes, often infiltrated near the capillaries, nerve fibers and striated muscles. Mast cells were often noticed during the first hours after PDT treatment.

Some of the nerve fibers showed injuries such as swelling of mitochondria and occurrence of vacuoles in the axons and Schwann cells as early as 10 min after PDT treatment. Severe changes widely occurred at 6-24 hr after PDT treatment, when the axons lost their neurofilaments and became empty in appearance, and the Schwann cells became seriously damaged or even necrotic (Fig. 5).

The injuries to striated muscle were mainly edema of the cytoplasm, separation or breakdown of myofibrils, as well as swelling and deformation of mitochondria. This could often be seen within the first hr after PDT treatment (Fig. 6). In comparison with the changes in endothelial cells, fibroblasts and nerve fibers, the damage to striated muscles occurred relatively late and less extensively.

In the epidermis, swelling and deformation of mitochondria, pyknosis of the nuclei and sometimes dilation of intercellular space could be observed, but this occurred infrequently. The tonofibrils and desmosomes were often well preserved and the cells of sebaceous glands unchanged. In comparison with the dermis, the epidermis was much less sensitive to the injurious effects of PDT treatment. Even in some severely damaged areas of the skin, the epidermis remained well preserved while many structures in the dermis were considerably damaged (Fig. 4). Chondrocytes were also often slightly changed after PDT treatment.

DISCUSSION

1. The mechanism of PDT action has been extensively investigated, both clinically and experimentally (8). Yet, very few have reported on the ultrastructural changes of tissues and cells caused by PDT in vivo. Coppola et al (9) examined the lymphoma cells in mice and observed changes in mitochondria as early as 10 min after PDT treatment. Lu (10) reported that mitochondria of mouse S-180 tumor cells showed considerable changes at 30 min after PDT. Both authors emphasized the significance of injury to mitochondria during PDT. Our observation of mitochondrial damage was consistent with these findings. From the

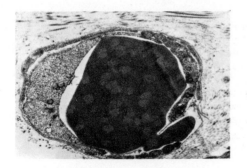

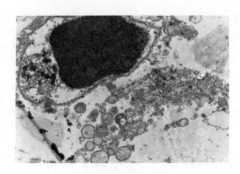

Figure 1

Figure 2

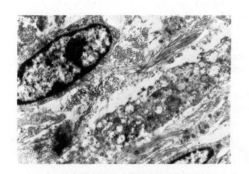

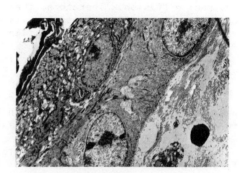

Figure 3

Figure 4

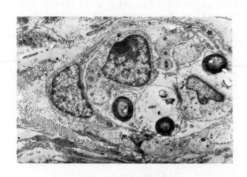

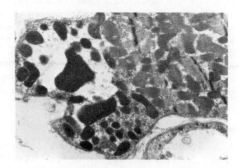

Figure 5

Figure 6

results of our examination and data from the literature (9,10) we found that the ultrastructural changes both in tumor cells and in normal tissue cells caused by PDT in vivo were quite similar.

2. As in the previous study (6) it was demonstrated in this experiment that in a given organ such as skin, various types of tissues or cells respond differently to the PDT treatment. The capillaries, fibroblasts and nerve fibers were much more sensitive to PDT action and showed earlier and more extensive changes than did the keratinocytes and chondrocytes, although the epidermis is the most external layer of the skin and is directly irradiated by light. To our knowledge, this phenomenon has not been described and emphasized in the literature. It is obvious that the mechanism of photodynamic action in vivo is much more complicated than that in some in vitro homogeneous systems. We believe that the following points may play an important role in the mechanism of photodynamic action:

a. The affinity of various cells to HpD may be different, which results in different concentrations of HpD in various cells.

b. The endothelial cells are directly contacted with blood, and fibroblasts as well as nerve fibers are well supplied with blood oxygen, but the epidermis and chondrocytes obtain their needed oxygen only by indirect diffusion. Therefore, the difference of actual oxygen concentration in various types of cells may influence the sensitivity of various cells to photodynamic action.

c. The difference of pH value in various cells may be also closely related to their sensitivity to PDT treatment.

d. The sensitivity of various cells to photodynamic treatment may also depend upon their difference in chemical composition, structure and function.

e. Early damage caused by PDT to endothelial cells and the nerve fibers innervating small vessels, as well as the early injury of mast cells, may cause serious disturbances in microcirculation, thrombosis, edema and hemorrhage, leading to hypoxia, degeneration and necrosis of surrounding tissues as secondary injuries.

SUMMARY

In the present study it has been demonstrated that the ultrastructural changes in mouse ear skin after PDT action occurred much more distinct and earlier than those observed in the histological study. Various types of cells in the skin respond to the photodynamic action differently. The endothelial cells, fibroblasts and nerve fibers are most sensitive. In contrast, the keratinocytes and chondrocytes often remain rather well preserved after PDT treatment. The possible mechanism of PDT action was briefly discussed.

REFERENCES
1.   Dougherty, TJ, JE Kaufman, A Goldfarb, KR Weishaupt, D Boyle and A Mittleman (1978) Cancer Res. 38:2628-2635.
2.   Ha, XW, XM Sun, JG XDie, XJ Fan and H Shen (1983) Chin. Med. J. 96:754-757.
3.   Lu, ZD (1983) Chin. Med. J. [in Chinese] 63:326-329.
4.   Pan, QQ, AL Ning and RZ Wang (1983) Chin. Med. J. [in Chinese] 63:330-335.

5. Dougherty, TJ, KR Weishaupt and DG Boyle (1982) In Cancer, Prin-
    ciples and Practice of Oncology. Ed. DeVita VT, Hellman and SA
    Rosenberg, pp.1836-1844. Lipincott, Philadelphia.
6. Zhou, CN, WZ Yang, XJ Fan, ZX Ding, YX Wang, H Shen and XW Ha
    (1985) [contained in this volume.]
7. Zelikson, AH /Edited/ (1967) Ultrastructure of normal and abnormal
    skin. Lea and Febiger, Philadelphia.
8. Spikes, JD (1982) In The Science of Photomedicine, Ed. JD Regan and
    JA Parrish, pp.113-114. Plenum Press, New York.
9. Coppola, A, E Viggiani, L Salzarulo, et al. (1980) Am. J. Pathol.
    997:175-192.
10. Lu, ZD (1983) Chin. Med. J. [in Chinese] 63:336-339.

STUDIES ON HEMATOPORPHYRIN-PHOTOSENSITIZED EFFECTS ON

HUMAN CANCER CELLS IN VITRO: TEM AND SEM OBSERVATIONS

Ning Ai-Lan and Pan Qiong-Qian

Cancer Institute, Chinese Academy of Medical Sciences
Beijing, People's Republic of China

INTRODUCTION

Hematoporphyrin derivative (HPD) is a dye with photodynamic action. The selective localization at tumor loci has been exploited for both tumor localization and tumor therapy (1-5). At present, the mechanism of photodynamic therapy has not been elucidated. In order to study the anti-tumor mechanisms at the subcellular level, the ultrastructural changes of ECA109 esophageal epithelial cancer cells were observed before and after treatment with HPD and irradiation.

MATERIALS AND METHODS

The ECA109 tumor is an epithelial cell line which was established by our laboratory (Department of Cell Biology, Cancer Institute, Chinese Academy of Medical Sciences, 1976). These cells are cultured in 80% 199 medium supplemented with 20% newborn calf serum, penicillin (100 U/ml) and streptomycin (100 μg/ml).

ECA109 cells were treated with 3 μg/ml HPD (Institute of Medical and Pharmaceutical Industry, Beijing) for 1 hr without light. The medium was changed after treatment. The ECA109 cells were exposed to a 20 watt UV light, maximum irradiation = 365 nm, at a distance of 2 cm for 10 min. Ultrastructural measurements were carried out immediately after treatment and 1, 3 or 5 days later.

After photodynamic therapy, specimens were rinsed twice in Palade buffer and centrifuged at 2500 rpm for 5 min. The supernatant fluid was discarded. The cell were fixed for 30 min in 2.5% glutaraldehyde and rinsed twice in Palade buffer, then fixed in 1% OsO_4 at 4°C for 30 min and rinsed in Palade buffer twice. Specimens were finally dehydrated by serial increments of 30-100% acetone before embedding in Chinese Epoxy resin 618 (5). Sections of 500-800mμ were prepared with a LKB ultratome and double stained by uranyl acetate and lead citrate before examination with a H600 electron microscope.

Scanning electron microscopy was carried out in ECA109 cells on a coverglass which had been rinsed with Palade buffer and fixed with 2.5% glutaraldehyde and 1% OsO_4. Specimens were then dehydrated with acetone, then iso-amyl acetate. The specimens were then dried in a critical-point apparatus and coated with carbon and gold. The specimens were examined with a JEM-35C scanning electron microscope.

RESULTS

Transmission electron microscopy showed that ECA109 cells could be subdivided into dark, light, and empty cells. These different cell types appeared to correspond to proliferating, degenerating, and dying cells. In the control group, dark and light cells with high and medium electron density were the major components, and only a few empty cells could be found (Fig. 1). After treatment with HPD and irradiation we observed mainly round or elliptical, low electron-density light or empty cells (Fig. 2). The decreased proportion of dark to light cells demonstrated the cytotoxic effects of HPD.

ECA109 cells in interphase were flat and irregular in shape. The microvilli were more abundant and often ruffled; some twisted with each other and some others showed round or funnel shaped dilation at their terminals. Spheroid cells having long microvilli, many ruffles, filopodias and blebs are mitotic. After treatment with HPD and irradiation, early effects were enlargement of cellular interspace, reduction of cell volume, diminution and shrinkage of microvilli and appearance of pseudopodia (Figs. 3 and 4). Some round holes or caves could be seen after irradiation (Fig. 5). All of these effects demonstrate extensive destruction of the cell membrane.

ECA109 cells have a large number of villi, few desmosomes and tonofilaments, large and irregular nuclei and large, distinct nucleoli. A large amount of aggregated RNP appears in ECA109 cells which is not found in normal esophageal epithelial cells. After HPD and irradiation, holes and caves could be seen by TEM. Some microvilli protruded from the membrane of the caves (Fig. 6). Part of the cell membrane was destroyed as indicated by the appearance of cytoplasmic shedding. Different sized vacuoles appeared in most cells, and the larger vacuoles occupied up to one half of the cell (Fig. 7). These were also found in dark cells. Loose atypical tonofilaments, instead of compact tonofilaments, appeared in the cytoplasm. Immediately after exposure to HPD and irradiation, the membranes of mitochondria were blurred. We also observed mitochondrial shrinkage and indistinct cristae filled with high density material and vacuoles. Some mitochondria had intact or partially intact cristae; others showed high electron density particles (Fig. 8). As the time after irradiation increased, fewer particles and more vacuoles were observed, and mitochondrial vasculization ultimately was seen (Fig. 9). Some rough endoplasmic reticulum filled with high density material was seen (Fig. 3). The number of Golgi complexes increased and circular shaped Golgi complexes appeared. These were connected with a large number of small vacuoles (Fig. 9). Lysosomes piled up in the concave part of the nucleus (Fig. 7). Most nuclei were round with lower density and granular euchromatin. Heterochromatin was attached to the nuclear membrane.

Mitosis was not difficult to find after HPD and irradiation. Up to seven mitotic cells were found in one field. Round mitotic cells appear different from the flat interphase cells under the scanning electron microscopic. However, no equator plate could be formed in cells during mitosis (Fig. 10). Some organelles such as degenerated mitochondria, dilated endoplasmic reticulum and vacuoles remained in mitotic cells. Even though centrioles existed, no spindle fibers appeared.

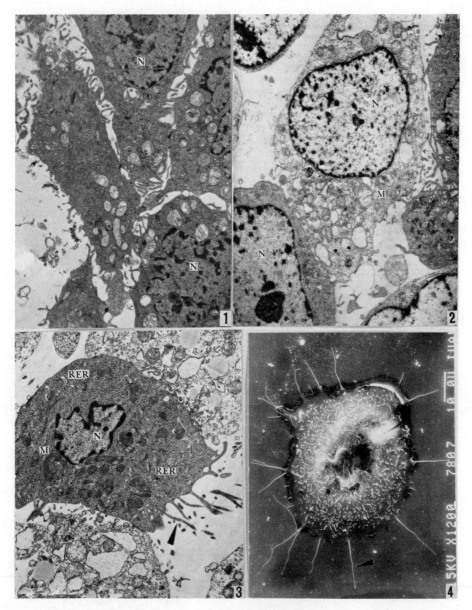

Plate 1.   Figures 1-4.

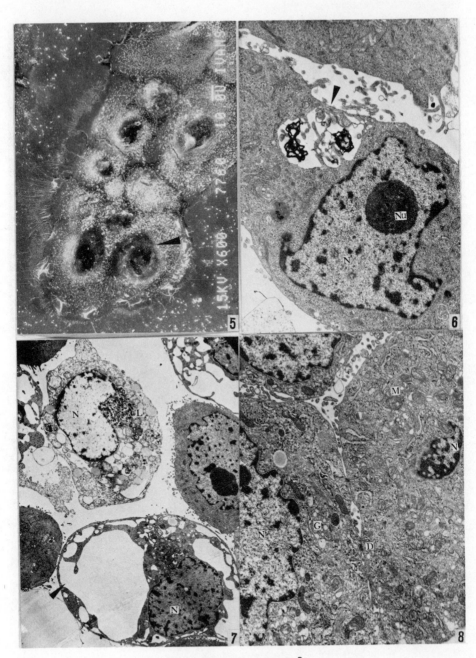

Plate 2. Figures 5-8.

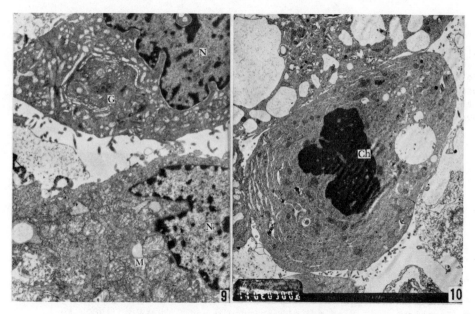

Plate III.   Figures 9 and 10.

DISCUSSION

Neither HPD nor irradiation alone were cytotoxic to the tumor
cells. But substantial photodynamic toxicity was observed. We observed
cytoplasmic vacuoles, destruction of membranes and degeneration of organelles including nuclei. This demonstrates that HPD and light is
highly effective in destroying ECA109 cells, and that these cells are an
effective model for the study of photoradiation therapy.

Despite the use of HPD and irradiation as a therapeutic method for
many years, the mechanism of this therapy remains obscure. It has been
suggested that membrane damage is the major cytotoxic effect, resulting
in loss of permeability, inhibition of membrane active transport systems, destruction of membrane protein and cytoskeleton systems, and alterations in membrane fluidity (6,7). It is also reported that the
cytotoxicity induced by HPD and irradiation results from formation of
the short-lived, highly-reactive singlet oxygen species (5,8). Malik
(9) noticed that the effects of photoirradiation corresponded with the
distribution of HPD. Membrane damage was observed when HPD localized in
erythroleukemia cells, but not in Burkitt's lymphoma, where localization
was not observed. Our results indicated photodynamic membrane destruction, including cell membrane, nuclear membrane, endoplasmic reticulum
membrane and mitochondrial membrane. These effects may derive from the
preferential affinity of HPD for tumor cell membranes (4). Nuclear
damage was also observed. The damage to euchromatin was much more than
to heterochromatin. This phenomenon could be explained by the destruction of DNA in the nucleus (10).

Effects of photochemotherapy on mitosis have never been reported.
The present study demonstrated that the number of metaphase cells in-

creased. Condensed chromosomes could be found, but no mitotic spindle or divided cells appeared. Similar characteristics were found in vincristine and Maytenus Hookerii Loas-treated ECA109 cells and hydroxyurea-treated cells (11). But vincristine and Maytenus Hookerii Loas caused formation of some multinuclear cells,; an effect not seen after photodynamic therapy. It has been reported that S-phase cells are more sensitive to photochemotherapy than are early G cells. This result may derive from DNA damage leading to metaphase arrest (12). In addition to cytotoxic effects on ECA109 cells, many desmosomes, tonofilaments and Golgi complexes appeared after HPD and irradiation The repair of tumor cell damage was initiated after photodamage. This phenomenon has not been reported before.

REFERENCES

1. Dougherty T.J., R.E. Thoma, D.G. Boyle and K.R. Weishaupt (1981) Cancer Res. 41, 401-404.
2. Dougherty T.J. (1980) J. Surgical Oncology 15, 209-210.
3. Kelly J.E. and M.E. Snell (1976) J. Urol. 115, 150-151.
4. Kessel D. and T.H. Chou (1982) Cancer Res. 43, 1994-1999.
5. Ning Ai-lan (1980) Acta Anat. Sinica (in Chinese) 11(1),69-75.
6. Coppola A., E. Viggiani, L. Salzarulo and G. Rasile (1980) Am. J. Path. 99, 175-192.
7. Moan J., J.V. Johannessen, T. Christensen, T. Espevikt and J.B. McGhie (1982) Am. J. Path. 109, 184-192.
8. Buettner G.R. and L.W. Oberley (1980) FEBE Lett 121, 161.
9. Malik Z. and M. Djaldetti (1980) Int. J. Cancer 26, 495-500.
10. Gomer C.J. (1980) Cancer Lett 11, 161-167.
11. Ning Ai-lan (1981) Acta Biol. Exp. Sinica (in Chinese) 14(3), 239-242.
12. Christensen T. (1981) Brit. J. Cancer 44, 433-439.

PHOTODYNAMIC EFFECT OF HEMATOPORPHYRIN ON HUMAN CANCER CELLS

AND NORMAL DIPLOID FIBROBLASTS IN VITRO

Qiong-qing Pan, Min Fu, and Xui-mai Zhao

Cancer Institute, Chinese Academy of Medical Sciences
Beijing, People's Republic of China

INTRODUCTION

Porphyrin is a photodynamic agent that renders cells vulnerable to light at wavelengths which correspond to the absorption spectra of porphyrin (1). This modality has been utilized for both diagnosis and therapy of human and animal neoplasia (2,3), but the mechanism of photodynamic damage is still not clear. Malignant tumors accumulate and retain more HPD than do normal tissues (2). Christensen et al. (4) concluded that the mouse embryo cell line C3H/10T1/2, as well as its DMBA-transformed counterpart bound equal amounts of HPD and had similar sensitivities to light. To study possible differences in uptake and retention of HPD by normal vs. malignant cells in vitro, we compared the photodynamic response of human esophageal tumor cells and diploid lung fibroblasts.

MATERIAL AND METHODS

Cell lines

1. ECA109 cells: An epithelial tumor cell line established from human esophageal carcinoma in our laboratory. The cells are grown in 199 medium supplemented with 20% new born calf serum and 100 μ/ml of penicillin and streptomycin. Monolayered cells were detached by addition of equal amounts of 0.25% trypsin and 0.2% EDTA.

2. 2BS cell line: Diploid fibroblasts from human embryo lung cells were purchased from the Beijing Institute of Biological Products. Cells in passage 28-35 were used in our experiments.

3. Co-cultivation of ECA109 and 2BS cells: 2BS cells (about 4 x 10^5/2 ml) were seeded in culture flasks with cover slips and grown at a 45° angle from the horizontal. Two days later, ECA109 cells (6 x 10^5) were added, and incubations continued at 37° with the flasks in a horizontal position. HPD and light were applied two days later.

Hematoporphyrin derivative

HPD was prepared by Beijing Institute of Medicine Manufacture, kept in the dark at -10°C to -20°C and diluted with 199 medium supplemented with 1% calf serum.

Light source

Using the method of Christensen et al. (5), a light source consisting of two UV lamps (Beijing Lamp Factory, major emission wavelength 365 nm) was employed. The distance between the light source and cells was adjusted by a movable glass platform on which culture vessels were placed. The total fluence was determined by the distance from the light to the platform and exposure time. At 2 cm of light distance and exposure time of 5 min, the temperature at the culture vessels was raised from 26°C to 27.5°C, and to 30° and 36°C after 10 min and 20 min of irradiation, respectively. All experiments described here were performed at temperatures below 37°C.

HPD uptake and irradiation

1. The effect of different concentrations of HPD was determined as follows. Cells were incubated with three different concentrations of HPD (5, 50 and 500 μg/ml) for 2,3,4,6 and 24 hr. The cell morphology was observed and the mitotic index determined after staining.

2. The effect of a 5 μg/ml concentration of HPD on the morphology of tumor cells as a function of irradiation was measured. After a 1 hr exposure of cells to 5 μg/ml HPD in the dark at 37°C, the cells were washed three times with PBS, and suspended in 199 medium containing 20% newborn calf serum, then irradiated with light for 5 to 30 minutes. The cells were fixed and stained after an additional 24 hr incubation in the dark.

3. The effect of low concentration of HPD on tumor cells using various light distances and exposure times was determined. Cells were incubated with 1, 3 or 5 μg/ml of HPD for 1 hour at 37°C in the dark, then irradiated for 5, 10 or 20 min with the light 1 or 2 cm from the cell layer. Cell numbers were determined after an additional 24 hr incubation in the dark.

4. The effect of HPD and/or light on growth of ECA109 and 2BS cells was measured. Four conditions were employed: neither HPD nor light, HPD, light, and HPD + light. Growth was assessed 1, 3 and 5 days after treatment.

5. The effect of varying the interval between HPD treatment and light exposure on cell growth was measured. Cells were exposed to light 1, 24, 48, 72 and 96 hr following HPD treatment, and counted after a further 24 hr incubation.

6. The recovery of cell growth was measured after treatment with different levels of HPD. The distance between the light and the cells was 2 cm, the irradiation time was 10 minutes; cells were incubated with 3 and 5 μg/ml HPD for 60 min before irradiation. Cells were counted on day 1, 3, 5, 7 and 9 after light exposure.

7. The effect of HPD (1-5 μg/ml) and light on the morphology of ECA109 and 2BS cells was measured with both cells present in the same flasks. One hr after addition of drug, the cells were irradiated for 10 min with the light 2 cm from the cell layer. The cells were stained and examined after an additional 24 hr of incubation in the dark. Additional studies were carried out to determine the effect of variation in time between drug exposure and irradiation. Co-cultured cells were exposed to 3 μg/ml HPD for 1 hr, then incubated in fresh medium for 1, 24, 48, 72 and 96 hr before exposure to light (10 min). The cells were stained and examined after an additional 24 hr incubation in the dark.

Table 1. The influence of different dosage and duration of HPD
treatment on morphology and mitosis of ECA109 cells

Time (hr)	500 μg/ml		50 μg/ml		5 μg/ml		control
	morph. change	mitotic index	morph. change	mitotic index	morph. change	mitotic index	mitotic index
2	no change	8	no change	14	no change	23	19
3	vacuoles	2	no change	18	no change	15	13
4	karyopyknosis vacuoles	0	vacuoles	12	vacuoles	11	22
6	condensed chromatin karyopyknosis	0	vacuoles	17	vacuoles	14	13
24	karyopyknosis cells split into fragments	0	vacuoles	6	vacuoles	10	24

RESULTS

1. Exposure of tumor cells to high HPD concentrations (500 μg/ml)
for > 4 hr led to cell damage, but low HPD concentrations (5 μg/ml)
caused no obvious morphological damage to cells during 2 or 3 hr incubations (in the dark). These results are shown in Table 1.

2. Exposure of ECA109 cells to light alone led to appearance of
cytoplasmic vacuoles as the exposure time was increased. After exposure
for 10 minutes (light distance at 1 or 2 cm) there were only a few small
vacuoles; these increased in number as the time was extended to 20 min.
Various stages of mitotic figures could also be seen. When the exposure
time was prolonged to 30 min, chromosomes in some mitotic cells ag-
glutinated. When cells were irradiated for 10 min after exposure to 5
μg/ml HPD (1 hr), many vacuoles appeared in the cytoplasm and nuclei
and chromosomal agglutination could be seen, but mitotic figures were
rare. With light exposure prolonged to 20 min, larger vacuoles appeared
in the cytoplasm. After 30 min, karyopyknosis appeared in all cells and
the cytoplasm stained faintly.

3. After exposure of cells to 1 μg/ml HPD, the cell number
decreased only when the light distance was 1 cm or less. When the distance between cells and light was >2 cm, no toxic effects were observed,
even after 10-20 min. When cells were treated with 3 or 5 μg/ml of HPD,
then irradiated, the extent of cell damage was proportional to the time
of light exposure.

4. When cells were treated with HPD and then exposed to light at
different time intervals, toxicity to both cell types examined increased
as compared to controls. With increasing time between drug exposure and

Table 2. The effect of different intervals between HPD
treatment and light exposure on the growth
of ECA109 and 2BS cells

| Duration between HPD and light (hours) | Cell number (10^4/flask) | | | |
| | ECA109 | | 2BS | |
	HPD+light	control	HPD+light	control
1	13.2	126.8	24	60
24	161.2	232	31.6	98.9
48	178.4	320.8	91.6	140.4
72	418.8	445.6	152.6	176.4
96	485.2	504	165.2	212

irradiation, the extent of cytotoxicity decreased, until a 4 day interval was reached. This result indicates that until the 4th day after incubation with HPD, a photosensitizing amount of HPD remained in the
cells examined here. The accumulation and retention of HPD was essentially identical in 2BS and tumor cells (Table 2).

5. The effect of HPD + light on cell growth during the subsequent
9 days was shown in Fig. 1. After exposure to 5 μg/ml of HPD the survival of ECA109 cells was only 1%. This indicates that no repair of
phototoxic damage had occurred. After treatment with 3 μg/ml + light
the growth rate of ECA109 cells was the same as that of a control
population, but the cell number was 30-60% less than that of control.
Partial repair of phototoxic damage is therefore indicated. The growth
curves of 2BS cells after exposure to 3 or 5 μg/ml HPD also declined.
Though the cell number on the 9th day after treatment had increased
slightly, complete repair had not occurred.

6. The degree of killing of ECA109 and 2BS cells was dependent on
the amount of HPD added. The extent of damage increased with the concentration of HPD. Morphological changes were not observed when either
cell line was treated with 1 or 2 μg/ml HPD. Such changes began to appear at a 3 μg/ml concentration of HPD, e.g., vacuoles in the cytoplasm
and karyopyknosis. Cells began to detach from culture flasks as the
dosage of HPD increased to 5 μg/ml. The extent of damage to both kinds
of cells was similar.

7. The most extensive damage to both cell lines occurred when irradiation occurred immediately after 1 hr of HPD treatment. The extent
of damage to 2BS cells appeared to be more extensive than occurred with
ECA109 cells. 2BS cells retracted as thin filaments with rod-like
necrotic nuclei. ECA109 cells retracted to a spheroid shape with large
vacuoles and little cytoplasm, condensed chromatin or necrotic nuclei.
The degree of injury gradually decreased as the interval between HPD
treatment and light exposure was increased. Irradiation 4 days after
HPD exposure caused no changes in the morphology of 2BS cells. Under
these conditions, ECA109 cells grew as compact sheets with more giant
mononucleated cells and some cells still having large cytoplasmic
vacuoles.

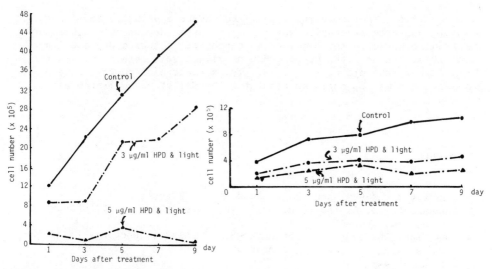

Figure 1. Effect of different concentrations of HPD and light (distance = 2 cm, time = 10 min) on recovery of ECA109 cells (left) and 2BS cells (right) from phototoxic damage.

DISCUSSION

The degree of photodynamic cellular damage is dependent on the quantity of HPD bound to the cells (6) and the interval between HPD exposure and irradiation (2). It is necessary to understand the effect of different dosages of HPD, light intensity and intervals on different kinds of cells before the mechanism of cytotoxicity of HPD and light can be elucidated. We observed no damage to cells treated with 5 - 50 μg/ml HPD alone. Light alone, with the source 1 or 2 cm from the cells had no effect after 10 min. But the combination of light + HPD exerted a lethal effect on both ECA109 cells and 2BS cells. A low dosage of HPD (1 μg/ml) could damage only a portion of the cell populations at the light distance of 1 cm. However, at HPD concentrations of 3-5 μg/ml, the extent of cell injury increased as the light distance decreased and exposure times were prolonged. Phototoxic effects of HPD depend upon the concentration of drug bound to cells, but high concentrations of HPD may be toxic even in the absence of light. Moan et al. (7) considered that the photodynamic effects of HPD at pH 6.7 and 7.2 were stronger than that at pH 7.8. The pH value of our culture medium was maintained at pH 7.0.

The target structure of photodynamic reaction on cells may include the cell surface (8) and membrane structures (7). Another morphological injury involves cytoplasmic vacuolization. The vacuoles increased in number and enlarged as the light intensity increased in the presence of HPD. Damage in cell nuclei was also observed. and the frequency of mitotic division decreased. It is not known if these nuclear injuries were influenced by cytoplasmic injury or were effected directly by HPD and light.

The difference between in vitro and in vivo tests was shown by Granelli et al. (9). Their results indicated that glioma cells in culture were destroyed in less than 8 minutes by HPD-photoradiation, but 40 min were required to eradicate the transplanted subcutaneous gliomas in rats. The monolayer cells in vitro are more sensitive than the solid

127

tumor in vivo. Therefore tests in vitro may be a valuable model for
studying photochemotherapy. Our in vitro system demonstrated that normal cells behaved in the same way as cancer cells toward HPD and light.
Tumor cells did not take up more HPD or selectively retain the HPD for
longer times than did normal cells. The structure of tumor tissue or
other factors may be more important for the selective anti-neoplastic
effect.

REFERENCES

1. Spikes JD. (1975) Ann. N. Y. Acad. Sci. 244:496.
2. Dougherty TJ, RE Thoma, DG Boyle and KR Weishaupt (1981) Cancer
   Res. 41:401-404.
3. Beijing HPD-Laser Study Group (1982) Chinese J. Oncology 4:122-126.
4. Christensen T, K Ferren, J Moan and E Pettersen (1981) Brit. J.
   Cancer 44: 717-724.
5. Christensen T, J Moan, E Wibe and R Oftebro (1979) Brit. J. Cancer
   39:64-68.
6. Bugelski PJ, CW Porter and TJ Dougherty (1981) Cancer Res. 41:4606-
   4612.
7. Moan J, EO Pettersen and T Christensen (1979) Brit. J. Cancer
   39:398-407.
8. Kessel D. (1977) Biochemistry 16:3443.
9. Granelli, S, I Diamond, A McDonagh, C Wilson, C and S Nielsen
   (1975) Cancer Res. 35:2567.

PHOTOTOXIC EFFECTS OF HEMATOPORPHYRIN DERIVATIVE AND ITS CHROMATOGRAPHIC

FRACTIONS ON HORMONE-PRODUCING HUMAN MALIGNANT TROPHOBLAST CELLS IN VITRO

Albert W. Girotti and Robert O. Hussa

Departments of Biochemistry, Gynecology and Obstetrics
Medical College of Wisconsin
Milwaukee, WI 53226

INTRODUCTION

There is a growing interest in developing model systems for studying
the phototoxic effects of hematoporphyrin derivative (HPD) (Bellnier and
Lin, 1984; Bernes et al., 1983; Christensen et al., 1983, Henderson et al.,
1983; Kessel and Chou, 1983; Kreimer-Birnbaum et al., 1984). Our studies
in this area involve cultured cells that secrete the hormone marker, human
chorionic gonadotropin (hCG). The availability of this marker protein opens
up new possibilities for examining HPD photodamage vis a vis cancer phototherapy. For example, (1) large numbers of samples can be evaluated
quickly and efficiently by sensitive radioimmunoassay procedures; (2) cell
cultures need not be sacrificed for counting cells or determining DNA
synthesis, since viability is directly related to the amount of hCG
secreted; (3) the system we describe is amenable to eventual in vivo
investigation using hCG-secreting cells transplanted into animals; and
(4) nonmalignant hCG-secreting cells are readily available for use as normal
counterparts of malignant cells.

While HPD exerts a high degree of selective photoxicity to malignant
tumors, it is not clearly established whether this selectivity can be
manifested under in vitro conditions (Andreoni et al., 1983; Gomer and
Smith, 1980; Henderson et al., 1983; Moan et al., 1981. With this
question in mind, we compared the photosensitivity of malignant and
nonmalignant cells in terms of cell survival and hormone production.

MATERIALS AND METHODS

Porphyrin Preparations

HPD (Photofrin I, 5 mg/ml in sterile saline) was purchased from
Oncology Research and Development, Inc. (Cheektowaga, NY) and stored in
the dark at 4°C. For some experiments HPD was used without further
purification. In other cases it was fractionated by gel exclusion
chromatography on a Bio-Gel P10 column (2 x 35 cm) at 25°C (Kessel and
Chou, 1983). All manipulations were carried out under minimal light.
The column was loaded with 1.5 ml of stock HPD and eluted with 150 mM
NaCl-5 mM HEPES (pH 7.2) at approximately 15 ml/h. The 2 ml fractions were
monitored for porphyrin absorbance at 500 nm. The material emerging
earlier (primarily in fractions 21-23) was designated HPD-F according to

Kessel and Chou (1983), while the material emerging later (fractions 37-40) was designated HPD-S. HPD-F is enriched in hydrophobic components, including dihematoporphyrin ether, the proposed tumor localizer (Dougherty et al., 1984). HPD-S is devoid of the ether and consists mainly of hematoporphyrin. Porphyrin concentrations in pooled fractions were determined spectrophotometrically at 399 nm, using 10 mM cetyltrimethyl-ammonium bromide, 50 mM HEPES (pH 7.0) as the solvent and an extinction coefficient of 220 cm^2/mg, as measured with hematoporphyrin obtained from Sigma Chemical Co. Porphyrin fractions were prepared as soon as possible before use and the Bio-Gel P-10 column was regenerated by successive washings with NaOH (5 mM), water, and NaCl-HEPES. For incubations with cells, stock porphyrin solutions were diluted in culture medium and filter sterilized (0.2 μm pore size) immediately before addition to culture flasks.

Cell Cultures

Two continuous cell lines of human malignant cells were employed, the BeWo line derived from choriocarcinoma (Pattillo and Gey, 1968) and the CaSki line derived from epidermoid carcinoma of the cervix (Pattillo et al., 1977). The BeWo cells secrete hCG and its free α-subunit, while the CaSki cells secrete low levels of hCGβ - like immunoreactive material. Both cell lines were subcultured at weekly intervals using 0.05% trypsin-0.02M EDTA dispersion. The culture medium (Medium 4510: 50% Waymouth's Special Medium M.B. 752/1, 40% Gey's Balanced Salt Solution, and 10% fetal calf serum, all purchased from Grand Island Biological Company, Grand Island, NY) was replenished daily unless stated otherwise.

Normal cell systems included placental cells, amniotic fluid cells, and human fibroblast cultures. The placental cell cultures were prepared by trypsinization of pieces of cotyledons from term placenta and plating into plastic culture flasks. Amniotic fluid cells, consisting of trophoblast cells and fibroblasts, were obtained by amniocentesis 16 weeks after the last menstrual period. Amniotic fluid was centrifuged and the pelleted cells were resuspended in Eagle's Medium and placed into plastic flasks. The cells were maintained at 37°C in a CO_2 incubator under an atmosphere of 95% air, 5% CO_2. Standard culturing procedures were followed, using Eagle's Medium with 20% fetal calf serum. Human foreskin fibroblast cultures were kindly provided by Dr. M.T. Story, Department of Urology, Medical College of Wisconsin. The fibroblasts were dispersed with 0.25% trypsin-EDTA for 5 min, and plated in Minimal Essential Medium containing 10% newborn calf serum. After allowing 3 h for the cells to attach to the culture flasks, HPD was added.

Procedures for Incubation, Washout, and Irradiation

All incubations were carried out at 37°C. Cells (1-2 x 10^5) were plated in plastic Leighton tubes (Corning #3393) and incubated for 3 to 24 h to allow attachment. The cultures were replenished with medium (2 ml per tube) containing 0, 10, 15, or 20% serum and up to 25 μg/ml of HPD starting material, HPD-S, or HPD-F (see RESULTS for details). The Leighton tubes were shielded from the light from this point to the time of irradiation. After standing for 1 h in serum-free or 20 h in serum-containing medium, the cells were rinsed and incubated in porphyrin-free, serum-containing medium, which was replenished daily. The "washout" time (typically 2, 3, or 4 days) was the interval between removal of porphyrin from the incubation medium and exposure of the cells to a light dose. Daily media were collected and frozen for later determination of hCG, hCGα, or hCGβ.

At varying times after removal of HPD (up to 4 days), the cells were irradiated by placing the Leighton tube on a translucent plastic platform and exposing to cool white fluorescent light for 5 min at 25°C. Incident light intensity was approximately 1 mW/cm^2, as measured with a Yellow Springs thermopile. The cells were then incubated in serum-containing medium for up to 4 days in the dark. Spent culture fluids were again stored for subsequent hormone analysis. At the indicated times, irradiated cells and controls were dispersed with trypsin-EDTA and counted by hemocytometer.

Controls

Experiments were controlled in three different ways. Procedural controls received neither HPD nor light, and were included to verify the reliability of the culturing techniques with respect to cell viability and growth, hormone production, and possible microbial contamination. Light controls, i.e., cells irradiated without porphyrin treatment, were prepared to assess possible phototoxic or thermal effects unrelated to HPD photo-activation. Dark controls were prepared to assess any light-independent toxicity of the porphyrin used. Controls for the specificity of HPD phototoxicity were also included; two malignant cell lines of different origin were compared: trophoblast cells (BeWo line) and epidermoid cells derived from carcinoma of the cervix cells (CaSki line). Normal cells from amniotic fluid, fibroblasts, and cells dispersed from term placenta were used as the closest possible counterparts of the malignant trophoblast.

Analyses

Cell counts were determined by hemocytometer following dispersion with 0.25% trypsin-EDTA and dilution in Ca^{2+}- and Mg^{2+}-free Earle's Balanced Salt Solution (Grand Island Biological Company). Cells taking up Trypan blue dye were counted as nonviable. The concentrations of hCG, hCGα, and hCGβ were measured by double antibody radioimmunoassay (Hussa, 1979).

RESULTS

Early experiments in this work focused on the relative in vitro sensitivity of normal and malignant cells to unfractionated HPD, whereas later experiments concentrated on the effects of different molecular excluded fractions of HPD on malignant trophoblast cells. In BeWo malignant trophoblast cells the onset of HPD phototoxicity was very rapid when cells were irradiated immediately after removal of HPD from the culture fluid, i.e., without a washout period (Table 1). The number of viable cells in irradiated cultures was only 8% that of dark controls, when cell counts were performed 2 h following exposure to light. In irradiated cultures, the level of hormone marker was 42% (hCG) or 37% (α-subunit), compared to dark controls. The lesser decrease in hormone marker could be explained by the difference in the nature of endpoints. Since the hormone markers measured all secretory activity that occurred during the entire period of incubation, a gradual decline in secretory activity would result in a cumulative concentration of hCG in the 2 h medium that appeared higher than the relative cell count. On the other hand, the cell counts reflected the number of viable cells at the end of the incubation period and would therefore maximize the effect of treatment.

Table 2 compares the relative phototoxicity of HPD (5 and 25 μg/ml) on two different malignant cell lines. Cultures were irradiated 4 h after removal of HPD at the lower concentration, or 24 h after removal of HPD at 25 μg/ml. Hormone levels were measured in the culture fluid 24 h and 48 h after removal of HPD, while viable cells were determined 48 h after its removal. When cells were irradiated 4 h after HPD removal, hCGα levels were 29 and 31% of dark control levels at the 24 and 48 h time points, and the number of viable cells was only 4% of dark control levels in the BeWo choriocarcinoma cell line. Under the same conditions in the CaSki cervical carcinoma cell line, hCGβ levels were less affected at both time points and viable cell counts at 48 h were 54% of dark control levels. Similar results were obtained when HPD was used at 25 μg/ml and irradiation was performed after 24 h of washout, except that no viable BeWo cells remained 24 h after irradiation, whereas 25% of viable CaSki cells still remained, relative to dark controls.

Table 3 shows the photokilling effects of HPD on malignant vs. nonmalignant cells, based on 11 separate experiments using 25 μg/ml HPD. Cell counts are shown as percentage of dark controls. As can be seen, malignant trophoblast cells were the most sensitive, decaying more rapidly than cervical carcinoma cells (as also shown in Table 2) and somewhat more rapidly than amniotic fluid cells. Placental cells and fibroblasts were the least sensitive and were killed at about the same rate (cf. 48 h washout point). The order of photosensitivity of the different cell types is summarized as follows: BeWo > amniotic fluid > CaSki > fibroblasts ≃ placenta. Hormone assays gave results which correspond to this same general relationship (results not shown). Comparison of several experiments suggested that the type and concentration of serum in the culture fluid during the washout period modulated the degree of phototoxicity. This is evident in the results with amniotic fluid cells (Table 3), where incubation with fetal calf serum at concentrations of 10, 15, and 20% and irradiation at 24 h resulted in cell counts of 7, 15, and 56%, respectively, relative to dark controls. This suggested that serum exerted a protective effect against phototoxicity in the in vitro protocol. In experiments 8-11 with fibroblasts, the phototoxic effect was less pronounced when 10% newborn calf serum was the source of serum rather than 10% fetal calf serum, for reasons that are not apparent.

Table 1. RAPID ONSET OF HPD PHOTOTOXICITY IN BEWO MALIGNANT TROPHOBLAST CELLS

	Analyses		
	Viable cells $(\times10^{-3})$	hCG (ng/ml)	α-subunit
Dark control	109	1.08	4.45
	134	0.53	5.74
	122	0.80	5.10
	(100%)	(100%)	(100%)
Irradiated	7	0.33	1.56
	13	0.35	2.20
	10	0.34	1.88
	(8.2%)	(42.2%)	(36.9%)

HPD (5 μg/ml) incubation was for 1 h in serum-free medium, followed by one rinse with balanced salt solution. Medium containing 10% serum was then added, followed immediately by irradiation for 5 min. Analyses on duplicate culture flasks were performed 2 h after irradiation.

Table 2. HPD IS MORE PHOTOTOXIC TO MALIGNANT TROPHOBLAST CELLS (BEWO) THAN TO CERVICAL CARCINOMA CELLS (CASKI)

| Cells | Incubation conditions | | | Analyses | | | | |
| | HPD inc. | | | 24 h [b] | | 48 h [b] | | |
	HPD (μg/ml)[a]	Washout Serum	Hours	hCGα (ng/ml)	hCGβ	hCGα (ng/ml)	hCGβ	Viable cells (x10^{-3})
BeWo-DC	5	10%	4	164 (100%)		331 (100%)		360 (100%)
BeWo-hν	5	10%	4	48 (29%)		102 (31%)		13 (4%)
CaSki-DC	5	10%	4		1.48 (100%)		1.56 (100%)	300 (100%)
CaSki-hν	5	10%	4		0.65 (44%)		0.63 (40%)	163 (54%)
BeWo-DC	25	10%	24	238 (100%)		382 (100%)		535 (100%)
BeWo-hν	25	10%	24	302[c] (127%)		30 (8%)		0 (0%)
CaSki-DC	25	10%	24		1.24 (100%)		1.58 (100%)	266 (100%)
CaSki-hν	25	10%	24		1.40[c] (113%)		0.18 (11%)	67 (25%)

DC, dark control; hν, irradiated (5 min). Mean of duplicates is shown.

[a]HPD incubation was for 1 h in serum-free medium; cells were then rinsed with serum-containing medium prior to washout incubation.

[b]Hours after removal of HPD.

[c]24 h hormone levels prior to irradiation.

Table 3.   HPD PHOTOTOXICITY TO NORMAL AND MALIGNANT CELLS[a]

Cell Type	Expt.	Serum	Cells (% dark control) at h washout		
			24 h	48 h	72 h
Malig. troph. (BeWo)	1	10% FCS	0		
	2	" "	0	3	
	3	" "		1	
Amniotic Fluid	1	10% FCS	2		
	2	" "	8	10	
	3	" "		22	
	4	" "	7		
	4	15% FCS	15		
	4	20% FCS	56		
Cerv. ca. (CaSki)	1	10% FCS	16		
	2	" "	14	26	
	3	" "		42	
Placenta (normal)	5	10% FCS	48	42	
	6	" "	56		
	7	" "	69		
Fibroblasts	8	10% FCS	2	40	
	9	" "		34	41
	10	10% NBC	73	92	
	11	" "	23	57	67
	9	" "		47	99
	9	" "		78	92

[a]Incubation with 25 µg/ml HPD was for 1 h in serum-free medium.
Cells were irradiated after the indicated periods of washout and counted
24 h later.   FCS = fetal calf serum; NBC = newborn calf serum.

Increasing the washout time from 0 to 2 h resulted in a substantial reduction in phototoxicity (Table 4). This implies that much of the damage immediately after removal of HPD was a result of readily diffusable components in the preparation. Based on previous observations (Kessel and Chou, 1982; Kessel and Chou, 1983), we suspect that these components would be poor tumor localizers, whereas the slowly diffusing components which persist after long washout periods would be enriched in aggregated localizers. Therefore we used washout times that were as long as possible in order to focus on the effects of the latter components. When no serum was present during washout, a greater phototoxic effect was observed at 4 h (Table 4: 0% vs 22% viability), suggesting that serum exerted a protective effect against phototoxicity. Similar observations have been made previously (Chang and Dougherty, 1978; Sery, 1979) and are consistent with the presence of porphyrin binding proteins (hemopexin, albumin) which will compete with cells for HPD components.

Table 4. PARTIAL PROTECTION AGAINST HPD PHOTOXICITY BY SERUM AND INCREASING WASHOUT TIME IN BEWO MALIGNANT TROPHOBLAST CELLS: EARLY WASHOUT TIMES

	Washout		h after irrad'n	Analyses viable cells $(\times 10^{-3})$		hCGα (ng/ml)
	h	serum[a]				
Dark control	–	10%	24	264 (100%)	5.45 (100%)	73.5 (100%)
Irradiated	0	10%	24	0 (0%)	0.55 (10%)	2.83 (3.9%)
	2	10%	22	56 (21%)	3.02 (55%)	22.6 (31%)
	4	10%	20	59 (22%)	2.70 (49%)	18.4 (25%)
Dark control	–	0%	20	250 (100%)	6.10 (100%)	62.3 (100%)
Irradiated	4	0%	20	0 (0%)	1.50 (24%)	7.56 (12%)

HPD (5 μg/ml) incubation was for 1 h in serum-free medium, followed by one rinse with balanced salt solution.

Irradiation (5 min) at 0, 2, or 4 h after removal of HPD; analyses were 24 h after HPD removal. Results represent mean of duplicate flasks.

[a]All flasks were replenished with medium containing 10% serum 20 h prior to analysis.

Photodamaging effects of HPD, HPD-F, and HPD-S on malignant trophoblast and amniotic fluid cells after 48 and 72 h of washout are summarized in Table 5. For both cell types, HPD-F was the most potent of the three porphyrin preparations. Like unfractionated HPD, HPD-F was less phototoxic at 72 h than at 48 h, suggesting that it still contained some rapidly effluxing components. Under the same conditions, both preparations were more phototoxic to malignant trophoblast cells than to normal amniotic fluid cells (cf. Table 3). Neither cell type was affected by HPD-S under the same experimental conditions.

Table 5. PHOTOTOXIC EFFECTS OF HPD, HPD-F, AND HPD-S ON AMNIOTIC FLUID CELLS AND BEWO MALIGNANT TROPHOBLAST CELLS

| | | Cells (% dark control) at h washout | | | | | |
| | | 48 h | | | 72 h | | |
Cell Type	Expt:	HPD	HPD-F	HPD-S	HPD	HPD-F	HPD-S
Malignant trophoblast		16	0	109	80	12	135
Amniotic fluid		38	5	150	59	47	189

The mean of duplicate flasks is shown.
All incubations contained HPD or derivative at 25 µg/ml in serum-free medium for 1 h. All washout incubations were in medium containing 10% fetal calf serum.

Experiments with BeWo malignant trophoblast cells were designed to assess the reversibility of the phototoxic effect of unfractionated HPD, HPD-F, and HPD-S, as well as potential dark effects of these porphyrins prior to the time of irradiation (Figs. 1-4). Cells (in duplicate) were plated into plastic Leighton tubes and incubated for 3 h in Medium 4510 to allow attachment. The culture medium was then replenished with Medium 4510 containing freshly diluted HPD (25 µg/ml), HPD-S (25µg/ml) or HPD-F (10 or 25µg/ml), and incubation was carried out in the dark for 20 h at 37°C. Light controls received Medium 4510 without porphyrin. Following the day of incubation with drug, all flasks were rinsed once with Medium 4510, and incubated in the dark in Medium 4510 for 6 days, with daily medium replenishment. All culture fluids were saved for measurement of hCG and hCGα. Separate duplicate sets of sister cultures (complete with light controls) were set up for irradiation at 2, 3, or 4 days of washout. This experimental design required 18 Leighton tubes for each washout time, or 54 tubes for the entire experiment. Eight culture fluids were saved from each Leighton tube, the total being 432 culture fluids, each analyzed for hCG and hCGα. In addition, viable cell counts were performed on all 18 flasks with 3 days of washout.

The results of these experiments are presented in Figs. 1-4, which show the hCGα and viable cell count data (mean values of duplicate flasks) for the light control, HPD, HPD-S, and HPD-F series, respectively. The dotted curve in Figs. 2, 3, and 4 represents the light control (cf. Fig. 1) which has been superimposed to facilitate comparison of the data. In each figure, the bottom, middle, and top panels depict 2, 3, and 4 days of washout, respectively. Although hCG itself was also determined in these experiments, the data are omitted, since the relative trend was similar to that shown for hCGα.

The pattern of hCGα secretion in light controls (Fig. 1) was identical to that of procedural controls, showing that irradiation per se was innocuous. Secretion of the tumor marker was biphasic, greatest levels being observed on days 3 and 4 after subculture, with a decline thereafter. This pattern of hormone secretion has been observed in previous studies with the malignant trophoblast cell lines (Story et al, 1974).

Dark incubation of the BeWo cells with HPD starting material (Fig. 2) caused a slight depression in hCGα secretion up to day 4 of washout, after which the levels increased (compare with HPD-F results below). This occurred despite a 2-fold reduction in cell count by day 6. Exposure to light after 2 or 3 days of washout caused a dramatic reduction in hCGα levels and viable cells, as had been observed in previous experiments (Tables 1-4). After 4 days of washout, irradiation had relatively little effect on hCGα secretion for the next 2 days.

Incubation with HPD-S had no significant effect on secretion of tumor marker or cell counts, regardless of whether cells were irradiated or not (Fig. 3). These results agree with earlier ones obtained on other malignant cell types (Kessel and Chou, 1983). In those studies HPD-S was shown to contain relatively polar constituents, the most abundant being hematoporphyrin, which is taken up poorly by cells.

HPD-F at 10 μg/ml was phototoxic, but measureably less so than at 25 μg/ml (Fig. 4). This was reflected in only a 50% reduction in hCGα secretion and cell viability after 3 days of washout (Fig. 4, middle panel). There was virtually no effect of irradiation after 4 days of washout, as was also seen with stock HPD. In addition, there was a moderate increase in hCGα levels in dark controls at 4-6 days of washout, and this was similar to that observed with HPD. At higher concentrations of HPD-F, cell killing and inhibition of hormone secretion was total, even when the light dose was given after 4 days of washout (Fig. 4). This contrasts with the negligible effects of stock HPD or the lower concentration of HPD-F under the same conditions. A remarkable dark effect was also observed with HPD-F at 25 μg/ml. This was manifested as a nearly complete ablation of hCGα secretion through day 2 of washout, followed by a recovery phase in which hormone levels continued to increase through at least day 6. The increase in hormone by day 6 (cf. level in procedural control) was considerably greater than that observed with HPD or HPD-F at 10 μg/ml despite a nearly 5-fold lower viable cell count. When expressed as hCGα per cell the HPD-F-treated cultures (25 μg/ml) produced at least 30 times more hormone than untreated cells (see Fig. 4, middle panel). The same trends were evident when hCG itself was determined. These results were reproduced in replicate experiments on BeWo cells, but similar experiments have not yet been performed with other cell types.

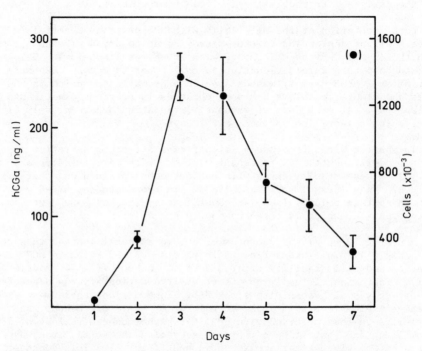

Fig. 1. Daily hCGα production in BeWo choriocarcinoma cell
cultures irradiated in the absence of HPD. Values shown are
means ± SD (n=6) and are identical to those obtained in
procedural controls. Cell counts (parentheses) were performed
on day 7. These data are also shown in Figs. 2-4.

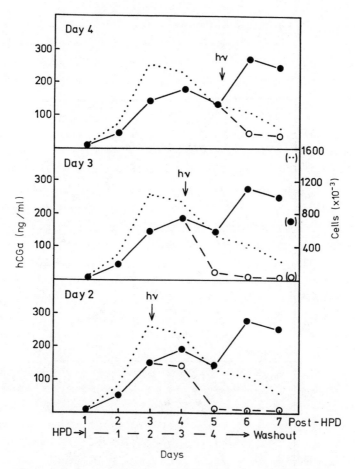

Fig. 2. Concentration of hCGα in culture fluids
of BeWo cells incubated for 20 h with stock HPD
(25 µg/ml) in Medium 4510. Closed symbols represent dark controls that received HPD but no
irradiation, and open symbols denote experimental
flasks subsequent to irradiation for 5 min. For
purposes of comparison, the hCGα levels in
light controls are also indicated (dotted line).
The times at which cultures were irradiated are
indicated by vertical arrows. Other details
were as for Fig. 1. Viable cell counts (on
washout day 6) were performed on cultures at
which the washout time was 3 days, and are
indicated in parenthesis. The symbols in
parenthesis correspond to those used to denote
hCGα levels.

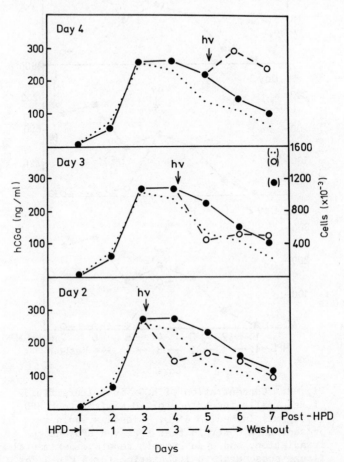

Fig. 3. Concentration of hCGα in the culture fluids of BeWo cells incubated for 20 h with HPD-S (25 µg/ml) in Medium 4510. Other conventions and details are as described in Fig. 2.

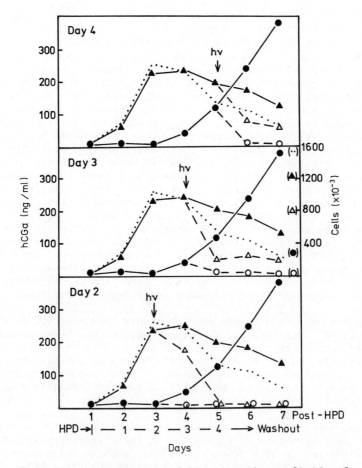

Fig. 4. Concentration of hCGα in culture fluids of
BeWo cells incubated for 20 h with HPD-F in Medium
4510. The HPD-F concentrations were 10 μg/ml
(triangles) and 25 μg/ml (circles). The open
symbols represent irradiated cultures. Other
conventions are as described in Fig. 2.

DISCUSSION

HPD-photodynamic therapy has been shown in clinical trials to be an effective means of selectively eradicating a wide variety of carcinomas (Dougherty et al., 1983). Two fundamental questions relating to this treatment are: (1) how are the localizing components of HPD retained by malignant tissues, and (2) when photoactivated, how do these components kill tumor cells? Various investigators have sought insights into the mechanism of localization by attempting to replicate the in vivo situation with cells in culture. Most of the efforts to date have yielded controversial results. For example, Henderson et al. (1983) compared the kinetics of HPD uptake and efflux in two transformed cell lines (HeLa and EMT-6) with two normal, albeit unrelated types (CHO cells and skin fibroblasts) and found no difference. On the other hand, Kreimer-Birnbaum et al. (1984) showed that AY-27 cells derived from bladder tumors in rats retained significantly more porphyrin than normal bladder cells after exposure to HPD. It should be noted that in the latter study porphyrin-treated cells were washed with serum-free media, whereas in the former study serum-containing media were used in order to more closely approximate in vivo conditions. In the present work we have demonstrated that HPD renders BeWo choriocarcinoma cells significantly more photosensitive than non-malignant close counterparts, viz. amniotic fluid cells and normal placental cells (Table 3). These effects were seen after long washout periods in the presence of serum and they presumably reflect greater retention of the true localizer, dihematoporphyrin ether (DHE), by the malignant cells (Kessel and Chou, 1983; Dougherty et al, 1984); however, this remains to be established. These results suggest that the BeWo system may be appropriate for studying the determinants of localization at the cellular level. Our results also indicate that HPD is more phototoxic to one malignant cell type, BeWo choriocarcinoma, than to another, CaSki cervical carcinoma. Whether this reflects a greater uptake or retention of porphyrin and/or oxygen by BeWo cells or an intrinsically greater photosensitivity is not known.

It is now well established that the tumor localizer DHE is a minor component in HPD preparations, the other constituents being hematoporphyrin itself (HP), protoporphyrin (PP) and the two isomers of hydroxyethyl-vinyl deuteroporphyrin (HVD). An aggregate of DHE can be purified by aqueous gel exclusion chromatography of HPD (Kessel and Chou, 1983; Dougherty et al. 1984), as used in this work. High molecular weight, rapidly eluting material (HPD-F; also designated as Photofrin II) consists of DHE and lesser amounts of PP and HVD. Slower eluting material (HPD-S) is more polar overall, and consists of HP and lesser amounts of HVD. Previous studies comparing the interaction of porphyrins in these fractions with Sarcoma-180 cells in vitro and in vivo demonstrated that substantially more fluorescent material from HPD-F was retained by these cells than from HPD-S (Kessel and Chou, 1984). Although cellular porphyrin levels were not determined in the present work, a comparison of the relative photosensitivities of HPD-S-treated and HPD-F-treated BeWo cells (Figs. 3 and 4) suggests a pattern of porphyrin uptake and retention that was similar to that observed with Sarcoma cells.

In reference to question 2 above, it is now well accepted that cell membranes are the principal target of HPD-mediated phototoxicity. Photolesions in plasma membranes as well as in various organelle membranes (mitochondrial, lysosomal, nuclear, and microsomal) have been reported (Deziel and Girotti, 1983; Girotti, 1983). It seems likely, therefore, that the indicators of photodamage in this study, viz. decreases in cell viability and in hormone production, are reflections of some type of membrane damage. It may also be possible that the remarkable dark effect of HPD-F on hCGα release by BeWo cells (Fig. 4) is membrane related. One might imagine, eg., that hormone synthesis and/or secretion if affected by the interaction of some hydrophobic component in HPD-F (possibly DHE) and a membrane component.

The use of hormone-secreting cells offers certain advantages in in vitro studies of HPD phototoxicity. Hormone measurements can be performed quickly and easily using radioimmunoassay procedures, and can thus be used to monitor daily changes in individual culture flasks. This contrasts with procedures involving measurements of DNA synthesis or cell number, in which the culture must be sacrificed at the time of measurement. The value of this approach was demonstrated in the present investigation, in which hormone determinations on daily culture fluids revealed an intriguing transient dark effect of the higher dose of HPD-F on BeWo cells (Fig. 4, circles). This effect might have been overlooked without the aid of hormone measurements. Preliminary experiments suggest that the late stimulation of hormone secretion is not related to a renewed logarithmic growth phase of the cells in these flasks, since early log BeWo cells did not secrete comparable high levels of hormone (unpublished observations). We are seeking to elucidate the nature of this dark effect and to compare its mechanism with that of phototoxicity. Another advantage of the model system we describe is that it provides a convenient biologic endpoint; this could be exploited in the form of a quality control test on HPD preparations destined for clinical use.

SUMMARY

The in vitro phototoxicity of HPD on malignant cells relative to normal cells has been examined. Two human malignant cell lines were studied: (1) the BeWo line of choriocarcinoma cells, which secrete the tumor marker human chorionic gonadotropin (hCG) and its α-subunit; and (2) the CaSki line of human cervical carcinoma cells, which secrete hCG and its β-subunit. Trophoblast-derived, hCG-secreting cells from human amniotic fluid were used as normal controls. In all experiments with HPD plus light, a close correlation was found 24 h after light between cell number and RIA-detectable marker concentration in the medium. Phototoxicity was greater when HPD was introduced in serum-free rather than serum-containing medium. No toxicity was observed in light and dark controls. Cells in Leighton tubes were incubated 1 h with HPD (25 µg/ml) in serum-free medium, then rinsed and incubated with medium containing 10% serum. At 2 and 3 days after contact with HPD, flasks were exposed to cool white fluorescent light (1 mW/cm^2) for 5 min. Viable cell counts taken 1 day after the light dose indicated (1) that HPD is significantly more phototoxic to BeWo than to CaSki cells; and (2) that both malignant cell types are more photosensitive than amniotic fluid cells, presumably because the latter retain HPD less effectively.

In another aspect of this work BeWo cells were used as a model system for comparing the phototoxic effects of the fast (F) and slow (S) eluting fractions of HPD obtained by Bio-Gel P-10 chromatography. Cells in light-shielded tubes were sensitized by incubating with porphyrins for 20 h in media containing 10% calf serum. At 2, 3, or 4 days after removal of porphyrin, with daily replacement of serum-containing medium, flasks were irradiated (see above), and then incubated in the dark for 2 to 4 additional days. Daily culture fluids were analyzed for hormone levels (hCGα), and cell counts were performed 2 or 3 days after the light dose. HPD-S (25 µg/ml) had no effect on either hormone secretion or cell viability in any of the flasks, whether exposed to light or not. HPD-F at low concentrations (0.25 or 2.5 µg/ml) had no detectable effect on cell count or hormone secretion in irradiated flasks. At 10 µg/ml, HPD-F was innocuous in dark controls, but caused a large decrease in cell count and hormone output in irradiated flasks. At 25 µg/ml, HPD-F, in addition to being strongly phototoxic, exhibited a dark toxicity, ie. it suppressed hCGα secretion for at least 3 days and caused an 80% decrease in cell count by day 6. The decline in hormone output (but not cell viability) showed a dramatic reversal during subsequent days of incubation such that hCGα per cell was about 30 times greater than in cultures lacking HPD-F by day 6. The dark- and light-mediated effects of HPD starting material (25 µg/ml) mimicked those of HPD-F (25 µg/ml) but, as expected, were less pronounced.

Summary of key findings: (1) the HPD fraction exerting greatest photodamage on trophoblast cells is HPD-F, which is enriched in dihematoporphyrin ether; (2) at high enough concentrations in the dark, HPD-F exhibits a reversible toxicity to the cells, the mechanism of which remains to be elucidated; (3) our model system provides a sensitive means of distinguishing between light-independent and light-dependent aspects of HPD-F toxicity.

ACKNOWLEDGEMENTS

This work was supported by a grant from the Elsa U. Pardee Foundation, and by grant #CA-23357 awarded by the National Cancer Institute, DHHS. We thank Dr. David Kessel for valuable advice on HPD fractionation; Dr. Roland Pattillo, Ms. Anna Ruckert, and Mr. James Kurtz for advice and assistance in the cell culturing; Ms. Martha Rinke, Mr. James Thomas and Mr. John Jordan for other technical assistance; and Ms. Christine Konczal for typing the paper.

REFERENCES

Andreoni, A., Cubeddu, R., De Silvestri, S., Laporta, P., Ambesi-Impiombato, F.S., Esposito, M., Mastrocinque, M., and Tramantano, D., 1983, Effects of laser irradiation on hematoporphyrin-treated normal and transformed thyroid cells in culture, Cancer Res. 43: 2976-2080.

Bellnier, D.A., and Linn, C-W., 1984, Photodynamic inactivation of cultured bladder tumor cells: A preliminary study of the effects of porphyrin aggregation, in: Porphyrin Localization and Treatment of Tumors, D.R. Doiron and C.J. Gomer, eds., pp. 361-371, Alan Liss, Inc., New York.

Bernes, M., Wile, A., Dehlman, A., Johnson, F., Burnes, R., Sperling, D., Guiltinan, M., Siemens, A., Walter, R., and Hammer-Wilson, M., 1983, Cellular uptake, excretion, and localization of hematoporphyrin derivative (HPD). Adv. Exper. Med. Biol. 160: 139-150.

Chang, C.T., and Dougherty, T.J., 1978, Photoradiation therapy: kinetics and thermodynamics of porphyrin uptake and loss in normal and malignant cells in culture. Radiat. Res. 74: 498.

Christensen, T., Moan, J., McGhie, J.B., Waksvik, H., and Stigum, H., 1983, Studies of HPD: chemical composition and in vitro photosensitization. Adv. Exper. Med. Biol. 160: 151-164.

Deziel, M.R., and Girotti, A.W., 1983, Photodynamic action of protoporphyrin on resealed erythrocyte ghosts: mechanisms of release of trapped markers, Adv. Exper. Med. Biol. 160: 213-225.

Dougherty, T.J., Boyle, D.G., Weishaupt, K.R., Henderson, B.A., Potter, W.R., Bellnier, D.A., and Wityk, K.E., 1983, Photoradiation therapy: clinical and drug advances, Adv. Exper. Med. Biol. 160: 3-13.

Dougherty, T.J., Potter, W.R., and Weishaupt, K.R., 1984, The structure of the active component of hematoporphyrin derivative, in: Porphyrin Localization and Treatment of Tumors, D.R. Dorion and C.J. Gomer, eds., pp. 301-314, Alan Liss, Inc. New York.

Girotti, A.W., 1983, Mechanisms of Photosensitization, Photochem. Photobiol. 38: 745-751.

Gomer, C.J., and Smith, D.M., 1980, Photoinactivation of Chinese hamster cells by hematoporphyrin derivative and red light, Photochem. Photobiol. 32: 341-348.

Henderson, B.W., Bellnier, D.A., Ziring, B., and Dougherty, T.J., 1983, Aspects of the cellular uptake and retention of hematoporphyrin derivative and their correlation with the biological response to PRT in vitro, Adv. Exper. Med. Biol. 160: 129-138.

Hussa, R.O., 1979, Effects of antimicrotubule agents, potassium, and inhibitors of energy production on hCG production, In Vitro 15: 237-245.

Kessel, D., and Chou, T.H., 1982, Components of hematoporphyrin derivatives and their tumor localizing capacity, Cancer Res. 42: 1703-1706.

Kessel, D., and Chou, T.H., 1983, Tumor localizing components of the porphyrin preparation hematoporphyrin derivative, Cancer Res. 43: 1994-1999.

Kreimer-Birnbaum, M., Baumann, J.L., Klaunig, J.E., Keck, R., Goldblatt, P.J., and Selman, S.H., 1984, Chemical studies with hematoporphyrin derivative in bladder cell lines, in: Porphyrin Localization and Treatment of Tumors, D.R. Doiron and C.J. Gomer, eds. pp. 335-350, Alan Liss, Inc., New York.

Moan, J., Steen, H.B., Feren, K., and Christensen, T., 1981, Uptake of hematoporphyrin derivative and sensitized photoinactivation of C3H cells with different oncogenic potential, Cancer Lett. 14: 291-296.

Pattillo, R.A., and Gey, G.O., 1968, The establishment of a cell line of human hormone-synthesizing trophoblastic cells in vitro, Cancer Res. 28: 1231-1236.

Pattillo, R.A., Hussa, R.O. Story, M.T., Ruckert, A.C.F., Shalaby, M.R., and Mattingly, R.F., 1977, Tumor antigen and human chorionic gonadotropin in CaSki cells: a new epidermoid cervical cancer cell line, Science 196: 1456-1458.

Sery, T.W., 1979, Photodynamic killing of retinoblastoma cells with hematoporphyrin and light, Cancer Res. 39: 96-100.

Story, M.T., Hussa, R.O., and Pattillo, R.A., 1974, Independent dibutyryl cyclic adenosine monophosphate stimulation of human chorionic gonadotropin and estrogen secretion by malignant trophoblast cells in vitro, J. Clin. Endocrinol. Metab. 39: 877-881.

EXAMINATION OF POTENTIALLY LETHAL DAMAGE IN CELLS TREATED WITH

HEMATOPORPHYRIN DERIVATIVE AND RED LIGHT

Charles J. Gomer, Natalie Rucker and A. Linn Murphree

Clayton Ocular Oncology Center
Childrens Hospital of Los Angeles
4650 Sunset Boulevard
Los Angeles, CA 90027

INTRODUCTION

Cell survival curves for mammalian cells treated with hematoporphyrin derivative (HpD) photodynamic therapy (PDT) exhibit a threshold (shoulder) region followed by exponential kill. This type of phenomenon indicates that there is accumulation of sublethal damage during in-vitro PDT, and also suggests that repair of damage may occur. Documentation of sublethal damage repair following HpD PDT (via standard split dose experiments) has recently been reported (1). Unfortunately, there are problems related to standardization of porphyrin concentrations and locations during fractionated treatments which may lead to unacceptable variables for any split dose experiment.

Photodynamic therapy at the clinical and preclinical level is normally performed in single treatment schedules (2). In these cases, the presence and repair of potentially lethal damage (PLD) may play a role in the ultimate effectiveness of treatment (3). It is therefore of both clinical as well as fundamental interest to examine PLD and repair of PLD as it relates to HpD PDT. We will define inhibition of PLD-repair as a post treatment effect which increases the cell killing that results from the expression of damage (normally only potentially lethal) that would have been repaired in the absence of the post treatment conditions (3).

In the current study, we have examined conditions which have been previously reported to inhibit the repair of PLD following exposure to ionizing radiation. Post treatment incubation at hypothermic temperatures ($0-4^{\circ}$ C), and incubations in growth medium containing either caffeine or 3-aminobenzamide were performed following HpD PDT. Repair of potentially lethal ionizing radiation damage is inhibited at $0-4^{\circ}$ C (4). The addition of either caffeine or 3-ABA to cells immediately following exposure to ionizing radiation also induces a potentiation of cell killing (5,6). In the current set of experiments, we have examined the phenomenon of PLD in V-79 cells incubated with HpD for either short (1 hour) or long (16 hours) time periods and then exposed to red light.

147

METHODS

Drugs

HpD was obtained from Photofrin Medical Inc., Cheektowaga, NY. Caffeine and 3-aminobenzamide (3-ABA) were obtained from Sigma Chemical Company, St. Louis, MO.

Light Source

A set of 14 parallel soft white 30 watt fluorescent bulbs (Sylvania, F30T12) were filtered with Plexiglass and Milar film (Rubylith SR-3; Ulano Corp., Brooklyn, NY) and were used as the light source for all HpD PDT experiments. The emission spectrum of this light source ranged from 570 nm to 650 nm (peak at 620 nm) as determined with a scanning monochromator (American ISA, Metuchen, NJ) (/). The power density at the treatment site was 0.35 mW/cm^2 as measured with a radiometer-photometer (EG&G, Model 550-1, Salem MA).

Cell Lines and Survival Assays

Chinese hamster lung fibroblasts (V-79) cells were used in all experiments. The cells were maintained in exponential growth in a monolayer culture in MEM medium supplemental with 10% fetal calf serum (FCS), and antibiotics (penicillin and streptomycin). Prior to treatments, appropriate numbers of V-79 cells were plated onto 60 mm plastic Petri dishes and incubated at 37 degrees for 4 hours to allow for cell attachment. Following cell attachment, the growth medium was removed, the cells were washed once with serum free MEM medium and then the cells were incubated in the dark with HpD. Two HpD incubation protocols were utilized: (1) a short incubation which consisted of a 1 hour HpD incubation (25 ug/ml) in MEM medium supplemented with 1% FCS; or, (2) a long incubation which consisted of a 16 hour HpD incubation (25 ug/ml) in MEM medium supplemented with 5% FCS (8). Immediately following the 16 hour incubation, the cells were incubated for an additional 30 minutes in MEM medium supplemented with 10% FCS (washout protocol). Following the various HpD incubation and washout procedures, the cells were rinsed once with serum free MEM. The dishes were then exposed to the red light source. The longest exposure time was 3 minutes.

The treated cells were refed immediately with complete growth medium and returned to a 37 degree CO2 incubation for standard clonigenic assays in control experiments. In PLD experiments, the treated cells were refed and incubated with; a) complete growth medium at 0-4° C for 6 hours; b) complete growth medium containing caffeine (3 mM) for 24 hours, or; c) complete growth medium containing 3-ABA (20 mM) for 5 hours. Following the various incubation protocols, the cells were refed with complete growth medium and returned to a 37° C 5% CO$_2$ incubator for clonogenic assays. All appropriate controls (HpD alone, light along, PLD agent alone, PLD agents plus HpD and PLD agent plus light) were examined. Plating efficiencies ranged from 60-90 percent. Surviving fractions of cells were determined from colony formation. Three dishes were treated at each dose point in each experiment and each experiment was repeated at least 3 times.

RESULTS

The effect of 6 hours of 0-4° C exposure on V-79 cells incubated for 1 or 16 hr with HpD and then exposed to light are shown in Figures 1 and 2 respectively. The incubation at 0-4° C potentiated the photosensitizing effect for cells in both HpD incubation protocols. Approximately a 3 fold increase in cytotoxicity was observed for cells incubated at 0-4° C for 6 hours.

Figures 3 and 4 illustrate the effect of post PDT treatment incubation with caffeine. We did not observe a statistically significant differential effect of caffeine for cells incubated with HpD for 1 hour (Figure 3). However, an approximate 4 fold enhancement in cell killing was observed for V-79 cells incubated with HpD for 16 hours prior to light exposure.

Figure 5 and 6 show survival curves for V-79 cells which received 3-ABA following HpD PDT. No statistically significant difference in resulting survival levels were observed by 3-ABA for cells incubated for either 1 hour (Figure 5) or 16 hours (Figure 6).

DISCUSSION

The major observation of this study is that under certain conditions it is possible to demonstrate the inhibition of PLD-repair in mammalian cells treated with HpD PDT. Incubation of V-79 cells for 6 hours at 0-4° C lead to a potentiation of HpD PDT phototoxicity for both short and extended HpD incubation protocols. Potentiation of phototoxicity was also demonstrated for cells in the extended HpD incubation group when caffeine was added for 24 hours following PDT. However, we have not observed any statistically significant potentiation (or protection) of HpD PDT for cells exposed to caffeine following a short HpD incubation protocol or for any HpD PDT treated cells exposed to 3-aminobenzamide.

The experimental protocols which were utilized in the current studies were chosen in order to minimize the number of variables which could effect our results. Variations in porphyrin localization, uptake, re-localization and photobleaching (which can occur in standard split-dose/sublethal damage repair experiments) were eliminated since all cellular manipulations (addition of repair inhibitors) were performed after PDT treatment. The 2 HpD incubation protocols (1 hour or 16 hours) were utilized in order to examine possible differences in damage/repair kinetics since subcellular localization of porphyrins can vary as a function of incubation time (9). The potential repair inhibitors (0-4° C, caffeine and 3-ABA) were utilized in concentrations and/or time intervals which were non-toxic by themselves.

The results of our studies do not allow for the determination of targets responsible for cell inactivation following HpD PDT. However, our results with 3-AbA indicate that DNA damage (which is induced by HpD PDT (10)) and repair of DNA damage do not play a significant role in whether a cell lives or dies following PDT. It has been shown in numerous studies that 3-ABA is a specific inhibition of poly (adenosine diphosphoribose) synthesis and that 3-ABA can potentiate the cytotoxic response of mammalian cells treated with ionizing radiation (6). There is also considerable information in the literature suggesting that a primary effect of 3-ABA is in inhibiting the repair of DNA damage. 3-ABA may also have multiple cellular effects which would complicate or limit the generalization that 3-ABA's mode of action in potentiating cell lethality is conclusively via inhibiting of DNA damage (11). This point is currently under debate (12). Nevertheless, the fact that we did not observe any potentiating effect when 3-ABA was added to cells following HpD PDT strongly indicates that DNA damage and repair does not play a major role in the ultimate expression of in vitro PDT lethality.

Caffeine has been reported to potentiate the lethal actions of

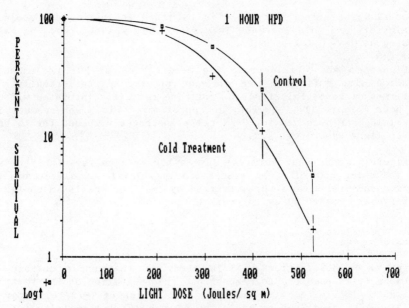

Figure 1. Survival curves for V-79 cells incubated for 1 hour with
HpD prior to light exposure. □ - control cells, + - cells
which received a 6 hr incubation at 0-4° C following
light exposure.

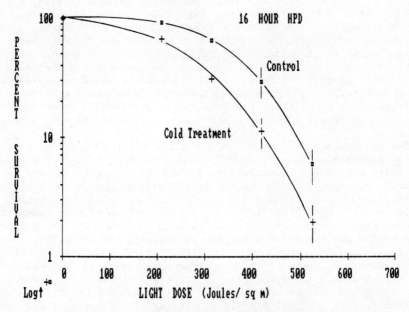

Figure 2. Survival curves for V-79 cells incubated for 16 hr with
HpD. □ - control cells, + cells which received a 6 hr
incubation at 0-4° C following light exposure.

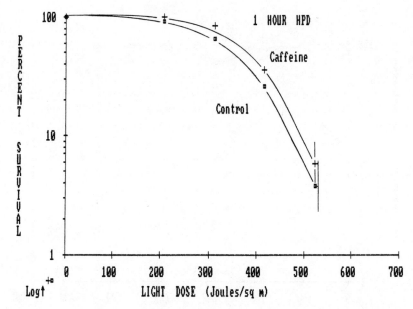

Figure 3.   Suvival curves for V-79 cells incubated for 1 hour with
HpD and exposed to red light. □- control cells, + - cells
which were incubated for 24 hr with 3 mM caffeine
after light treatment.

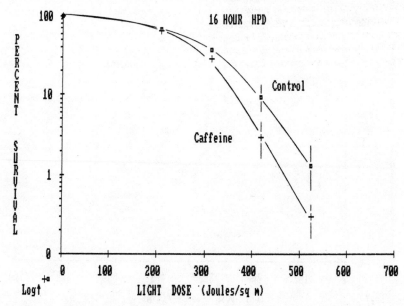

Figure 4.   Same as in Figure 4 except V-79 cells received a 16 hr
HpD incubation. □- control cells, + - caffeine
treated cells.

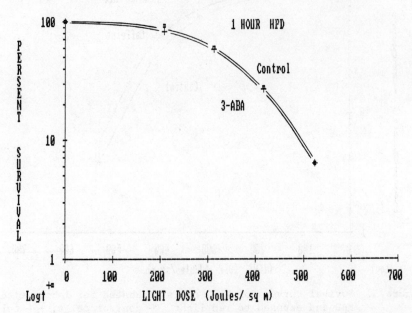

Figure 5. Survival curves for V-79 cells incubated for 1 hr with HPD and then exposed to red light. □ - control cells, + - cells which were incubated for 5 hr with 3-ABA (20 mM) following light treatment.

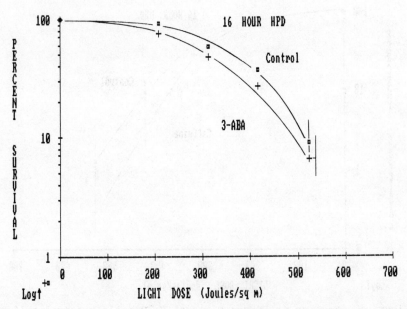

Figure 6. Same as in Figure 5 except V-79 cells were incubated with HpD for 16 hr. □ - control cells, + - 3-ABA treated cells

ionizing and ultraviolet radiation (5). This compound has many subcellular effects and the exact mechanism of potentiation (inhibition of PLD-repair) is still unclear. In our experiments, caffeine was shown to potentiate the cytotoxicity of PDT for extended (16 hr) HpD incubations but not for PDT using short (1 hr) HpD incubations. This suggests that different targets may be damaged for the 2 incubation conditions and that caffeine acts only on one of these targets. The protocol of HpD PDT followed by 6 hours at 0-4° C was shown to potentiate cell lethality for both HpD incubations conditions. While the mechanism of action of post treatment cold incubation is not known, it can be assumed that this procedure may act by inhibiting a wide variety of repair enyzmes (4).

In conclusion, our study has demonstrated that the phenomenon of PLD-repair can be observed in cells treated with HpD PDT. The sites of potentially lethal damage and the repair mechanisms associated with this damage are probably different from that which has been reported for cells treated with ionizing radiation. These studies suggest that post treatment manipulation of tumors with hypothermia could be advantageous in clinical PDT in that it may result in the inhibition of repair of potentially lethal damage.

REFERENCES

1. Bellnier, D.A. and Lin, C.W., Photosensitization and split dose recovery in cultured human urinary bladder carcinoma cells containing nonexchangable hematoporphyrin derivative. Cancer Res. 45: 2507-2511, 1985.

2. Dougherty, T.J., Photodynamic therapy (PDT) of malignant tumors, CRC Crit. Rev, pp. 83-116, CRC Press, Boca Raton, FL, 1984.

3. Utsumi, H. and Elkind, M.M., Two forms of potentially lethal damage have similar repair kinetics in plateau and in log phase cells., Radiat. Res.: 47, 569-580, 1985.

4. Henle, K.J. and Leeper, D.B., Interaction of sublethal and potentially lethal 45° hyperthermia and radiation damage at 0, 20, 37 or 40° C., European J. Cancer, 15, 1387-1394, 1979.

5. Waldren, C.A. and Rusko, I., Caffeine enhancement of x-ray killing in cultured human and rodent cells., Radiat. Res., 73, 95-110, 1978.

6. Ben-Hur, E. and Elkind, M.M., Poly (ADP-ribose) metabolism in x-irradiated Chinese hamster cells: Its relationship to repair of potentially lethal damage., Int. J. Radiat. Res., 45, 515-523, 1984.

7. Gomer, C.J., Rucker, N., Banerjee, A. and Benedict, W.F., Comparison of mutagenicity and induction of SCE's in Chinese hamster cells exposed to HpD photoradiation, ionizing radiation of ultraviolet radiation., Cancer Res. 43, 2622-2627, 1983.

8. Gomer, C.J., Rucker, N., Razum, N.J. and Murphree, A.L. In-vitro and in-vivo light dose rate effects related to HpD PDT. Cancer Res. 45, 1973-1977, 1985.

9. Moan, J., McGhie, J. and Jacobsen, P.B., Photodynamic effects on cells in-vitro exposed to HpD and light. Photochem. Photobiol., 37, 599-604, 1983.

10. Gomer, C.J., DNA damage and repair in CHO cells following HpD photoradiation. Cancer Lett. 11, 161-167, 1980.

11. Milam, K.M. and Cleaver, J.E., Inhibitors of poly (ADP-ribose) synthesis: effect on other metabolic processes. Science, 223, 589-591, 1984.

12. Ben-Hur, E., Chen,C.C. and Elkind, M.M. Inhibitors of poly (ADP-ribose) synthetase, examination of metabolic perturbations and enhancement of radiation response in Chinese hamster cells., Cancer Res., 45, 2123-2127, 1985

PORPHYRIN PHOTOSENSITIZATION OF BACTERIA

Berit Kjeldstad, Terje Christensen* and Anders Johnsson

Institute of Physics, University of Trondheim
*7055 Dragvoll, Norway
Department of Toxicology, National Institute of Health
Geitmyrsv. 75, 0462 Oslo 4, Norway

ABBREVIATIONS USED

CTAB   : Cetyltrimethylammonium bromide
Hpd   : Hematoporphyrin derivative (Photofrin I, Photofrin
          Inc, USA)
HPLC   : High pressure liquid chromatography
P.acnes : Propionibacterium acnes

INTRODUCTION

In the present study we focus our attention on the gram positive
bacterium P.acnes. It is located on the human skin and normally
produces large amounts of porphyrins, both copro- and protoporphyrin
(Fanta et al. 1981, Kjeldstad et al. 1984, Lee et al. 1978). The
bacteria can be visualized on the skin by their strong, red porphyrin
fluorescence, in near UV-light (Mc Ginley et al. 1980). It is of
interest to study possible photosensitivity of P.acnes due to these
internally produced porphyrins. Investigations of the possible effects
of externally added porphyrins, e.g. Hpd, could also be useful in a
broader context since Hpd is used as a sensitizer in photodynamic
therapy of certain cancer forms (Dougherty et al. 1978). Therefore
it is desirable to study the detailed light reactions involved in Hpd
photosensitization in a variety of cellular systems. Studies of
bacterial photosensitization may also adress the question on whether
phototherapy of infections is feasible.

Since P.acnes accumulates larger amounts of porphyrins than many
other bacterial cells, we have started a study of its photoreactions.
The following presentation is deliberately very short and details are
omitted.

PHOTOINACTIVATION OF P.ACNES DUE TO ENDOGENOUS PORPHYRINS

In order to investigate the influence of internally produced
porphyrins on photoinactivation of the cells, P.acnes cells with

Table 1. Concentration of porphyrins pr. mg protein in
P.acnes grown on Eagles medium with different pH.
The parentheses show standard deviation of the mean.

	pH in the growth medium			
	5.3	6.1	6.7	7.2
nmol porphyrin/ mg protein	0.7 (±0.1)	1.4 (±0.2)	1.2 (±0.1)	0.4 (±0.1)

different porphyrin concentration were produced. The porphyrin production in these cells is influenced by the chosen growth conditions (Fanta et al. 1981, Kjeldstad et al. 1984). When P.acnes (serotype I, CN 6278) ·was grown in anaerobic jars ($H_2:CO_2$/95:5 v/v) on Eagles medium at different pH, the amount of porphyrin pr. mg protein varied. The medium was phosphate buffered (60mM), and cells were incubated for four days when the porphyrin concentration was determined, see Table 1. The porphyrins were quantified by fluorescence measurements (ex 410nm and em 610 nm) of porphyrin extracts from the bacteria. The extractions were carried out by ethylacetate/acetic acid (3:1 v/v), followed by a transfer to 3M HCl solution. Protoporphyrin IX was used as a standard. The amount of protein was determined by a commercially available kit, using bovine serum albumin as a standard.

The porphyrins were identified as copro- and protoporphyrin by HPLC analysis (formed as described in details by Husby and Romslo 1980, Sandberg and Romslo 1981).

The changes in the photosensitivity of P.acnes at different porphyrin contents were investigated by determining the ability of the cell to form colonies after light irradiation. The cells were suspended in PBS (pH 6.7) at a concentration giving a clear solution. Illumination was carried out at 415 nm by a Xenon lamp, 450W and a double monochromator (bandwidth 15 nm). The irradiance at the sample position was 52 W/m^2. During irradiation two samples, containing 200 - 600 cells, were taken out at different times and spread on plates with a yeast extract containing agarmedium (Kjeldstad 1984). After four days of anaerobic incubation, the number of colonies was counted. Surviving fractions were determined and sensitivity was defined as the inverse value of the fluence which gave 37% survival.

The results (Fig. 1) showed that cells with different concentration of porphyrin were inactivated to different extent, under otherwise equal conditions. The sensitivity increased with increasing porphyrin content. A threefold increase in the porphyrin content resulted in a corresponding sensitivity increase.

Light at 415 nm corresponded to the maximum absorbance of the fluorescing porphyrins in the cells, measured from the fluorescence excitation spectrum (em 635nm) of a cell suspension (Kjeldstad and Johnsson, to be published).

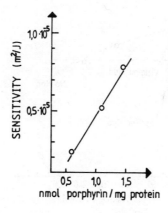

Figure 1. Sensitivity ($1/F_{37}$) of P.acnes for 415nm irradiation
versus the amount of porphyrins pr. mg. protein.
The curves represent the means of four experiments.

PHOTOSENSITIZATION OF P.ACNES DUE TO EXTERNALLY ADDED HPD

The effect of externally added Hpd on the blue light sensitivity of
P.acnes was investigated.

The cells were incubated in PBS with Hpd at different concentrations
(0.1, 0.5, 1.0, 2.0 and 5.0 µg/ml). The incubation time was 30 min and
the whole procedure was performed in darkness. The cells were then
quickly sentrifuged and resuspended in Hpd-free PBS. The uptake of Hpd
was calculated by suspending the cells in 1% of the cationic detergent
CTAB, and by measuring the porphyrin fluorescence (ex 398nm and
em 623nm). CTAB was added to dissolve porphyrin aggregates (Christensen
et al. 1983). The presence of the aggregates would have led to an
underestimate of the porphyrin concentration since they are known to
fluoresce with a low quantum yield (Moan and Sommer 1981). The uptake
of Hpd in non-aggregated form could therefore be compared at different
incubation concentrations. Fluorescence was also measured without
adding CTAB. This gave an estimate of the concentration of fluorescing
porphyrins in the bacteria.

Irradiation and assessment of viability were carried out exactly
the same way as described above. The irradiation was, however, performed
at 400nm (bandwidth 15nm, irradiance 62 W/m^2), corresponding more closely
to the absorbance maximum of Hpd in protein bound state or in organic
solvents (Moan and Christensen 1981, Jori et al. 1980).

The results showed that the light sensitivity of P.acnes increased
with increasing incubation concentration of Hpd. In Fig. 2 survival
curves from one typical experiment are shown. The Hpd uptake did not
increase linearly with the incubation concentration.

A comparison based on relative values was made between the
sensitivity of bacteria incubated with different concentrations of Hpd
and the uptake of Hpd. It showed, as seen in Table 2, that the relative
uptake of fluorescing Hpd varied exactly as the photosensitivity of the
bacteria. The total uptake of Hpd (measured after CTAB addition) was,
however, consistently higher than the relative photosensitivity of the
bacteria in the concentration range tested.

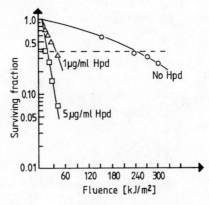

Figure 2'. Dose-response curves for P.acnes incubated in PBS with Hpd concentrations 0(o), 1.0(Δ), 5.0(□) μg/ml, washed and irradiated at 400nm (bandwidth 15nm, 65 W/m^2).

It can therefore be concluded that the non-aggregated porphyrins in Hpd are most important in the photodestruction of the bacteria. This is also found for human cells (Moan and Sommer 1984).

CONCLUSION AND PERSPECTIVES

The procaryotic cells of P.acnes are sensitive to porphyrin photodamage when illuminated in PBS-solution. Both naturally occuring porphyrins and externally added porphyrins that can be taken up by the cells, are involved in the destruction after a treatment with blue light. In the P.acnes system both the type and the concentration of endogenous as well as exogenous porphyrins can be varied in a controlled manner. This might be of importance in studies of the mechanism behind porphyrin induced photodamage in bacteria. Investigations of such mechanisms in procaryotic cells is of value in itself, and might also highlight our knowledge about corresponding processes in eucaryotic cells.

In clinical context it is relevant to point to the possible treatment of acne vulgaris with light. If light treatment of acne should be of any value it is necessary to understand the porphyrin

Table 2. Relative photosensitivity ($1/F_{37}$) of P.acnes and relative uptake, measured by fluorescence with and without 1% CTAB added. The data are normalized to 1 for Hpd concentration 5.0 μg/ml.

Rel. extracellular Hpd conc. (5.0 μg/ml corresponds to 1)	Rel. sensitivity	Rel. uptake of fluorescing porphyrins	Rel. total uptake of porphyrins
0.02	0.10	0.07	0.12
0.1	0.16	0.18	0.26
0.2	0.27	0.30	0.41
0.4	0.46	0.50	0.58
1	1	1	1

sensitization in these bacteria compared to the sensitization of the human cells in order to optimize the light conditions during the treatment.

We have found that P.acnes has a higher uptake of Hpd than other bacteria (cf. Bertoloni et al. 1984). Studies to understand uptake mechanisms and photosensitization in different bacteria will certainly be important if treatment of bacterial infections with light and porphyrins should be a reality. A preliminary investigation in this direction in now underways in our laboratory.

REFERENCES

Bertoloni,G., Salvato, B., Dall'Acqua, M., Vazzoler, M. and Jori G. 1984. Photochem. Photobiol. 39, 811-816.
Christensen, T., Sandquist, T., Feren, K., Waksvik, H. and Moan, J. 1983. Br. J. Cancer 48, 35-43.
Dougherty, T.J., Kaufman, J.E., Goldfarb, A., Weishaupt, K.R., Boyle, D. and Mittelman, A. 1978. Cancer Res. 38, 2628-2635.
Fanta, D., Formanek, I., Poitscek, Ch., Thurner, J. 1981. Arch. Dermatol. Res. 271, 127-133.
Husby, P. and Romslo, I. 1980. Biochem. J. 188, 459-465.
Jori, G., Reddi, E., Rossi, E., Cozzani, I., Tomio, L., Zorat, P.L., Pizzi, G.B. and Calzavara, F. 1980. In Medecine Biologie Environnement, Guglielmo et al. eds., vol. 8, INEC. 141-154.
Kjeldstad, B. 1984. Z. Naturforsch. 39c, 300-302.
Kjeldstad, B., A. Johnsson and S. Sandberg. 1984. Arch. Derm. Res. 276, 396-400.
Lee, W.S., A.R. Shalita and M.B. Poh-Fitzpatrick. 1978. J. Bacteriol. 133, 811-815.
McGinley, K.J., Webster, G.F. and Leyden, J.J. 1980. Br. J. Dermatol. 102, 437-441.
Moan, J. and Christensen, T. 1981. Cancer Lett. 11, 209-214.
Moan, J. and Sommer, S. 1981. Photobiochem. Photobiophys. 3, 93-103.
Moan, J. and Sommer, S. 1984. Photochem. Photobiol. 40, 631-634.
Sandberg, S. and Romslo, I. 1981. Biochem. J. 198, 67-74.

PHOTOSENSITIZATION OF MITOCHONDRIAL ADENOSINE-TRIPHOSPHATASE

AND ADENYLATE KINASE BY HEMATOPORPHYRIN DERIVATIVE IN VITRO

Nai-Wu Fu, Shu-Yong Yeh, Chin Chang
Xiu-Hua Zhao, and Li-Sung Chang

Department of Pharmacology, Cancer Institute
Chinese Academy of Medical Sciences, Beijing

INTRODUCTION

In study of the site(s) of action of HPD photoirradiation, our interest centered on the biomembranes and subcellular organelles. Our previous work showed that nucleoside transport was significantly inhibited by HPD photosensitization (1,2). Incorporation of radioactive nucleosides into acid-soluble and acid-insoluble fractions of cell extracts was also reduced, and this reduction was not due to impaired nucleoside or nucleotide phosphorylation. It is inferred that this inhibition may be the result of an alteration in the plasma membrane induced by HPD photoirradiation. In the present study, we have investigated the photodynamic action of HPD on mitochondrial enzymes and on cytochrome P-450 of mouse liver.

MATERIALS AND METHODS

Animals and Tumor

Female Kunming white mice were supplied by the animal breeding room in our institute. Ascitic hepatoma cells were routinely maintained in the laboratory by weekly intraperitoneal passage in Kunming white mice of 18 to 22 gm.

Chemicals

HPD was obtained from Institute of Pharmaceutical Industry, Beijing. Vials containing 5 mg HPD per ml of 0.9% NaCl solution were stored in the dark until used. All the other chemicals were purchased from Beijing Chemical Co.

Preparation of Mitochondria

Mitochondria were prepared according to Gibson (3). Briefly, this consisted of sonication of ascitic tumor cells, a low-speed centrifugation to remove debris, a 15,000 x g centrifugation to obtain a pellet containing mitochondria, resuspension and centrifugation at 30,000 x g for 30 min at 4° to obtain the mitochondria pellet. This pellet was suspended in PBS: final concentration = 1 to 2 mg of protein/ml.

Table 1. Photoradiation effect of various concentrations of
HPD on mitochondrial ATPase of hepatoma cells.

| HPD (μg/ml) | ATPase activity (Pi nmole/mg protein/min)[a] | | | |
	Control	Irradiation	HPD	HPD+light
10	51.0±6.5[b]	45.0±2.8	42.0±3.1	5.9±0.01[c]
5	90.0±17.0	74.0±11.0	74.0±73.0	7.9±1.7[d]
2.5	95.0±9.6	92.0±25.0	99.0±21.0	27.0±1.1[c]
1.25	95.0±9.6	92.0±25.0	95.0±11.0	57.0±32.0

[a] Control: no photoradiation and no HPD. HPD was added into the mitochondrial suspension and incubated at 37°C for 30 min, then irradiated
for 10 min.

[b] Mean±S.E. of 3 separate experiments.

[c] $p < 0.01$ by the Student t test.

[d] $p < 0.001$ by the Student t test.

Cytochrome P-450 from liver microsomes was assayed according to
Omura (4). The difference spectrum (base line) over the range 490-400
nm was measured in a Shimadzu UV-200 double beam spectrophotometer, and
the P-450 content determined by the value of spectral difference divided
by molar extinction coefficient 91 $cm^{-1}nm^{-1}$. G-6-Pase activity was
measured as described by Harris (5). Mitochondrial ATPase was assayed
according to Emmelot (6). Release of orthophosphate from ATP was
measured with the Fiske-Subbarow procedure (7). Adenylate kinase activity was assayed by the method of Criss (8). Malonaldehyde was
measured as described by Placer (9). TG was assayed according to Felix
(10). Protein was measured by the method of Lowry (11).

Photoradiation studies in vitro

HPD solutions were added into the 2 ml aliquots of mitochondrial
suspensions. The mitochondria-HPD mixtures were incubated at 37° in the
dark for 30 min, then placed 1 cm from two 20-watt fluorescent bulbs at
room temperature. The total photoradiation dose was 0.2 mW/cm². After
photoirradiation, the mixtures were centrifuged and the pellet was
resuspended in PBS and used for enzyme assay.

RESULTS

Table 1 shows inhibition of ATPase of mitochondria by various concentrations of HPD in the presence of light. When irradiation time was
10 min, doses lower than 2.5 μg/ml showed no effect on ATPase activity,
and neither light alone nor HPD in the absence of light had a significant effect on ATPase. As seen from the data in Table 2, when HPD
and Photofrin II were used at the same dose, Photofrin II showed much
less effect than did HPD. If irradiation time was increased, a lower

Table 2. Inhibition of mitochondrial ATPase of hepatoma cells
by HPD- and Photofrin II-induced photosensitization

Drug	Percent of Inhibition (%) [a] HPD Concentration (μg/ml)			
	1.25	2.5	5.0	10.0
HPD	39.3[b]	71.8	91.2	88.5
Photofrin II	No effect	57.6	53.8	68.8

[a]Values are represented as percentage of control (no HPD and no ir-
radiation=100%).

[b]Mean of 2 separate experiments.

Condition of experiments was the same as described in Table 1; irradiation time was 10 min.

Table 3. HPD plus various time of irradiation on
mitochondrial ATPase of hepatoma cells

Group	ATPase activity (Pi nmole/mg Protein/min)
Control	128.0±23.0[a]
Irradiation 5 min	126.0±1.1
Irradiation 30 min	137.0±11.0
HPD 0.625 μg/ml	131.0±12.0
HPD 0.625 μg/ml + Irradiation 5 min	106.0±20
HPD 0.625 μg/ml + Irradiation 30 min	83.0±17.0[b]

[a]Mean±S.E. of 3 separate experiments

[b]$p < 0.05$ by the Student t test.

Experimental conditions were the same as described in Table 1.

Table 4. HPD-induced photosensitization of mitochondrial
adenylate kinase of hepatoma cells in Vitro

Group	Adenylate kinase activity (ADP μmole/mg protein/min)
Control	1.01±0.13[a]
Irradiation 10 min	1.01±0.42
HPD 10 μg/ml	0.78±0.38
HPD 10 μg/ml + Irradiation 10 min	0.41±0.27[b]

[a]Mean±S.E. of 4 separate experiments.

[b]$p < 0.05$ by the Student t test.

Experimental conditions were the same as described in Table 1.

dose of HPD (0.625μg/ml) could also photosensitize ATPase (Table 3).

Adenylate kinase activity of hepatoma mitochondria was greatly inhibited by HPD at the dose of 10 μg/ml after a 10 min irradiation (Table
4).

Measurement of malonaldehyde (MDA) is one of the most useful
methods (9,12) for determination of lipid-peroxidation products in
studies of lipid-peroxidation damage. As shown in Table 5, the mitochondrial MDA content was greatly elevated when HPD was used at the dose
of 10 μg/ml and irradiation time was 30 min. Neither irradiation alone
nor HPD in the dark affected the MDA content (Table 5).

The data presented in Table 6 demonstrate that mitochondrial G-6-
phosphatase activity in hepatoma cells was not sensitive to HPD photosensitization, using an HPD at the dose of 10 μg/ml and irradiation for
30 min.

The optical density (450 nm) representing P-450 of mouse liver
microsomes was increased 2.5-fold after an intraperitoneal injection of
HPD for 4 days at the dose of 5 mg/kg per day. The liver weight, G-6-
Pase, MDA and TG content showed no difference with that of the control
(Table 7).

DISCUSSION

The data presented here demonstrate that mitochondria displayed a
HPD-induced photosensitization, as evidenced by a significant inhibition
of ATPase and adenylate kinase activity. The inhibition of ATPase that
HPD was a more effective agent for the activity was related to the concentration of HPD in vitro, as well as to the length of time of photoirradiation. The inhibitory effect of Photofrin II on ATPase was much
less than that of HPD.

This result was in accordance with our previous work (1). In this

Table 5. Lipid peroxidation of mitochondrial membrane of hepatoma cells by HPD photosensitization[a]

Group	MDA content (nmol/mg protein) Incubation time at 37°C after irradiation		
	30 min	60 min	90 min
Control	0.42 ± 0.14[b]	0.44 ± 0.06	0.38 ± 0.11
Irradiation 30 min	0.38 ± 0.06	0.28 ± 0.12	0.34 ± 0.08
HPD 10 μg/ml	0.34 ± 0.04	0.28 ± 0.08	0.36 ± 0.29
HPD 10 μg/ml + Irradiation 30 min	1.29 ± 0.15[c]	1.75 ± 0.39[d]	1.51 ± 0.08[e]

[a] HPD was added in the mitochondrial suspension and incubated at 37°C in the dark for 30 min. Then the suspension was irradiated for 30 min and incubated at 37°C in the dark for specified periods.

[b] Mean±S.E. of 4 separate experiments

[c] $p<0.01$ by the Student t test

[d] $p<0.05$ by the Student t test

[e] $p<0.02$ by the Student t test

Table 6. Photodynamic action of HPD on mitochondrial G-6-Pase activity of hepatoma cells

Group	G-6-Pase activity (Pi μmol/mg protein/min)[a]
Control	1.45 ± 0.85[b]
Irradiation 30 min	1.49 ± 0.95
HPD 10 μg/ml	1.80 ± 1.14
HPD 10 μg/ml + Irradiation 30 min	1.90 ± 1.41

[a] The results of tested groups showed no significant difference with that of the control by the Student t test

[b] Mean±S.E. of 4 separate experiments

Table 7. Influence of HPD on cytochrome P-450, G-6-Pase, MDA and TG content of mouse liver microsome

Drug	Liver weight (g/100g Bd.Wt)	P-450 (nmol/mg protein)	MDA (nmol/mg protein)	TG (mg/g liver wet Wt)	G-6-Pase (Pi mol/mg protein/min)
---	5.58±0.63	1.08±0.59	1.31±0.40	19.78±17.90	0.45±0.24
HPD	6.0±0.90	1.78±0.48	1.35±0.27	14.97±4.47	0.59±0.12

HPD was injected intraperitoneally at the dose of 5 mg/Kg per day for 4 days. 24 hours later the mice were killed and the liver microsomes were prepared according to Omura (2).

Mean±S.E. of 5 separate experiments
p<0.01 by the Student t test

work we demonstrated that HPD was a more effective agent for photo-dynamic inhibition of nucleoside transport than was Photofrin II. It appears that different photosensitizing activities of different HPD species could be detected by measuring photodynamic inhibition of membrane transport functions and of mitochondrial enzyme activities. Mito-chondrial and microsomal G-6-Pase activity was not influenced by HPD photoirradiation, even when the irradiation time was extended to 30 min. ATPase is a critical mitochondrial respiratory enzyme and adenylate kinase is a central component in the adenylate energy system of all biological cells. The adenylate energy system plays an important role in the regulation of various metabolic pathways. Inactivation of these enzymes may be a significant contributory factor to photodynamic cytotoxicity. Microsomal cytochrome P-450 of mouse liver was markedly induced by HPD, but liver weight was not increased. These results suggest that the photodynamic effect of HPD on cytochrome P-450 is not the same as phenobarbital which is a P-450 inducer. The biological significance of this inducing effect by HPD is not clear. Two possibilities should be considered: (a) transformation and elimination of HPD in the liver may occur via cytochrome P-450. (b) toxic products of metabolic activation may be related to cytochrome P-450.

The mitochondrial MDA content was markedly increased by HPD photo-sensitization. Since mitochondrial membrane phospholipids contain relatively high concentrations of polyunsaturated fatty acids, they are major sites of lipid peroxidation damage, with a resulting profound effect on the cell (13). Inactivation of mitochondrial enzymes may be related to this lipid peroxidation, but the mechanism of such inactivation is not clear.

SUMMARY

Mitochondrial ATPase and adenylate kinase activity of hepatoma cells were inhibited by hematoporphyrin derivative (HPD) followed by photoirradiation. Inhibition of ATPase activity was a dose- and time-related event. Malonaldehyde (MDA) content of mitochondrial membranes was markedly increased by HPD plus light. The content of mouse liver microsomal cytochrome P-450 was greatly increased after intraperitoneal injection of HPD for 4 days (5 mg/kg/day). The liver weight, and levels of liver microsomal G-6-phosphatase, MDA and triglyceride (TG) showed no difference in treated vs. control animals. The data presented here demonstrate that mitochondria may be a sensitive site of action of HPD photosensitization, and inactivation of ATPase and adenylate kinase may be an important contributing factor to tumor cell damage and death.

REFERENCES

1. Nai-Wu, Fu, et al. (1984) Chinese Oncology. 6, 232.
2. Nai-Wu, Fu, et al. (In Press) Acta Academiae Medicinae Sinicae.
3. Gibson, SL, et al. (1983) Cancer Research, 43, 4191.
4. Omura, T, et al. (1964) J. Biol. Chem., 239, 2370.
5. Harris, RC, et al. (1954) J. Clin. Invest. 33, 1204.
6. Emmelot, P, et al. (1954) Brit. J. Cancer 13, 348.
7. Fiske, CH, et al. (1925) J. Biol. Chem. 66, 375.
8. Criss, WE, et al. (1978) Methods in Enzymol. 51, 459.
9. Placer, ZA, et al. (1966) Anal. Biochem. 16, 358.
10. Felix, GS (1971) Clin. Chem. 17, 529.
11. Lowry, OH, et al. (1951) J. Biol. Chem. 193, 265.
12. Sawicki, E, et al. (1963) Anal. Chem. 35, 199.
13. Tappel, AL (1973) Fed. Proc. 32, 1870.

PORPHYRIN-MEMBRANE INTERACTIONS: STRUCTURAL, KINETIC AND THERMODYNAMIC ASPECTS STUDIED USING FLUORESCENCE TECHNIQUES

R.C. Chatelier,* W.H. Sawyer,*
A.G. Swincer,[†] and A.D. Ward[†]

* The Russell Grimwade School of Biochemistry,[§]
University of Melbourne
Parkville, Victoria, 3052

† The Department of Organic Chemistry
University of Adelaide
Adelaide, South Australia, 5001

INTRODUCTION

"Haematoporphyrin derivative" (HPD) is being used increasingly in the photochemical treatment of malignant tumours. Its mode of action has yet to be elucidated. It is believed that the exposure of HPD to red light results in the production of singlet oxygen which reacts with surrounding bimolecules and kills the cell.[1] The primary site of action is thought to be at the membrane surfaces,[2] where porphyrins collect at hydrophobic loci[3] and catalyse the cross-linking of membrane proteins[4,5] which leads to cellular necrosis. We have investigated the structural, kinetic and thermodynamic aspects of HPD association with synthetic membranes as an aid to understanding its efficacy in photo-irradiation therapy. Proto-porphyrin (PP) and haematoporphyrin (HP) have also been examined, and the results are presented here for comparison.

The rate and extent of porphyrin uptake into lipid bilayers was monitored using the technique of fluorescence enhancement. The transverse organization of the porphyrins in membranes was determined by quenching their fluorescence with a set of n-doxyl stearates (n-NS). The quenching (doxyl) moiety of the n-NS probes is attached to various carbon atoms of a stearic acid chain. It therefore locates at a graded series of depths

---

§ Address correspondence to Dr. Sawyer at this address.

in the membrane and is capable of sensing the transverse location of the fluorescent porphyrins.[6,7]

MATERIALS AND METHODS

The preparation of HP, HPD and PP is described elsewhere.[8] The aggregate fraction of HPD from a Bio-Gel P-10 column was used; this corresponds to commercially available Photofrin II. Egg yolk phosphatidyl-choline (eggPC) was purchased from Lipid Products (Nutfield Nurseries, England). n-Doxyl stearates (n-NS; n = 5,7,10,12,16) were purchased from Molecular Probes (Junction City, Oregon) and their concentrations determined by ESR spectroscopy. All chemicals were of analytical grade.

Fluorescence was measured and lipid vesicles were prepared in 0.1 M Tris-HCl buffer, pH 7.5 as previously described.[9] Porphyrins were added as 5-20 µl aliquots of a 0.1 mg/ml aqueous stock solution to a 3 ml vesicle dispersion. n-Doxyl stearates were added as 5 µl aliquots of 2 mM methanolic stock solutions. There was no time-dependence of the fluore-scence signal after quencher addition.

THEORY
Partition Coefficient of a Fluorophore

Consider a partitioning fluorophore whose quantum yield of fluore-scence is higher in the membrane than in the aqueous phase. When the total concentration of fluorophore ($[F]_T$) is fixed and the concentration of membrane ($[M]_T$) is varied:

$$I_{obs} = I_M f + I_A (1-f) \qquad (1)$$

where f is the fraction of fluorophore associated with the membrane, I_{obs} is the observed intensity of fluorescence, and I_M and I_A are the fluore-scence intensities when the fluorophore is fully associated with the membrane and aqueous phase, respectively. I_M may be obtained by extra-polating a plot of $1/I_{obs}$ vs $1/[M]_T$ to $1/[M]_T = 0$.[10] Rearranging equation 1:

$$f = \frac{(I_{obs} - I_A)}{(I_M - I_A)} \qquad (2)$$

The concentrations of fluorophore in the membrane and aqueous phases may be calculated:

$$[F']_M = \frac{f[F]_T}{\bar{v}[M]_T} \qquad (3)$$

170

$$[F']_A = [F]_T (1-f) \tag{4}$$

where the primed symbols, $[F']_M$ and $[F']_A$, refer to concentrations relative to the volume of the membrane (V_M) and aqueous phase (V_A), respectively, and $\bar{V}$ is the volume of one mole of membranes. V_M is small compared to the total volume, V_T, and thus $V_A \simeq V_T$. If the fluorophore exhibits "ideal partitioning", then a plot of $[F']_M$ vs $[F']_A$ will be linear with slope equal to the partition coefficient $(K_p = [F']_M/[F']_A)$.

Analysis of Fluorescence Quenching

When quenching takes place by a mixture of static and dynamic mechanisms, the degree of quenching of the membrane-bound fluorophore by the partitioning quencher is dependent on the lifetime of the fluorophore, the partition coefficient of the quencher, the charges on the two molecules, the viscosities of their environments, and their proximity to one another. An effective way to obtain an index of fluorophore-quencher proximity from fluorescence quenching measurements is as follows:

Consider a series of Stern-Volmer plots obtained at various membrane concentrations. As the volume of the membrane phase is increased, the degree of quenching decreases because the quencher is diluted in the membrane phase. A given level of quenching corresponds to a particular average number of quenchers in the membrane.[11] Invoking the law of conservation of mass:

$$[Q]_T = [Q']_M \frac{V_M}{V_T} + [Q']_A \tag{5}$$

where V_M/V_T is the ratio of the membrane volume to the total volume, and $[Q]_T$, $[Q']_M$ and $[Q']_A$ are the quencher concentrations defined analogously to $[F]_T$, $[F']_M$ and $[F']_A$ above. Thus a plot of $[Q]_T$ vs V_M/V_T, at a particular quenching efficiency, is linear with slope $[Q']_M$ and ordinate intercept $[Q']_A$. This approach yields the "absolute" partition coefficient of the quencher $(K_{p,abs} = [Q']_M/[Q']_A)$, that is, a value of K_p which is independent of the organization of fluorophore and quencher within the membrane.

The same set of data can be used to obtain a "local" partition coefficient $(K_{p,local})$ which is dependent on the proximity of fluorophore and quencher.[11] In this case, the quenching process may be described in terms of a rearranged Stern-Volmer equation[12,13]

$$\frac{[Q]_T \, I}{I_o - I} = \frac{1}{k_q \tau_o} \frac{V_M}{V_T} + \frac{1}{k_q \tau_o \, K_{p,local}} \tag{6}$$

171

where I_o and I are the fluorescence intensities in the absence and presence of quencher, respectively, k_q is the bimolecular rate constant, and I_o is the excited-state lifetime of the fluorophore. A plot of the left-hand side of equation 6 vs V_M/V_T is linear with slope = $1/k_q\tau_o$ and slope/ordinate intercept = $K_{p,local}$.

A proximity index (V_R) can be defined in terms of the local and absolute partition coefficients:[13]

$$V_R = \frac{K_{p,abs}}{K_{p,local}} \tag{7}$$

As the fluorophore and quencher approach one another, $K_{p,local}$ increases relative to $K_{p,abs}$ and therefore the value of V_R decreases.[13]

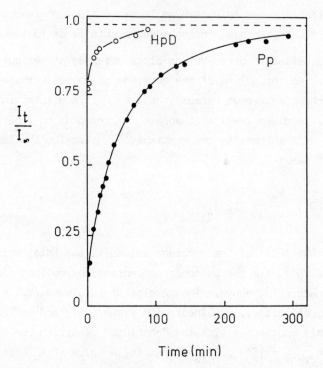

Figure 1. Time course of uptake of HPD and PP into eggPC vesicles. I_t and I_∞ refer to the fluorescence intensity at time = t and time = "infinity", respectively. Excitation and emission wavelengths were 410/634 nm and 398/625 nm for PP and HPD, respectively. Slit widths were 6/10 nm. Temperature = 20°C. [Lipid] = 1 mM.

RESULTS

Kinetics of Porphyrin Uptake into Membranes

Figure 1 shows the time course of fluorescence enhancement after PP and HPD were added to eggPC vesicles. The uptake of porphyrins was very slow and occurred on the time scale of hours. An attempt to analyse the data in Figure 1 according to first-order kinetics was unsuccessful since a plot of $\ln (F_{\infty}-F_t)$ *versus* time was non-linear (data not shown). This non-linearity was especially evident when HPD was used as the fluorophore, possibly because of its heterogeneous nature. We can make a qualitative statement, however, that the uptake of HPD was more rapid than PP (90% uptake in < 20 min compared with > 3 h).

Partition Coefficients of Porphyrins

The fluorescence of PP and HPD was enhanced 60- and 9-fold, respectively, when they were incorporated into membranes (see Figure 2a). The fluorescence of HP was not changed significantly by the presence of membrane. When the data for PP and HPD are treated using equations 2-4 the plot of $[F']_M$ *vs* $[F']_A$ is linear (Figure 2b) indicating that the uptake is adequately described by a partitioning process. The partition coefficients of PP and HPD were 19,000 and 43,000, respectively. The absence of any saturation effect in Figure 2b indicates that the uptake does not occur by a binding process although a wide range of fluorophore concentrations needs to be examined to establish this conclusively. The possible association of HP with membranes was investigated by an alternative procedure. Multilamellar eggPC liposomes were mixed with HP and equilibration was allowed to occur overnight. The liposomes were pelleted in an Airfuge and the amount of HP remaining in the supernatant was determined. The partition coefficient was < 200 even when no corrections were made for occlusion of HP between vesicles and for entrapment of HP within the aqueous interior of the vesicles.

The Transverse Organization of Protoporphyrin in Lipid Bilayers

Figure 3 shows Stern-Volmer plots of PP quenched by 7-NS at various lipid concentrations. The data were analysed according to equations 5 and 6 to obtain $K_{p,abs}$ and $K_{p,local}$, respectively.[13] The value of V_R was determined from the two partition coefficients (equation 7). Figure 4 shows a plot of V_R *versus* the carbon number of the n-NS quenchers. Consider the deepest quencher (16-NS) first. Its value of V_R is close to unity indicating that it is far away from the fluorophore and that quenching occurs throughout the membrane volume.[13] As the quencher is moved closer to the membrane surface, V_R decreases. This indicates that the quencher is getting closer to the fluorophore and the quenching inter-

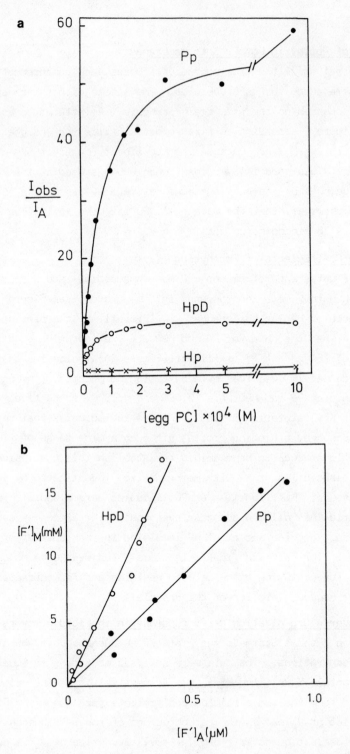

Figure 2. (a) Uptake of HP, HPD and PP as a function of lipid concentration. Excitation and emission wavelengths were 398/615 for HP. Other conditions as in Figure 1.
(b) Data in Figure 2a treated using equations 2-4 and replotted as $[F']_M$ vs $[F']_A$.

actions are confined to a smaller volume. 5-Doxyl stearate (5-NS) was also used but its partition coefficient was so large that reliable estimates of V_R could not be obtained.

Stern-Volmer plots obtained by quenching HPD with the n-NS probes tended to plateau (data not shown). The possible implications of this phenomenon will be explored below.

DISCUSSION

The slow uptake of PP and HPD into membranes (Figure 1) indicates that the rate was not diffusion controlled. The dissociation of the large aggregates of PP and HPD[14] or the intercalation of the bulky porphyrin ring into the membrane may have resulted in such a slow rate-limiting step for the uptake. A similar, slow rate of uptake into tissues occurs *in vivo*.[15]

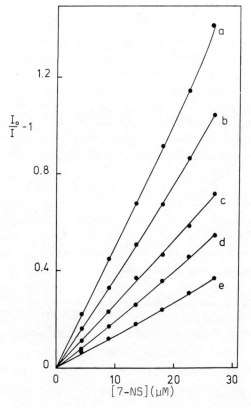

Figure 3. Stern-Volmer plots for the quenching of PP by 7-NS at various concentrations of eggPC (a) 0.075 mM, (b) 0.1 mM, (c) 0.15 mM, (d) 0.2 mM, (e) 0.3 mM. Other conditions as in Figure 1.

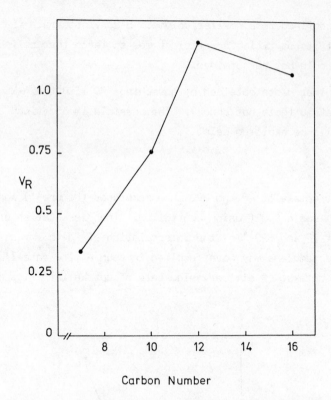

Figure 4. A plot of V_R *versus* the carbon number of
the n-NS quenchers for the quenching of
PP in eggPC vesicles.

The fluorescence of PP and HPD (but not HP) was considerably enhanced
in the presence of membranes (Figure 2). This may be because PP and HPD
exist as self-quenched aggregates in solution but as fluorescent species
in the membrane. Haematoporphyrin, on the other hand, probably exists as
a fluorescent monomer in the aqueous phase and does not interact with the
membrane. The ability of porphyrins to form aggregates (possibly in the
form of micelles) in aqueous solution seems to be related to their
membrane/buffer partition coefficients. This is because the "hydrophobic
effect"[16] influences both micelle formation and partitioning between
phases.

The partition coefficients of drugs and anaesthetics between the
aqueous and membrane phases have pharmacological significance.[3,17] Haema-

toporphyrin, which is the least cytotoxic of the three species studied, has the lowest value of K_p (< 200) while HPD and PP have high cytotoxicities[8] and K_p values of 43,000 and 19,000, respectively. This correlation does not give any indication of relative *in vivo* tumour localising ability, however, as PP does not give any tumour fluorescence, whilst that for HPD is quite strong.[18,19] Values of K_p influence the concentration of porphyrin in the membrane, their ability to bind to membrane proteins[20] as well as their rate of transport across the membrane.[21,22] Hence these partition coefficients do not tell us whether HPD acts on the plasma membrane or the components within the cell.

The data in Figure 4 imply that PP is located near the membrane surface. Its two carboxyl groups probably prevent the hydrophobic ring from sinking too deep into the membrane. Fluorescence quenching can yield information on the location of a fluorophore (or quencher) in the membrane but not on the orientation. We can predict the orientation of PP in lipid bilayers, however, on the basis of thermodynamic principles. PP probably behaves like a typical amphiphile with its charged carboxyl groups located at the membrane/buffer interface and its non-polar section in contact with the hydrophobic chains of the lipids (see Figure 5). In-spection of Figure 5 reveals that the addition of hydroxyl groups at the positions on the porphyrin ring marked by arrows would change the inter-action of the porphyrin with the membrane because the placement of polar groups in the hydrophobic interior of the bilayer is energetically un-favourable. This is probably why HP has a lower partition coefficient than PP.

Consideration is now given to the behaviour of HPD in lipid bilayers. When the n-(doxyl) stearates are used to quench the fluorescence of HPD in eggPC vesicles, the slopes of the Stern-Volmer plots decrease with increasing n-NS concentration. There are three factors which may contri-bute to such behaviour. Firstly, the quencher may bind to a limited number of sites on the membrane. This is not true for the n-NS/eggPC system (Blatt, Chatelier and Sawyer, manuscript in preparation). Secondly, two or more fluorophore sub-populations with different lifetimes may contribute to the fluorescence;[23] a likely factor in view of the hetero-geneous composition of HPD. Thirdly, there may be differences in the accessibility of the fluorophore sub-populations to the quencher;[24,25] such a situation may arise if the HPD components are located at various depths in the membrane.

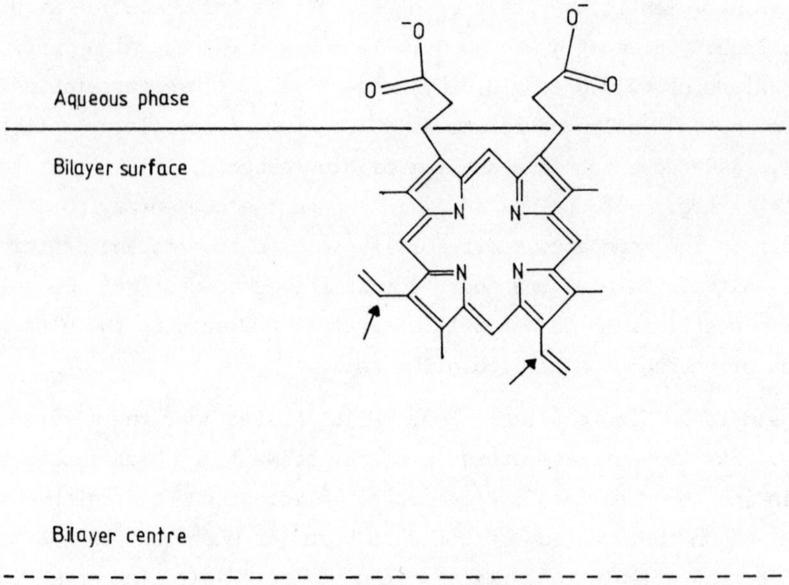

Figure 5. Schematic diagram of the transverse organization of PP in an eggPC leaflet. Addition of a hydroxyl group at the two positions marked by arrows would generate the structure of HP.

The ability of HPD to locate at various depths within the membrane cannot be explained by the simple polarity arguments used for PP and HP. The uptake kinetics and quenching data indicate considerable hetero-geneity in the fluorescent constituents of HPD with respect to their organization in the lipid bilayer. Hence at least some of the components of HPD are able to move freely into and, by implication, through membrane bilayers. This may be an important factor in the unique ability of HPD to cause tumour cell necrosis *in vivo*.

ACKNOWLEDGEMENTS

This research was supported by the Australian Research Grants Committee and the Australian National Health and Medical Research Council. We also thank Miss E. Laubman for a generous donation to the University of Adelaide Phototherapy Fund.

REFERENCES

1. K.R. Weishaupt, C.J. Gomer, and T.J. Dougherty, Identification of singlet oxygen as the cytotoxic agent in photo-inactivation of a murine tumor, Cancer Res. 36:2326 (1976).
2. D. Kessel, Effects of photoactivated porphyrins at the cell surface of leukemia L1210 cells, Biochemistry 16:3443 (1977).
3. K. Kohn, and D. Kessel, On the mode of cytotoxic action of photo-activated porphyrins, Biochem.Pharmacol. 28:2465 (1979).
4. A.W. Girotti, Photodynamic action of protoporphyrin IX on human erythrocytes: cross-linking of membrane proteins, Biochem.Biophys. Res.Commun. 72:1367 (1976).
5. T.M.A.R. Dubbleman, A.F.P.M. De Goeij, and J. Van Steveninck, Photodynamic effects of protoporphyrin on human erythrocytes, Biochim. Biophys.Acta 511:141 (1978).
6. V.G. Bieri, and D.F.H. Wallach, Fluorescence quenching in lecithin and lecithin/cholesterol liposomes by paramagnetic lipid analogues. Introduction of a new probe approach, Biochim.Biophys.Acta 389:413 (1975).
7. J. Luisetti, H. Möhwald, and H.J. Galla, Monitoring the location profile of fluorophores in phosphatidylcholine bilayers by the use of paramagnetic quenching, Biochim.Biophys.Acta 552:519 (1979).
8. P.A. Cowled, I.J. Forbes, A.G. Swincer, V.C. Trenerry, and A.D. Ward, Separation and phototoxicity *in vitro* of some of the components of haematoporphyrin derivative, Photochem.Photobiol. 41:445 (1985).
9. K.R. Thulborn, and W.H. Sawyer, Properties and the locations of a set of fluorescent probes sensitive to the fluidity gradient of the lipid bilayer, Biochim.Biophys.Acta 511:125 (1978).
10. E.A. Haigh, and W.H. Sawyer, Interpretation of double reciprocal plots used to determine the spectroscopic parameters of bound ligand for binding assays, Aust.J.Biol.Sci. 31:1 (1978).
11. M.V. Encinas, and E.A. Lissi, Evaluation of partition constants in compartmentalised systems from fluorescence quenching data, Chem. Phys.Lett. 91:55 (1982).
12. K.A. Sikaris, K.R. Thulborn, and W.H. Sawyer, Resolution of partition coefficients in the transverse plane of the lipid bilayer, Chem. Phys.Lipids 29:23 (1981).
13. E. Blatt, R.C. Chatelier, and W.H. Sawyer, The transverse location of fluorophores in lipid bilayers and micelles as determined by fluorescence quenching techniques, Photochem.Photobiol. 39: 477 (1984).
14. A.G. Swincer, A.D. Ward and G.J. Howlett, The molecular weight of haematoporphyrin derivative, its gel column fractions and some of its components in aqueous solution, Photochem.Photobiol. 41:47 (1985).
15. C.J. Gomer, and T.J. Dougherty, Determination of [^{3}H]- and [^{14}C]-haematoporphyrin derivative distribution in malignant and normal tissue, Cancer Res. 39:146 (1979).
16. C. Tanford, "The Hydrophobic Effect: formation of micelles and biological membranes", J. Wiley and Sons, New York (1973).
17. J.C. Skou, Local anaesthetic potency and penetration of monomolecular layers of nerve lipoids, Nature 174:318 (1954).
18. D. Kessel, and T. Chou, Tumor-localizing components of the porphyrin preparation haematoporphyrin derivative, Cancer Res. 43:1994 (1983).
19. T.J. Dougherty, D.G. Boyle, K.R. Weishaupt, B.A. Henderson, W.R. Potter, D.A. Bellnier, and K.E. Wityk, Photoradiation therapy - clinical and drug advances, in "Advances in Experimental Medicine and Biology. 160. Porphyrin Photosensitization" D. Kessel and T.J. Dougherty, eds., Plenum Press, New York (1983).

20. C.D. Richards, K. Martin, S. Gregory, C.A. Keightley, T.R. Hesketh, G.A. Smith, G.B. Warren, and J.C. Metcalfe, Degenerate perturbations of protein structure as the mechanism of anaesthetic action, Nature 276:775 (1978).
21. W.R. Lieb, and W.D. Stein, Biological membranes behave as non-porous polymeric sheets with respect to the diffusion of non-electrolytes, Nature, 224:240 (1969).
22. T. Wang, G.T. Rich, W.R. Galey, and A.K. Solomon, Relation between adsorption at an oil/water interface and membrane permeability, Biochim.Biophys.Acta 255:691 (1972).
23. J.C. Dalton, and N.J. Turro, Kinetic analyses of photochemical reactions which involve quenching of more than one excited state, Mol.Photochem. 2:133 (1970).
24. S.S. Lehrer, Solute perturbation of protein fluorescence. The quenching of the tryptophyl fluorescence of model compounds and of lysozyme by iodide ion, Biochemistry 10:3254 (1971).
25. M.R. Eftink, and C.A. Ghiron, Fluorescence quenching studies with proteins, Anal.Biochem. 114:199 (1981).

PHOTOPHYSICS AND PHOTOCHEMISTRY OF HEMATOPORPHYRIN, HEMATOPORPHYRIN DERIVATIVE AND UROPORPHYRIN I

L. I. Grossweiner, A. Blum and G. C. Goyal

Physics Department
Illinois Institute of Technology
Chicago, Illinois 60616

INTRODUCTION

The putative action mechanism in photodynamic therapy (PDT) involves serum transport of HPD to tumor tissue, localization and retention of the active constituent, generation of singlet molecular oxygen (Δ) by the action of visible light and the attack of Δ on the cellular targets. Early workers postulated that tumor tissue membranes are key targets in PDT (Dougherty et al., 1978), which is consistent with evidence that photosensitization of red blood cell membranes by protoporphyrin is mediated by Δ (Lamola et al., 1973). Subsequent studies on photosensitization of model membranes by hematoporphyrin (HP) and HPD are consistent with this hypothesis. However, the specific targets have not been identified and the involvement of non-membrane targets has not been ruled out. HPD is a porphyrin mixture leading to a complicated dependence of the photophysical and photochemical properties on the medium. The active constituent has not been adequately characterized, although there is evidence that it is a covalent dimer or oligomer of porphyrin units (Berenbaum et al., 1982; Dougherty et al., 1984; Swincer et al., 1985). Dougherty et al. (1984) refer to this material as dihematoporphyrin ether (DHE), which will be used for convenience in this paper, although the structure implied by the terminology has not been proven. The present results were obtained with an enriched HPD material prepared by polyacrylamide gel filtration of HPD referred to as HpD-A (Grossweiner and Goyal, 1983). According to Dougherty et al. (1984) a similar preparation contained 80-90% DHE. Results are reported for uroporphyrin I (Uro-I), which was proposed by El Far and Pimstone (1984) as a superior tumor photosensitizer than HPD. This work emphasizes photosensitization of two substrates that represent targets accessible to HPD in vivo. Egg phosphatidylcholine (EPC) liposomes were employed as models of biomembranes. Human serum albumin (HSA) was the other model substrate investigated in this work. The experimental methods described include isolation of HpD-A, liposome preparation, sizing and photosensitization, porphyrin binding and photosensitization of HSA, and measurement of singlet oxygen generation with the "RNO" method. The results demonstrate the role of the microenvironment on porphyrin photophysics and initial photochemistry and support, on the whole, the early speculation that singlet oxygen attack on membrane targets is a key factor in the molecular action mechanism of PDT.

MATERIALS AND METHODS

This section summarizes and updates experimental methods reported in
recent publications. In carrying out spectroscopic measurements on porphyrins, care should be taken to minimize exposure to room light and photolysis
in measuring instruments. Aqueous solutions of many porphyrins do not obey
Beer's law because of aggregations effects. The use of short path cuvettes is
preferable to dilution when spectrophotometric measurements on the Soret
band exceed the range of the instrument. Free base porphyrins may extract
metals from glassware, buffers and chromatography media, especially at low
concentrations. This contamination can be minimized by using fresh solutions
and washing glassware with glacial acetic acid. The growth of a fluorescence
band at 580 nm in aqueous porphyrins provides a sensitive test for metal
uptake (Ricchelli and Grossweiner, 1984).

Crude HPD (the brown powder) was obtained from Porphyrin Products
(Provo, UT) as Hematoporphyrin D and prepared from hematoporphyrin dihydrochloride (Sigma Chemical Co., St. Louis, MO) with the method of Lipson et al.
(1961) as modified by Dougherty et al. (1978). This material was dissolved in
0.1 M NaOH (2.5 mg/ml) and stirred for 30 min in the dark, neutralized with
1 M HCl and adjusted to pH 7.0 or 7.4 in 10 mM phosphate buffer (PB). The
resulting HPD had the same absorption spectrum and extinction coefficient
as reported by Doiron et al. (1984) for HPD in saline and by Poletti et al.
(1984) for Photofrin in neat water. The Soret band maximum was 370±2 nm and
the maximum absorbance was 0.50±0.02/cm at 0.05 mg/ml. HpD-A was prepared by
polyacrylamide gel filtration of HPD on Bio-Gel P-10 (Bio-Rad Laboratories,
Richmond, CA) using PB as the eluant (Grossweiner and Goyal, 1983; Goyal et
al., 1983). A 2 mg sample of HPD in PB was loaded on a 27 cm x 1 cm column
and 1.3 ml fractions were collected at a flow rate of 9.6 ml/hr. The fastest
moving material was HpD-A with the Soret peak at 363±2 nm and the absorbance
ratio at 365 nm to 390 nm equal to 1.50±0.1. The separation was faster and
confined to smaller volumes when the column was run at 50 °C, providing a
more concentrated HpD-A solution. HpD-A comprised approximately 35% of the
initltial HPD. The molar extinction coefficient of HpD-A determined by carrying out the gel separation in distilled water and evaporating to dryness was
66,000 at 365 nm and 42,00 at 390 nm for nominal 600 dalton molecular weight.
Uroporphyrin I was obtained from Porphyrin Products (98% purity) and used as
received. Human serum albumin was obtained from Sigma Chemical Co. (fraction
V, < 0.005% fatty acid) and used as received. The other chemicals were the
best available biochemical grades.

Liposome Preparation

Liposomes were prepared from egg yolk L-α-phosphatidylcholine (Sigma
Chemical Co., type VII) in chloroform. The material was handled in syringes
and vessels filled with nitrogen to minimize autooxidation. A 15 mg EPC
aliquot was evaporated to dryness with a nitrogen stream at the bottom
of a 10 mm test tube, the dry lipid film was dispersed in 2 ml of PB by
vortexing for 3 min and allowed to swell for at least 3 hr at 3 °C. The
resultant liposomes had a wide size distribution, the largest of which were
visible with the light microscope. Small unilammelar vesicles (SUV) were
obtained by probe sonicating for 2 hr with a Heat Systems Ultrasonics Model
W-10 sonicator using a 2.4 mm tungsten alloy tip. This operation was carried
out inside a polyethylene glove bag that had been filled with nitrogen and
the liposome suspension was saturated with nitrogen by bubbling. The resultant SUV suspension was clear to the eye. The size distribution was estimated
by gel filtration on Sephacryl S-1000 (Pharmacia Fine Chemicals, Inc.,
Piscataway, NJ) following the procedures of Nozaki et al. (1983). A 30 cm x
14 mm column was eluted with PB at 3 ml/hr. This flow rate required a high
hydrodynamic head and is best attained with a pump. The SUV eluted in a
symmetrical light scattering band at ratio of elution volume to void volume

equal to 1.80. The column calibration based on the data of Nozaki et al.
(1983) and specifications from Pharmacia indicate that the average particle
diameter was 40 nm. The uptake of HpD-A by the SUV was complete when
0.1 mg/ml HpD-A was mixed 1:1 with a 4 mg/ml SUV suspension and incubated for
several hours at 38 °C. Complete uptake was shown by passing the dyed SUV
through a Sephadex G-50 (fine) column. A single band at the void volume was
identified by light scattering at 335 nm and absorption at 398 nm, with no
evidence of porphyrins in the slower eluting fractions.

Irradiations

The light source was a 200 watt mercury-xenon arc (Hanovia 901-B1) in
an Oriel Corp. (Stratford, CT) Model 6137 lamp housing plus appropriate
filters. The solutions were irradiated in 4 x 1 x 1 cm polystyrene cuvettes
located on a water-cooled sample holder with temperature control to ± 0.1 °C.
They were bubbled with oxygen or nitrogen prior to and during the irradia-
tions. The incident intensity levels were estimated with an Eppley thermopile as the difference between the readings with sharp-cut glass filters.
This measurement provides only a rough estimate of the fluence levels and was
employed to correct for aging of the arc lamps. Most experiments were done
with Corning C.S. No. 0-52 (> 350 nm) and 3-70 (> 500 nm) filters. The Δ
quantum yields were measured at 546 nm with a filter combination consisting a
Corning C.S. No. 3-69 filter (> 510 nm), 1 cm of 60% (w/w) $NdNO_3$ and 1 cm of
a solution consisting of 20% $CuCl_2.H_2O$ and 27% $CaCl_2$.

Serum Albumin Binding Measurements

Binding of HPD-A to HSA was measured with the dialysis method using
Spectrapor-I (Spectrum Medical Industries Inc., Los Angeles, CA) membrane
tubing. Mixtures of HpD-A and HSA (2 ml) in pH 7.4 phosphate buffer (10 mM)
plus 0.9 % NaCl (PBS) were loaded into pre-boiled measured lengths of tubing
and placed inside sealed flasks containing 20 ml of PBS at 38 °C. The HpD-A
concentration in the outer solution was measured after 22 and 42 hrs. In
prior experiments with HP (Richard et al., 1983) the dialysate was diluted
to 1-3 µM in order to employ a constant extinction coefficient, but this
was not necessary with HpD-A which obeyed Beer's law from 3 to 300 µM
(Grossweiner and Goyal, 1984). Porphyrin holdup by the tubing was measured at
different HpD-A concentrations without HSA. The dialysis measurements with
with HSA were corrected by assuming two limiting assumptions: (a) the non-
dialysing ligand was proportional to the total ligand; (b) the non-dialysing
ligand was proportional to the free ligand. The details are given by
Grossweiner and Goyal (1984).

Singlet Oxygen Quantum Yields

The quantum yield of Δ generation by prophyrins in phosphate buffer
and in the presence of SUV was determined with the N,N-dimethyl-4-nitroso-
aniline (RNO) method (Kraljic and El Mohsni, 1978). Porphyrin solutions of
OD/cm ≈ 0.2 at 546 nm were prepared in PB containing 0.01 M histidine plus
100 mM RNO. The solutions were irradiated at 546 nm with continuous
bubbling by oxygen or argon. RNO bleaching was measured at 440 nm at 2 min
intervals for at least 20 minutes. The fractional absorption by the sensitizer (f_s) was calculated with the relationship: $f_s = [A_s/A_{tot}]$ x
$[1 - Exp-2.303\ A_{tot}]$, where A_s is the initial optical density of the
sensitizer alone in PB at 546 nm and A_{tot} is the optical density of the
sensitizer-RNO solution, plus light scattering by SUV when present. The
bleaching rates were corrected for aging of the lamps by measuring the
incident fluence before each run with an Eppley thermopile. The relative
singlet oxygen quantum yields (Φ_Δ) were obtained by dividing the corrected
bleaching rates by f_s. The absolute values of Φ_Δ were obtained by scaling
to $\Phi_\Delta = 0.75$ for rose bengal (RB) reported by Gandin et al. (1983) in neat

water based on the RNO method, and by M. A. J. Rodgers (private communication) for RB in pH 7.4 PB based on flash photolysis bleaching of 1,3-diphenylisobenzofuran. The RNO method showed an inherent dependence of the bleaching rate on pH from pH 8.0 to pH 5.7 (Blum and Grossweiner, 1985) and all measurements were made in pH 7.0 and pH 7.4 PB.

RESULTS

Initial Photophysics and Structure of Hematoporphyrin Derivative

The spectral and initial photophysical properties of HPD and related porphyrins are summarized in Table 1. It is generally assumed that HP is "aggregated" in aqueous media, as evidenced by the large blue shift of the Soret band with increasing concentration accompanied by fluorescence quenching. However, the intersystem crossing efficiency (Φ_T) and singlet oxygen quantum yield (Φ_Δ) remain high and independent of concentration. As discussed below, the available evidence indicates that the van der Waals dimeric state dominates in aqueous media and the large blue shift of the Soret band is associated with weak aggregates that do not affect the triplet state photochemistry. The shift of the Soret band to 400 nm and high values of Φ_Δ for HP in non-polar media and aqueous Uro-I are characteristic of porphyrin monomers. HpD-A appears to resemble HP in that the Soret peak shifts from 365 nm in aqueous media to 398 nm in SUV, accompanied by a dramatic increase of from 0.06 to 0.87. However, the blue-shifted Soret peak was unchanged from 3-300 μM (Grossweiner and Goyal, 1984), which can be explained either by unusually strong aggregation or its complete absence. Alternatively, the apparent aggregation of HpD-A may result from folding of the dimeric components to a "clam shell" conformation (Grossweiner and Goyal, 1984; Kessel and Cheng, 1985). It is difficult to draw specific conclusions about the properties of HPD which contains approximately equal amounts of DHE and HP plus other porphyrins. The Soret band of HPD in PB has two components, probably corresponding to DHE plus porphyrin aggregates (370 nm) and porphyrin monomers (400 nm). Partial disaggregation occurs in methanol with the shoulder appearing near 375 nm and the main band near 400 nm.

Photosensitization of Liposomal Membranes

Earlier work on photosensitization of large EPC liposomes by HP showed that the mechanism depends on the sensitizer concentration (Grossweiner et al., 1982). At low concentrations of external or incorporated HP, membrane lysis was promoted by oxygen with indirect tests for the involvement of Δ. The rate of membrane lysis did not depend on oxygen at high HP concentrations. Similar concentration effects were found for photosensitized lysis of large EPC liposomes by incorporated HPD (Grossweiner et al., 1982). Oxygen had no effect at 3% HPD and it was strongly sensitizing at 0.1% HPD. Type I photosensitization of liposomes was less efficient with HpD-A. For example, the lytic rate under oxygen was 4.3-fold faster than under nitrogen for large EPC liposomes containing 1.8% HpD-A (Goyal et al., 1983). Measurements of lipid peroxidation in EPC liposomes provide direct evidence of membrane photosensitization by HpD-A (Table 2). Membrane lysis was measured by the optical density decrease at 750 nm (A_{750}). Lipid peroxidation (A_{530}) was assayed by the formation of malonyldialdehyde (MDA) based on the thiobarbituric acid test (Asakawa and Matshushita, 1980). Photobleaching of the membrane bound HpD-A was measured at 365 nm (A_{365}). The dependence of damage on dose indicates lysis was delayed by an incubation period when MDA attained its maximum value. Similar delays prior to rapid lysis were observed with large EPC liposomes sensitized by methylene blue (Grossweiner and Grossweiner, 1982), HP (Grossweiner et al., 1983) and external HpD-A (Grossweiner and Goyal, 1983) and probably resulted from damage to the inner membranes of the multilamellar liposomes prior to lysis of the outer membrane. The rate of membrane lysis was almost completely inhibited by the addition of 0.1 M

Table 1. Effect of Medium and Concentration on Porphyrin Photophysics.

Porphyrin	Conc. (μM)	λ_s * (nm)	λ_{fl} # (nm)	Φ_{fl} $	Φ_T @	Φ_Δ &
HP/PB	2-3	391 (a)	613 (b)			0.45 (c)
"	17	376 (a)			0.63 (d,e)	0.42 (c)
"	50	372 (a)	613 (b)			0.43 (c)
"	70-80	371 (a)		0.02 (j)	0.58 (j)	0.40 (c)
HP/SDS	20	400 (k)	620 (b)		0.87 (d)	0.55 (d)
HP/MeOH	2	400 (b)	625 (b)	0.27 (b)		
HP/9:1 MeOH/H$_2$O	17				0.83 (d)	0.65 (d)
	83			0.09 (j)	0.91 (j)	
HP/SUV	30	398 (c)				0.77 (c)
Uro-I/PB	18	398 (c)				0.71 (c)
HpD-A/PB	2	367 (f)	618 (f)			
"	20	364 (c)		0.005 (h)**		0.06 (c)
"	300	364 (g)				
Hpd-A/SUV	2	399 (f)	627 (f)			
"	20	398 (c)				0.87 (c)
HPD/PB	2	375,sh400 (b)	614 (b)	0.24 (b)		
"	50		614 (b)	0.12 (b)		
"	100		614 (b)	0.09 (b)		
HPD/MeOH	2	400,sh375 (b)	624 (b)	0.24 (b)		
"	50		624 (b)	0.28 (b)		
"	100		624 (b)	0.29 (b)		
HPD/SUV	25	398 (g)	625 (g)			

*   Soret band maximum
#   first fluorescence maximum
$ fluorescence efficiency
@   triplet state quantum yield
&   singlet oxygen quantum yield
** high MW fraction for Bio-Gel P-60 preparation
(a) Brown et al. (1976)
(b) Andreoni et al. (1982)
(c) Blum and Grossweiner (1985)
(d) Reddi et al. (1983)
(e) Reddi et al. (1985)
(f) Ricchelli and Grossweiner (1984)
(g) Grossweiner (1984)
(h) Bellnier and Lin (1984)
(j) Smith (1985)
(k) Sconfienza et al. (1980)

azide or 1 M 1,4-diazabicyclo[2.2.2]octane (DABCO), providing indirect
evidence for the key role of Δ. The photobleaching rate of HpD-A was the same
under nitrogen and oxygen ruling out the involvement of Δ in this process.
Photobleaching of HPD should be considered in PDT dosimetry at high light
dose levels. Moan et al. (1984) reported photobleaching of HPD in the
presence of serum and after uptake by human cancer cells. Lipid peroxidation
was observed for small EPC liposomes incorporating 0.3% HpD-A, based on MDA
formation and the absorbance increase at 235 nm attributed to conjugated
diene hydroperoxides (Klein, 1970). The rate of lipid peroxidation was
unchanged from 25 °C to 50 °C and inhibited by 10 mM azide (Goyal et al.,
1983). Photobleaching of HpD-A in small EPC liposomes was oxygen-independent
and faster at higher temperatures.

Uro-I (8 μM) was photobleached at the same rate under oxygen and
nitrogen. The addition of 10 mM azide or 0.1 M DABCO led to partial inhibition of photobleaching, also at the same rate under oxygen and nitrogen. The
Uro-I Soret band at 400 nm was not affected by the addition of large EPC
liposomes and there was no evidence of lipid peroxidation when 70 μM Uro-I
was irradiated in the presence of 6-fold excess liposomes. However, 70 μM
Uro-I sensitized lipid peroxidation of 100-fold excess SUV at a rate comparable to external HpD-A and this damage was completely inhibited by 10 mM
azide. The RNO measurements show that Φ_Δ = 0.71 for Uro-I in PB (Blum and
Grossweiner, 1985). The addition of 5 mM azide prevented RNO bleaching
induced by Uro-I. The present results are consistent with the work of Spikes
et al. (1984) showing that aqueous Uro-I and Uro-I covalently bound to
agarose beads was an efficient photosensitizer of furfural alcohol and amino
acids with evidence for Δ involvement based on oxygen uptake. The general
conclusion from these results is that Uro-I is an efficient generator of Δ
and an inefficient sensitizer of lipid peroxidation because of its strong
hydrophilic properties and rapid photobleaching in the absence of substrates.

Photosensitization of Human Serum Albumin

Photosensitization of HSA provides information about the photosensitization of systemic protein in PDT. Reddi et al. (1981) showed that HSA has
one strong binding site for HP. The association constant (K_a) was 3.6 x 10^6
liters/mol at pH 6.8-7.5. Photochemical studies by Richard et al. (1983)
showed that exposure of HP in the presence of excess HSA to visible/near-UV
radiation led to oxidation of tryptophan (TRP) in the HSA and photobleaching
of the complexed HP. The major involvement of Δ in TRP oxidation was demonstrated with indirect tests including 8-fold enhancement in D O PB. The
photobleaching of complexed HP was 30-fold faster than free HP and at the
same rate in oxygen and nitrogen. Laser flash photolysis measurements of

Table 2.   Photosensitization of Large EPC liposomes by HpD-A

| Dose * | A750 | A530 | A365 |
10^{-4} x kJ/m^2			
0.00	1.00	0.00	2.05
0.44	0.86	0.35	1.43
0.71	0.83	0.42	1.43
1.06	0.67	0.38	0.96
1.42	0.59	0.35	0.76

*   irradiated at 38 °C in pH 7.4 PBS at > 360 nm under oxygen

Table 3. Photosensitization of HSA by Uro-I

System *	R_{340} #
Control	0.0016
N_2	0.0087
O_2	0.044
O_2 /1 mM N_3^-	0.024
O_2 /5 mM N_3^-	0.012

* 50 μM HSA + 8 μM Uro-I, pH 7.4 PB
# initial loss of TRP fluorescence,
  irradiated at > 510 nm

Reddi et al. (1984) showed that HP binding to HSA did not affect Φ_T which
was 0.66 at pH 7.4. However, the triplet decay constant was 10-fold slower
for complexed HP and the rate constant for oxygen quenching of the triplet
state was 10-fold slower for complexed HP. These workers confirmed that Δ
oxidizes TRP in HSA and obtained 7×10^8 liters/mol-sec for the chemical
reactions of Δ with HSA.

Binding measurements HpD-A to HSA are complicated by aggregation in
aqueous media. Grossweiner et al. (1984) obtained approximate equilibrium
conditions after 42 hr dialysis in pH 7.4 PBS at 38 °C and found that each
HSA molecule binds 5 to 8 porphyrin units with $K_a = 2 \times 10^5$ liters/mole.
Moan et al. (1985) made similar measurments on Photofrin II with differential
absorption spectroscopy in PBS at room temperature. This method indicates
that HSA binds up to 8 porphyrin units with $K_a \simeq 3 \times 10^5$ liters/mole. New
data on photosensitization of HSA by Uro-I is summarized in Table 3. Solutions of 8 μM Uro-I plus 50 μM HSA in pH 7.4 PB were irradiated with visible
light (> 510 nm) and HSA damage was assayed by the loss of 340 nm TRP fluorescence excited at 295 nm. Binding of Uro-I to HSA was indicated by red shifts
of the Uro-I Soret peak and first fluorescence band. Photooxidation of HSA
was accompanied by rapid photobleaching of Uro-I at the same rate under
oxygen and nitrogen. However, the presence of azide led to a slower rate of
Uro-I photobleaching with no difference in oxygen and nitrogen.

DISCUSSION

Photophysics of HP

The Soret band of HP is located near 375 nm in concentrated aqueous
solutions and near 400 nm in aqueous surfactants, polar solvents and liposomes. The Soret band corresponds to the second excited singlet state (B
transitions) with exceptionally high oscillator strength that can magnify the
effects of relatively weak exciton interactions. Conversely, the properties
orginating in first excited singlet state should show weak exciton effects
because the Q transitions (visible bands) are weakly allowed. In fact, the
visible absorption bands in aqueous media show relatively small changes from
10-1000 μM (Gallagher and Elliott, 1973). The HP fluorescence data reported
by Andreoni et al. (1982) show a decrease in Φ_{f1} from 0.32 to 0.11 from 2 to
100 μM in PB accompanied by a decrease of the 15 ns decay component and an
increase of the 4 ns decay component. The relatively high values of Φ_{f1} are
rejected by Smith (1975) both on technical grounds and inconsistency with Φ_T.
His lower values give $\Phi_{f1} + \Phi_T \simeq 1$ in PB and methanol (Table 1). The values
of Φ_{f1} and Φ_T strongly suggest that the dimer state dominates in PB and the
monomer state dominates in less polar environments. The more weakly emitting,

shorter decay lifetime component measured by Andreoni et al. (1982) probably corresponds to the dimer. The weak aggregates responsible for the large Soret band blue shift, which does not affect the triplet state photochemistry, may involve a micellar structure as suggested by Brown et al. (1976). The unusual nature of these HP "aggregates" was noted also by Swincer et al. (19875) who found that the molecular weight could not be measured by vapor phase osmometry or sedimentation. Table 4 summarizes the approximate properties of the HP aggregation states.

Photophysics of HPD

HPD contains interconvertable porphyrins and therefore has a variable composition that depends on the preparation, history and method of analysis. The analysis of HPD by HPLC led to 35% HP, 20% hydroxyethylvinyldeuteroporphyrin (HVD), 5% protoporphyrin (PP) and 35% DHE (Kessel and Cheng, 1985). Photophysical measurements on HPD provide limited information about the DHE component because of interference from high concentrations of the other porphyrins. These is certainly the case for the fluorescence of HPD in hydrophilic cellular environments because aggregated DHE is essentially non-fluorescent. Results with HpD-A (Photofrin II) are more useful than HPD because the DHE content is approximately 85% with HVD as the major contaminant (Dougherty et al., 1984). A major difference between DHE and HP is the apparently much greater strength of the aggregation interactions in aqueous DHE. In one type of measurement, the molecular weight of HPD constituents was evaluated with chromatographic methods (Moan and Sommer, 1982; Kessel, 1982; Dougherty et al., 1983) and sedimentation (Swincer et al., 1985). These procedures consistently identified a stable, high molecular weight component comprising approximately 35% of HPD (Kessel and Cheng, 1985). The porphyrin concentrations in this type of measurement were high, in the millimolar range. Optical measurements are usually made at much lower concentrations, typically from 0.1-300 μM. It is possible that much tighter stacking occurs with covalent dimers compared to HP monomers, such that van der Waals dimers are stable throughout the concentration range where optical measurements have been made and true aggregates are stable in the high concentration range where physical measurements were made. An alternative possibility is that the covalent dimers are not self-associated in the optical range and the apparent effects of aggregation are caused by folding around the covalent bond. This explanation does not require the occurrence of aggregation at sub-micromolar DHE concentrations and it is consistent with the ease of apparent disaggregation in non-polar media such as liposomes. The values of Φ_Δ for HpD-A correlate with the dependence of the Soret band peak on the ratio of porphyrin to lipid (Blum and Grossweiner, 1985). The limiting values of Φ_Δ = 0.87 (±17%) and λ_s = 398 nm were attained below 1% (w/w) HpD-A. These effects were accompanied by a shift of λ_{fl} from 618 nm in PB to 627 nm in the SUV with an approximate three-fold increase of Φ_{fl}. The fluorescence of HpD-A in SUV is

Table 4.   Aggregation States of Aqueous Hematoporphyrin.

	λ_s (nm)	λ_{fl} (nm)	Φ_{fl}	τ_{fl} (ns)	Φ_T	Φ_Δ
monomer	400	625	0.09	15	0.8-0.9	0.6-0.7
dimer	395	615	0.02	4	0.6	0.4-0.5
aggregate	375	-	0.00	-	?	?

consistent with the work of Kessel and Cheng (1985) in which the emission peak of DHE varied from 624 nm in formamide, 628 nm in methanol and 637 nm in tetrahydrofuran, corresponding to dielectric constants of 109, 32 and 4, respectively. The intermediate value of λ_{fl} in SUV suggests that DHE partially intercalates into the lipid bilayer in a planar conformation.

Membrane Photosensitization

The high value of Φ_Δ for HpD-A in SUV is consistent with the assays of photosensitized lipid peroxidation, with protection under nitrogen and in the presence of DABCO and azide. Recent work of Dearden et al. (1985) showed that Δ generated by RB does not oxidize the saturated lipids in egg lecithin and the oxidation rate of the unsaturated lipids correlated with the degree of unsaturation. The photositization of lysis in large liposomes indicates that peroxidation can lead to membrane disruption. Similar photosensitization of membrane lysis was observed at low HP concentrations. However, Type I photosensitization dominated at high HP concentrations. Subsequent efforts to observe Type I photosensitization by HPD in biological systems have been been unsuccessful (See et al., 1984; Gomer and Razum, 1984; Salet et al., 1984) and may require much higher HPD concentrations than achieved in cellular targets. Uro-I was an ineffective photosensitizer of lipid peroxidation in EPC liposomes. In view of the high Φ_Δ, the low efficiency must be attributed to its inability to enter the membrane. Emiliani and Delmelle (1983) estimated that the partition coefficient of Uro-I between Tris buffer and EPC liposomes was 51, compared to 490 for HP and 1330 for PP. The same workers found that Uro-I photosensitized the production of the 5α-hydroperoxide from cholesterol in EPC liposomes with 7-fold lower efficiency than HP, indicating that Δ generated in the external aqueous medium is far less effective than when generated within the membrane bilayer.

Photosensitization of Human Serum Albumin

HSA has one TRP residue that dominates in the fluorescence excited at 295 nm. It forms a 1:1 complex with HP leading to fluorescence quenching, which was employed for measurements of the binding constants at HP to HSA concentration ratios up to unity (Reddi et al., 1981). HP binding shifted the Soret peak to 401 nm and the HP fluorescence from 614 nm to 624 nm. Exposure of the complex to near-UV/visible radiation led to TRP oxidation mediated by Δ (Richard et al., 1983). The experiments at HSA to HP ratios from 10-50 led to oxidation of a high fraction of HSA, limited by the photobleaching of the complexed HP. It is concluded that Δ generated by complexed HP diffused into the medium and oxidized TRP in the excess free HSA. The addition of HSA to Photofrin II led to absorption decrease at 365 nm and the formation of a shoulder at 402 nm attributed to the bound component (Moan et al., 1985). The calculation of the binding constants by means of differential absorption spectroscopy indicated that each HSA molecule binds up to 8 porphyrin units. Measurements of HpD-A binding to HSA with dialysis at 38 °C led to 5-8 porphyrin units per HSA, depending on the assumptions made in correcting for retention of HpD-A by the membrane tubing (Grossweiner and Goyal, 1984). No evidence for cooperativity was found with differential absorption spectroscopy, whereas the results of the dialysis method followed the Hill equation with α = 3.7. The discrepancy may result from the higher temperature employed in the dialysis method, in which case the HSA molecule may be more sensitive to binding-induced conformation changes. The fractional binding of HP by 500 μM HSA (the approximate concentration in human serum) was > 99% compared to 85% for HpD-A. The weaker binding of HpD-A may be involved in the more effective uptake of DHE by tumor tissue. This hypothesis is supported by measurements of HpD-A uptake by small EPC liposomes, in which three-fold excess HSA increased the uptake time from hours to several days (Goyal et al., 1983). Competitive inhibition of DHE binding by HP may explain the relatively high effectiveness of PDT by impure HP containing 15-20% DHE

(Dougherty, 1983). The photosensitization of TRP oxidation in HSA by HpD-A and Uro-I was similar to HP, with indirect evidence for the involvement of Δ based on protection by azide and under nitrogen. The generation of systemic Δ by serum-bound HPD may be involved in skin photosensitization accompanying PDT. A similar mechanism was proposed by Pathak (1984) in connection with PUVA phototherapy of psoriasis.

Conclusions

The photophysics and photochemistry of HPD are markedly different from HP and Uro-I in aqueous media. The DHE-enriched form of HPD, HpD-A, is a poor photosensitizer with spectral properties characteristic of strong exciton interactions induced by aggregation and/or folding around the covalent bonds. Aqueous HP is a moderately good photosensitizer with spectral properties indicating the prevalence of van der Waals dimers. Aqueous Uro-I is a good photosensitizer and weakly aggregated, at most. Binding of HP and HpD-A to human serum albumin or liposomal membranes induces a significant increase of the Type II photosensitizing ability, attributed to "monomerization" and higher singlet oxygen generation efficiencies. Uro-I weakly binds to liposomes and human serum albumin, if at all, and is less effective as a photosensitizer than HP and HpD-A. Photosensitization of liposomes and serum albumin was accompanied by photobleaching of the porphyrins. This process does not require oxygen and was much faster with Uro-I than HP and HpD-A.

ACKNOWLEDGMENT

This work was supported by NIH Grant GM 20117 and DOE Contract No. AC02-76EV02217. This is publication DOE/EV02217-57.

REFERENCES

Andreoni, A., Cubeddu, R., De Silvestri, S. and Laporta, P., 1982, Hematoporphyrin derivative: Experimental evidence for aggregated species, Chem. Phys. Lett., 88:33.

Asakawa, T., and Matsushita, S., 1980, Coloring conditions of thiobarbituric acid test for detecting lipid hydroperoxides, Lipids, 15:137.

Bellnier, D. A., and Lin, C.-W., 1984, Photodynamic inactivation of cultured bladder tumor cells: A preliminary study of the effects of porphyrin aggregation, in: "Porphyrin Localization and Treatment of Tumors," D. R. Doiron and C. J. Gomer, eds., Alan R. Liss, Inc., New York.

Berenbaum, M. C., Bonnett, R., and Scourides, P. A., 1982, In vivo biological activity of the components of hematoporphyrin derivative, Br. J. Cancer, 45:571.

Blum, A., and Grossweiner, L. I., 1985, Singlet oxygen generation by hematoporphyrin IX, uroporphyrin I and hematoporphyrin derivative at 546 nm in phosphate buffer and in the presence of egg photophatidylcholine liposomes, Photochem. Photobiol., 41:27.

Brown, S. B., Shillcock, M., and Jones, P., 1976, Equilibrium and kinetic studies of the aggregation of porphyrins in aqueous solution, Biochem. J., 153:279.

Dearden, S. J., Hunter, T. F., and Philip, J., 1985, Fatty acid analysis as a function of photo-oxidation in egg yolk lecithin vesicles, Photochem. Photobiol., 41:213.

Doiron D. R., Gomer, C. J., Fountain, S. W., and Razum, N. J., 1984, Photophysics and dosimetry of photoradiation therapy, in "Porphyrins in Tumor Phototherapy", A. Andreoni and R. Cubeddu, eds., Plenum Press, New York.

Dougherty, T. J., 1983, Hematoporphyrin as a photosensitizer of tumors, Photochem. Photobiol., 38:377.

Dougherty, T. J., Boyle, D. G., Weishaupt, K. R., Henderson, B. A., Potter, W. R., Bellnier, D. A. and Wityk, K. E., 1983, Photoradiation therapy - Clinical and drug advances, in: "Porphyrin Photosensitization", D. Kessel and T. J. Dougherty, eds., Plenum Press, New York.

Dougherty, T. J., Potter, W. R., and Weishaupt, K. R., 1984, The structure of the active component of hematoporphyrin derivative, in: "Porphyrins in Tumor Therapy", A. Andreoni and R. Cubeddu, eds., Plenum Press, New York.

Dougherty, T. J., Kaufman, J. E., Goldfarb, A., Weishaupt, K. R., Boyle, D., and Mittleman, A., 1978, Photoradiation therapy for the treatment of malignant tumors, Cancer Res., 38:2628.

El-Far, M., and Pimstone N.,1984, A comparative study of 28 porphyrins and their abilities to localize in mammary mouse carcinoma: Uroporphyrin I superior to hematoporphyrin derivative, in: "Porphyrin Localization and Treatment of Tumors," D. R. Doiron and C. J. Gomer, eds., Alan R. Liss, Inc., New York.

Emiliani, C., and Delmelle, M., 1983, The lipid solubility of porphyrins modulates their phototoxicity in membrane models, Photochem. Photobiol., 37:487.

Gandin, E., Lion, Y., and Van de Vorst, A., 1983, Quantum yield of singlet oxygen production by xanthene derivatives. Photochem. Photobiol., 37:271.

Gomer, C. J., and Razum, N. J., 1984, Acute skin response in albino mice following porphyrin photosensitization under oxic and anoxic conditions, Photochem. Photobiol., 40:435.

Goyal, G. C., Blum, A., and Grossweiner, L. I., 1983, Photosensitization of liposomal membranes by hematoporphyrin derivative, Cancer Res., 43:5826.

Grossweiner, L. I., 1984, Membrane photosensitization by hematoporphyrin and hematoporphyrin derivative, in: "Porphyrin Localization and Treatment of Tumors", D. R. Doiron and C. J. Gomer, eds., Alan R. Liss, Inc., New York.

Grossweiner, L. I., and Grossweiner, J. B., 1982, Hydrodynamic effects in the photosensitized lysis of liposomes, Photochem. Photobiol., 35:583.

Grossweiner, L. I., Patel, A. S., and Grossweiner, J. B., 1982, Type I and type II mechanisms in the photosensitized lysis of phosphatidylcholine liposomes by hematoporphyrin, Photochem. Photobiol., 36:159.

Grossweiner L. I., and Goyal, G. C., 1983, Photosensitized lysis of liposomes by hematoporphyrin derivative, Photochem. Photobiol., 37:529.

Grossweiner, L. I., and Goyal, G. C., 1984, Binding of hematoporphyrin derivative to human serum albumin, Photochem. Photobiol., 40:1.

Kessel, D., 1982, Components of hematoporphyrin derivatives and their tumor-localizing capacity, Cancer Res., 42:1703.

Kessel, D., and Cheng, M.-L., 1985, On the preparation and properties of dihematoporphyrin ether, the tumor-localizing component of HPD, Photochem. Photobiol., 41:277.

Klein, R. A., 1970, The detection of oxidation in liposome preparations, Biochim. Biophys. Acta, 210:486.

Kraljic, I., and El Mohsni, S., 1978, A new method for the detection of singlet oxygen in aqueous solution. Photochem. Photobiol., 28:577.

Lamola A. A., Yamane, T., and Trozzolo, A. M., 1973, Cholesterol hydroperoxide formation in red blood cell membranes and photohemolysis associated with erythropoietic protoporphyria, Science, 179:1131.

Lipson, R., Baldes, E., and Olsen, A., 1961, The use of a derivative of hematoporphyrin in tumor detection, J. Natl. Cancer Inst., 26:1.

Moan, J., and Sommer, S., Fluorescence and absorption properties of the components of hematoporphyrin derivative, Photobiochem. Photobiophys., 3:93.

Moan, J., Christensen, T., and Jacobsen, P. B., 1984, Porphyrin-sensitized photoinactivation of cells in vitro, in: "Porphyrin Localization and Treatment of Tumors," D. R. Doiron and C. J. Gomer, eds., Alan R. Liss, Inc., New York.

Moan,J., Rimington, C., and Western, A., 1985, The binding of dihematoporphyrin ether (Photofrin II) to human serum albumin, Clinica Chim. Acta, 145:227.

Nozaki, Y., Lasic, D. D., Tanford, C., and Reynolds, J. A., 1982, Size Analysis of phospholipid vesicle preparations, Science, 217:366.

Pathak, M. A., 1984, Mechanisms of psoralen photosensitization reactions, in: "Photobiologic, Toxicologic, and Pharmacologic Aspects of Psoralens," M. A. Pathak and J. K. Dunnick, eds., U.S. Government Printing Office, Washington, D.C.

Poletti, A., Murgia, S. A., Pasqua, A., Reddi, E., and Jori, G., 1984, "Photophysical and photosensitizing properties of Photofrin II, in: Porphyrins in Tumor Phototherapy," A. Andreoni and R. Cubeddu, Plenum Press, New York.

Reddi, E., Ricchelli, F., and Jori, G., 1981, Interaction of human serum albumin with hematoporphyrin and its Zn - and Fe -derivatives, Int. J. Peptide Protein Res., 18:402.

Reddi, E., Jori, G., Rodgers, M. A. J., and Spikes, J. D., 1983, Flash Photolysis of hemato- and copro-porphyrins in homogeneous and microheterogeneous aqueous dispersions, Photochem. Photobiol., 38:639.

Reddi, E., Rodgers, M. A. J., Spikes, J. D., and Jori, G., 1984, The Effect of medium polarity on the hematoporphyrin-sensitizized photooxidation of L-tryptophan, Photochem. Photobiol., 40:415.

Ricchelli, F. and Grossweiner, L. I., 1984, Properties of a new state of hematoporphyrin in dilute aqueous solution, Photochem. Photobiol., 40:599.

Richard, P., Blum, A., and Grossweiner, L. I., Hematoporphyrin photosensitization of serum albumin and subtilisin BPN', Photochem. Photobiol., 37:287.

Salet, C., Moreno, G., Vever-Bizet, C., and Brault, D., Anoxic photodamage in the presence of porphyrins: Evidence for the lack of effects on mitochondrial membranes, Photochem. Photobiol., 40:145.

See, K. L., Forbes, I. J., and Betts, W. H., Oxygen dependency of photocytotoxicity with hematoporphyrin derivative, Photochem. Photobiol., 39:631.

Smith, G. J., 1985, The effects of aggregation on the fluorescence and and the triplet state yield of hematoporphyrin, Photochem. Photobiol., 41:123.

Spikes, J. D., Burnham, B. F. and Bommer, J. C.,1984, Photosensitizing properties of free and bound uroporphyrin I, in "Porphyrins in Tumor Phototherapy", A. Andreoni and R. Cubeddu, eds., Plenum Press, New York.

Swincer, A. G., Ward, A. D., and Howlett, G. J., 1985, The molecular weight of hematoporphyrin derivative, its gel column fractionations and some of its components in aqueous solution, Photochem. Photobiol., 41:47.

BINDING OF PORPHYRINS TO SERUM PROTEINS

Johan Moan, Claude Rimington, Jan F. Evensen and
André Western

Norsk Hydro's Institute for Cancer Research
The Norwegian Radium Hospital
Montebello, 0310 Oslo 3, Norway

INTRODUCTION

It is of importance to determine how porphyrins are transported in blood, for two main reasons: (i) porphyrins are used for fluorescence detection and photodynamic therapy of tumors, and (ii) certain diseases such as the porphyrias and disorders such as lead intoxiation lead to an increased level of porphyrins in the blood. Until recently, albumin was thought to be the main porphyrin binding protein in serum. However, since most porphyrins are cleared through the liver, and since albumin is not taken up by hepatocytes, it has been conjectured that other porphyrin-binding elements exist in serum[1]. Attention has been drawn to hemopexin, a protein known to carry heme into hepatocytes, and to lipoproteins. Already in 1956, it was shown that porphyrins have a marked affinity for tissues with a high lipoprotein content[2].

If it is assumed that a significant fraction of the porphyrins when transported in the blood is bound to lipoproteins, their binding to certain tissues, such as liver and adrenals, may be related to the high concentration of lipoprotein receptors in these tissues. Recent investigations seem in fact to show extensive binding of protoporphyrin and hematoporphyrin to lipoproteins in serum[3,4]. Such binding was not found in earlier work, although the investigators were obviously aware of the possibility, since they compared the electrophoretic migration pattern of porphyrin-protein complexes in lipoprotein-containing sera with that in lipoprotein-depleted sera[5].

A number of methods have been used to study porphyrin-protein binding. Several of these methods will be discussed below. However, porphyrins possess some properties that introduce errors into the results obtained by practically all the methods. One should be aware of: (i) porphyrin aggregation that leads to retention in dialysis bags, fluorescence quenching and altered binding properties, (ii) porphyrin binding to dialysis bags, columns and glassware, and (iii) chelation of porphyrins (notably when present in low concentrations) with metal ions present as impurities in polar solvents, glassware and column materials[6]. Such chelation leads to altered absorption spectra, altered fluorescence spectra (Zn^{2+}, Mg^{2+}), or complete fluorescence quenching (Fe^{2+}, Cu^{2+}, etc.).

MATERIALS AND METHODS

Chemicals

Photofrin II (PII), a somewhat purified version of hematoporphyrin
derivative (Hpd), was bought from Photofrin Medical, Cheektowaga, NY.
Hematoporphyrin (Hp), protoporphyrin (Pp), uroporphyrin I (Up) and
tetraphenylporphine-suplphonate (TPPS) was bought from Porphyrin
Products, Logan, UT. Human serum albumin (HSA) fraction V and bovine
serum albumin (BSA) was bought from Sigma, St. Louis, MO. Low density
lipoproteins (LDL) obtained by ultracentrifugation, were a kind gift
from Dr. Rune Blomholt, Oslo University. Lipid vesicles (95% egg yolk
phosphatidylcholine + 5% L-α-dipalmitoylphosphatidic acid) were
prepared by Dr. Arnt I. Vistnes, Oslo University. All chemicals used
were obtained commercially and were of the purest grade available.

Dialysis and Ultrafiltration

Dialysis tubings with a cut-off at about 12000 D were used (A.H.
Thomas Co., PA). Hp solutions of low concentrations (<1μM) equilibrated
completely within 24 h at room temperature when 5 ml Hp solution was
dialysed against 10 ml PBS.

Ultrafiltration was carried out by means of the Centricon centrifugation system (cut-off at 10 000 and 30 000 D, respectively), Amicon,
MA. The tubes were loaded with 2 ml porphyrin solution and centrifuged
at 1000 g for 30 min.

Chromatography

To check the purity of porphyrin solutions, reversed phase HPLC was
carried out with a water/methanol gradient system as described elsewhere .

For gel permeation chromatography a Bio-Gel P10 column (Bio-Rad,
CA) eluted with PBS was used. The flow rate through the 260x10 mm
column was 0.5 ml/min.

Electrophoresis

To accord with the physiological situation, gel electrophoresis of
50 μl samples of diluted serum was carried out at pH 7.3. The buffer
consisted of 5.52 g diethylbarbituric acid + 1.0 g Tris in 1 l water.
The separating gels (7.5% polyacrylamide, 75x5 mm) were made by mixing
stock solutions A, B, C and D in volume ratios 6:1:0.005:1, where A
consisted of 10 g acrylamide + 0.37 g N, N'-methylene-bis-acrylamide +
100 ml water, B consisted of 6.85 g Tris + 100 ml water adjusted to
pH 7.3 by adding 1 N HCl, C was N, N, N', N'-tetramethylethylene-
diamide and D was 60 mg amoniumpersulphate in 10 ml water. The time of
electrophoresis was adjusted, so that albumin migrated 2/3 of the total
gel length. The current was 2-2.5 mA per gel. After elecotrophoresis,
the gels were stained with coomassie brilliant blue (0.62 g in 545 ml
50% methanol + 46 ml acetic acid) and rinsed in a mixture of acetic
acid, methanol and water (3:2:35). The gels were squeezed between two
quartz plates to a thickness of about 3 mm and scanned by a gel scanner
mounted in an absorption spectrometer. Unstained gels were scanned at
280 nm (proteins) and at 405 nm (porphyrins) while stained gels were
scanned either at 570 or at 660 nm.

Ultracentrifugation

Human serum was collected in EDTA tubes and porphyrins were added either to whole blood to estimate the fraction of the porphyrins that was bound to cells or to plasma after removal of the cells by centrifugation. For the analysis of lipoproteins a Beckman L2-65B ultracentrifuge with a SW40Tl rotor run at 40 000 rpm was used. A step-wise gradient with densities d=1.006 (equal to that of plasma), d=1.063 and d=1.21 was applied. Each density zone contained 4 ml solution. 14x95 mm polyallomer tubes were used. The desired densities were obtained by mixing a solution of 153 g/l NaCl+354 g/l KBr with 0.15 M NaCl. The plasma samples (1-4 ml) were either layered directly on top of the gradient or adjusted to d=1.21 by addition of 0.325 mg KBr per ml plasma and placed at the bottom of the tube. The centrifugation time was 24 h. The samples were fractionated and studied by absorption and fluorescence spectroscopy.

Photobleaching

Photobleaching experiments were carried out by exposing the protein-porphyrin mixtures to the light from a bank of fluorescent tubes: Philips TL 20W/09, emitting light mainly in the wavelength region 330-440 nm, and delivering a fluence rate of approximately 15 W/m^2 to the solutions which usually had an optical path length of 1-2 mm.

Absorption- and Fluorescence Spectroscopy

A Cary 118 spectrophotometer was used for the absorption measurements. Difference spectra were recorded in the following way (Fig. 1): The reference beam passed through two cuvettes, one with a given porphyrin solution and the other containing PBS, to which graded amounts (5-20 μl) of a concentrated HSA solution were added. The sample beam was passed through a solution containing pure PBS, and then another containing the same porphyrin solution as that in one of the reference cuvettes. To the porphyrin solution in the sample beam, graded amounts of HSA were added to give the same concentrations as those in the PBS reference cuvette. Thus, the changes in the absorption spectrum of the porphyrin resulting from HSA binding could be accurately registered. The data were corrected for the small reduction in absorbance resulting from the volume increase due to HSA addition.

Fluorescence spectra were registered by means of a Perkin Elmer LS-5 spectrofluorimeter. This instrument gives corrected excitation spectra. A red-sensitive Hammamatsu photomultiplier was used. The quenching of tryptophan fluorescence from the proteins by porphyrins is one of the most frequently used parameters employed to estimate the binding of porphyrins to proteins. Since porphyrins absorb slightly at the absorption maximum of proteins ($\lambda^\sim$ 280 nm) and significantly more at their fluorescence emission maximum ($\lambda^\sim$ 350 nm), one must be aware of inner filter effects. To minimize inner filter effects, a special cuvette holder was constructed with 0.2 mm slits for excitation and emission close to (0.1 mm) the edge of the cuvette. In this system no optical path length was larger than 0.3 mm. Nevertheless, we had to correct our data for inner filter effects at the highest porphyrin concentrations. It seems that such inner filter effects have been neglected in several experimental investigations of this type.

Energy transfer from tryptophan in the proteins to bound porphyrins may be studied by fluorescence excitation spectroscopy, and may give valuable information about the binding sites. Because of the overlapping of the protein fluorescence emission spectra with the porphyrin fluorescence excitation spectra, one should check if trivial absorption of protein fluorescence by the porphyrins takes place to an extent that may compete with energy transfer. We checked for this possibility by mixing the porphyrin solutions with a tryptophan solution whose fluorescence was equally intense as that of the relevant protein solutions. The excitation spectra of such mixtures, studied at the wavelengths of porphyrin emission, had no peaks or shoulders at 280 nm, indicating that the above mentioned trivial absorption-emission process does not contribute in the present case.

RESULTS AND DISCUSSION

Absorption Spectroscopy

Using the described experimental set up (Fig. 1), difference spectra such as those shown in Fig. 1 were registered. It can be seen that for PII no well defined isosbestic point is found. Thus, more than two species, bound and free PII, were involved. This is in agreement with the HPLC chromatograms which also indicate that several components are present in PII. However, one component seems to dominate. This is indicated both by the HPLC finding and by the facts: (i) that the absorbance was found to increase linearly with the concentration over the wide concentration range 0.6 - 12 μM porphyrin rings and: (ii) that the shape of the spectrum remained unchanged over the same concentration range. Assuming that one species dominates, one can use data such as those shown in Fig. 1 to calculate binding parameters. ΔA at 402 nm plotted as a function of $[HSA]^{-1}$ yields a straight line. Extrapolation to $[HSA]^{-1} = 0$ gives the assymtotic value ΔA_{∞} which is the absorbance change in solutions where the total amount of PII is bound. Assuming that the bound fraction of PII is $f_b = \Delta A / \Delta A_{\infty}$, the concentration of bound PII is given by $[PII]_b = f_b [PII]$. When [PII] is the total amount of PII and $[PII]_f$ is the concentration of free PII, $[PII] = [PII]_b + [PII]_f$. A plot of $[PII]_b^{-1}$ as a function of $[PII]_f^{-1}$ gives a straight line:

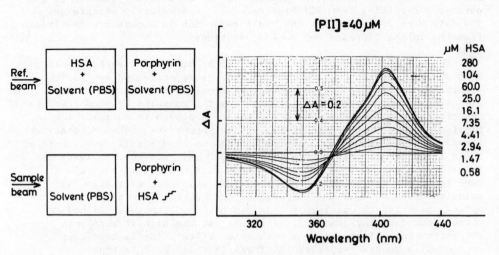

Fig. 1. Experimental set-up for measurements of the changes in the absorption spectrum of PII upon the addition of HSA (left part of the figure). A typical family of difference spectra is shown in the right part of the figure.

196

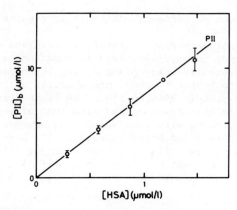

Fig. 2.   The maximum concentration of binding sites for PII plotted
          as a function of the HSA concentration.

$$\frac{1}{[PII]_b} = \frac{1}{[PII]_{bm}} + \frac{1}{k_a[PII]_{bm}} \cdot \frac{1}{[PII]_f}$$

$[PII]_{bm}$ represents the concentration of binding sites on the HSA
molecules and can be determined from the intercepts between the lines
and the ordinate axis. The binding constant k_a can also be determined
from the plot. For PII $k_a \approx 0.3 \ \mu M^{-1}$. A Hill plot of the binding
data, (log Y/(1-Y) as a function of log $[PII]_f$, where
$Y = [PII]_b/[PII]_{bm}$), is a straight line with a slope close to unity
(1.00 ± 0.15)[8], indicating that the binding is not of a cooperative
nature[8]. $[PII]_{bm}$ plotted as a function of HSA gave a straight
line, as expected (Fig. 2). The same was true for $[Hp]_{bm}$. From this
figure one can see that each HSA molecule can bind maximally 8 porphyrin
rings of PII. Correspondingly, 2 Hp molecules can be bound per HSA
molecule[8].

Dialysis, Ultrafiltration and Gel Permeation Chromatography

        PII was completely retained by the dialysis bags as well as by the
filters. Thus, it seems that in aqueous solutions the aggregates of PII
are of a greater equivalent size than 30 000 D. Porphyrin binding to
the filters and to the bags was also observed. This cannot alone
explain why PII was retained, since small molecules (H_2O) were not
retained. When PII was applied to a P10 column and its retention time
was compared with that of HSA (66 000 D), it was found that the PII
aggregates were larger than about 20 000 D. These findings are in
agreement with the work of Swincer et al.[9], who studied Hpd and some
of its components by ultracentrifugation, and found aggregate sizes
larger than 20 000 D in aqueous solutions. Neither dialysis nor ultrafiltration nor gel permeation chromatography is well suited to study PII
binding to macromolecules. On the other hand, Hp when applied in low
concentrations, is not retained by filters or by the dialysis bags.
However, one should also be aware of binding of Hp to the bags and to
the filters when quantitative data are sought.

        The finding that maximally 8 porphyrin rings of PII may bind to
each HSA molecule, is not at first sight in agreement with an aggregate
size of 20 000 - 30 000 D. Binding may possibly result in a break-down
of aggregates. However, at least two possible sources of error make
this suggestion at best, only tentative. Firstly, large aggregates
might bind to HSA without any change in the absorption spectrum.
Secondly, the spectral shifts seen might be due to the binding of only

a fraction of PII, such as monomers and small aggregates.

Gel permeation chromatography, ultrafiltration and electrophoresis are not suited to study porphyrin-protein interactions quantitatively since dissociation certainly takes place during the separation process. Neither can one use porphyrin fluorescence spectroscopy for quantitative studies of porphyrin-protein binding since the fluorescence quantum yield of aggregates is very low compared with that of monomers. This leads to a very complicated situation where free monomers, free aggregates, bound monomers and bound aggregates all have different fluorescence quantum yields. Finally, dimers and/or oligomers have an intermediate quantum yield which is about one half of that of monomers[10].

Fluorescence Spectroscopy

It is well known that when proteins contain tryptophan, their fluorescence is emitted mainly by the tryptophan residues[11], emission from other aromatic amino acids being of much lower intensity. Energy transfer to tryptophan from other chromophores in the proteins has been demonstrated. Furthermore, it is known that when porphyrin or hemin bind to proteins, the protein fluorescence is quenched[12, 13]. This has been used to determine binding constants. The main limitation of this method is that it gives information mainly about binding sites close to the emitting chromophores. When PII or Hp was added to HSA solutions, we found that the fluorescence was quenched[8]. When the fluorescence was monitored at 350 nm, where emission from the tryptophan residue of HSA dominates (HSA contains 1 tryptophan residue and 18 tyrosine residues[14]), the kinetics of the quenching was nearly

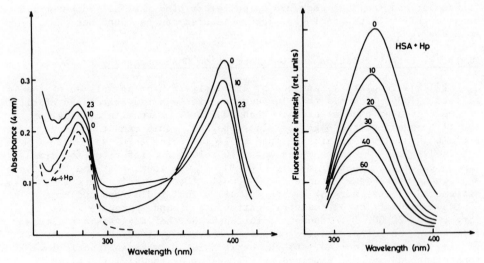

Fig. 3. Left part: Absorption spectra 15 µM HSA + 8.3 µM Hp exposed to near UV light (see Materials and Methods). Exposure times in minutes are given on the figure. Right part: Fluorescence emission spectra (λ_{exc} = 280 nm) of 15 µM HSA + graded amounts of Hp. The concentration of Hp in µM are given on the figure. A very similar family of emission spectra was obtained when a solution of 15 µM HSA + 8.3 µM Hp was exposed to near UV-light. An exposure of 23 min reduced the fluoresence by 50% at 350 nm, but only by 30% at 310 nm.

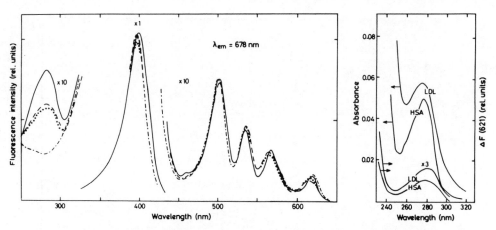

Fig. 4. Left part: Fluorescence excitation spectra of 2 μM Hp
bound to: (————) LDL (0.8 mg prot/ml, (————) HSA (15 μM),
(—·—·—) liposomes and (·····) human serum (2% in PBS).
Right part: Absorption spectra (1 cm) of HSA (1.5 μM) and
LDL (0.05 mg prot/ml) and energy transfer spectra of the
same solutions mixed with 0.84 μM Hp. The emission
monochromator was set at 621 nm.

exponential. At increasing porphyrin concentrations, the fluorescence
emission maximum was shifted towards lower wavelengths (Fig. 3). This
indicates that the tyrosine fluorescence is not quenched to the same
extent as the tryptophan fluorescence. Similar observations have been
reported for hemin-HSA binding[13]. Thus, strong porphyrin binding
seems to take place close to the tryptophan locus. Photobleaching
experiments were in agreement with this (Fig. 3). During exposure to
light, the tryptophan fluorescence was selectively bleached. Absorption
spectroscopy also indicated that the porphyrin, as well as the protein
was degraded (Fig. 3). The fine structure in the HSA absorption
spectrum was not significantly changed, indicating that some chromophores are insensitive to the treatment.

As already mentioned, the overlapping of the protein fluorescence
spectra with the porphyrin absorption spectra makes energy transfer
possible, provided that the distance between the bound porphyrins and
the protein chromophores is not too great. Such energy transfer has
been demonstrated for Hp bound to HSA[15]. The excitation spectra of
Hp-HSA-, TPPS-HSA-, PII-HSA- and Pp-HSA-complexes show well defined
peaks at about 280 nm when the emission monochromator is set in the
region of porphyrin fluorescence, as examplified by Fig. 4. No such
peak is found for mixtures of Up and HSA indicating that this porphyrin
does not bind to HSA.

The energy transfer peak at 280 nm is relatively higher for
BSA-porphyrin-complexes than for HSA-porphyrin-complexes. This is in
agreement with the fact that HSA contains one tryptophan residue[14]
while BSA contains two such residues[16]. As expected, porphyrins bound
to liposomes show no peak in·the 280 nm region (Fig. 4). LDL-Hp-com-
plexes show a marked energy transfer peak, larger than that of HSA-Hp-
complexes (Fig. 4). It is to be expected that porphyrins bound to
apolipoprotein B-100 of LDL would give strong energy transfer bands,
since this protein contains 13 ± 2 tryptophan residues[17]. From similar
data the distance between bound Hpd and the tryptophan locus of HSA was

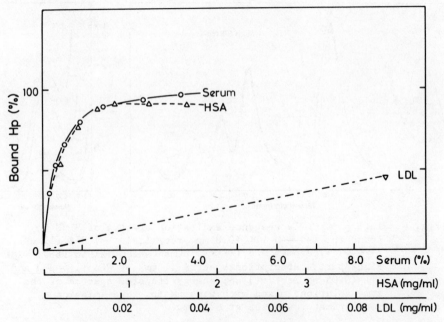

Fig. 5. Binding of Hp to HSA, LDL and human serum as measured by energy transfer from proteins to Hp. The scales for HSA and LDL along the abscissa correspond roughly to the concentrations of these components in the given concentrations of serum. [Hp] = 0.4 μM.

calculated[15]. Such calculations cannot be performed for porphyrins bound to LDL since its apoprotein contains several tryptophan residues. One cannot even conclude that the energy transfer spectra prove that the porphyrin is bound to the protein. Energy transfer may also occur if the porphyrin is able to diffuse in the lipid part of LDL and collide with the apolipoprotein. Such an energy transfer, dependent on diffusion of the energy acceptor in the lipid region of LDL, was suggested for pyrene in LDL[17]. TPPS also showed an energy transfer peak when mixed with fresh LDL. This porphyrin, which is very polar, is hardly dissolved in the lipid fraction of LDL, and is therefore most probably bound to the apolipoprotein. This observation also indicates that the apolipoprotein is exposed to water. The energy transfer peak for LDL-bound porphyrins increase in the order TPPS<Hp≈Hpd<Pp. This may be explained in terms of the lipid solubility of the porphyrins and probably offers a method to determine the location of the protein in the LDL particle, which is a matter of dispute[17].

By monitoring the height of the energy transfer peak during titration of a porphyrin solution with a protein, one can determine binding curves (Fig. 5). This analysis seems to indicate that Hp is mainly bound to albumin in human serum. However, one should remember that such measurements only give information about porphyrin binding close to the protein, and in the case of LDL, the porphyrin might in principle bind or localize also at other regions.

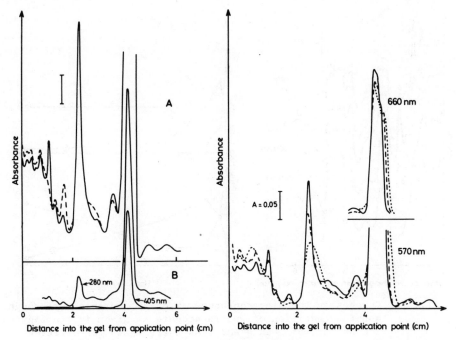

Fig. 6. Electrophoreograms of human serum.
Left part: A) (————) 20% serum without porphyrins,
stained with coomassie blue, (-----) 20% serum + 100 μg/ml
Hpd, stained with coomassie blue. The bar corresponds to
A=0.2 at 570 nm. B) The same gel as that corresponding
with the stippled electrophoreogram in A), but here scanned
at 280 nm and at 405 nm before staining. The bar
corresponds to A=0.02. Applied amount equivalent to 2.5 μl
serum.
Right part: Electrophoreograms of 20% human serum exposed
to light in the presence of 12.5 μg/ml Hpd. Exposure
times: (————) 0 min, (-----) 2 min, and (·····) 10 min.
The gels were stained with coomassie blue. The absorbance
of the stained bands is about a factor 4 larger at 570 than
at 660 nm.

Electrophoresis

Electrophoreograms of human serum with and without Hpd are shown in
Fig. 6. According to the 405 nm scan (absorption max. of protein bound
porphyrin), Hpd seems to migrate mainly with the albumin band. However,
the addition of Hpd as well as PII to human serum, clearly changed its
elecrophoreogram also in other regions than that of albumin. These
changes are shown more detailed in Fig. 7. It can be seen that at
intermediate porphyrin concentrations peak 4 is diminished while peak 6
is increased. At higher concentrations these changes are reversed.
From the present data it is impossible to explain this unexpected
behaviour, except that the changes must be brought about by the interaction of porphyrins with serum proteins, and that they are found in the
region where IgA, haptoglobins and lipoproteins migrate[18].

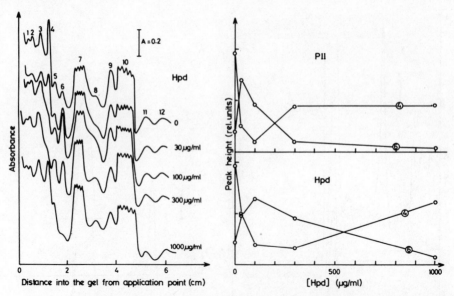

Fig. 7. Changes in the electrophoreogram of 20% human serum brought
about by addition of Hpd in amounts as given in the figure.
The heights of peak 4 and peak 6 are shown for different
concentrations of Hpd and PII in the right part of the
figure.

The β-lipoproteins (HDL) are probably spread over a large fraction of the
gel [18] so this method is not well suited to study their ability to bind
porphyrins.

The electrophoreograms of serum exposed to light in the presence of
Hpd indicate that moderate light exposures lead to changes which are
more marked in the regions of transferrin, hemopexin (peak 7, Fig. 7)
and of proteins with low R_F values than in the region of HSA. This
may indicate porphyrin binding to these proteins, although a porphyrin
molecule bound to one protein molecule may act as sensitizer of another
protein molecule.

Ultracentrifugation

Analysis of the plasma of fresh, whole EDTA-blood incubated for 1 h
at 37°C with TPPS, Hp, Hpd or PII (all in concentrations of 70 μg/ml),
indicated that the major fraction of the porphyrin remained in the
plasma fraction. The uptake into cells should not be completely
neglected, however, since about 20% of Hp, and somewhat less of the
other porphyrins, was associated with the cells. We have no reason to
believe that equilibrium was reached after 1 h incubation. Thus, when
porphyrins are circulating in the blood for several hours, the cell-
bound fraction may be significant.

Ultracentrifugation of human plasma containing PII or Hpd gave three
main porphyrin bands in the gradient, one strong at the expected position of HDL, and weaker bands at the positions of LDL and HSA (Fig. 8).
Similar observations have been made by others (D. Kessel, personal

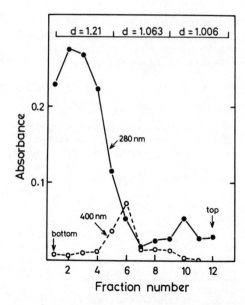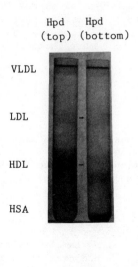

Fig. 8. Absorbance profiles at 280 nm (proteins) and 400 nm (porphyrins) of a KBr/NaCl-gradient after ultracentrifugation.
1 ml of human plasma with 70 µg/ml Hpd was layered at the
top of the gradient. Similar results were obtained when
the sample was placed at the bottom of the centrifuge
tube, although in this case more porphyrin was associated
with the protein fraction at the bottom of the tube (right
part of the figure). Dilution 1:40 before analysis.

communication). A relatively larger fraction of TPPS was found in the
bottom of the tube, although some of this polar porphyrin was also found
in the HDL fraction.

This observation of porphyrin binding to lipoproteins is in
agreement with other recent reports[3,4]. However, there seems to be
some conflict between these data and the electrophoresis and fluorescence data. Several explanations are possible:

1) Albumin binds fatty acids and lipoproteins, and may in this way
carry the porphyrins during the electrophoresis.
2) Some lipoproteins, mainly HDL are found close to albumin in
electrophoreograms[18].
3) The high salt concentrations used to achieve the gradients for
ultracentrifugation may change the affinity of albumin for
porphyrins.
4) It is well known that the solubility of porphyrins, and therefore
probably also their binding properties, changes with the salt
concentration.
5) It has been shown that HDL is contaminated by HSA when analysed in a
similar way as we have done[19]. Furthermore, lipoproteins may
aggregate and/or change configuration during the centrifugation[19]
6) During electrophoresis, porphyrins may dissociate from the lipoproteins. Albumin, which itself moves fast in the electric field
may be able to keep the porphyrins bound.

An argument in favour of the role of lipoproteins for porphyrin
transport in blood is that tissues that are known to take up large

amount of injected porphyrins (liver, adrenals) also have large concentrations of lipoprotein receptors. The transport processes of porphyrins in blood and their binding to blood components, should be studied much more thoroughly in the future.

ACKNOWLEDGEMENTS

The present work was supported by The Norwegian Cancer Society (Landsforeningen mot Kreft) and by The Association for International Cancer Research. The authors got valuable assistance from dr. Arnt I. Vistnes and dr. Rune Blomholt at Oslo University, dr. Kristian Drevon at The National Institute for Public Health and Elisabeth Paus at The Norwegian Radium Hospital.

REFERENCES

1. A. A. Lamola, Fluorescence studies of protoporphyrin transport and clearance, Acta Dermatovener (Stockholm) Suppl. 100:57 (1982).

2. T. Kosaki and T. Saka, Studies on the affinities of cells and their formative elements for porphyrin bodies, Mie Med. J. VI:55 (1956).

3. J. P. Reyftmann, P. Morliere, S. Goldstein, R. Santus, L. Dubertret, and D. Lagrange, Interaction of human serum low density lipoproteins with porphyrins: A spectroscopic and photochemical study, Photochem. Photobiol. 40:721 (1984).

4. G. Jori, M. Beltramini, E. Reddi, B. Salvato, A. Pagnan, L. Ziron, L. Tomio, and T. Tsanov, Evidence for a major role of plasma lipoproteins as hematoporphyrin carriers in vivo, Cancer Lett. 24:291 (1984).

5. P. Koskelo, I. Toivonen, and P. Rintola, The binding of ^{14}C-labelled porphyrins by plasma proteins, Clin, Chim. Acta 29:559 (1970).

6. S. Sommer, C. Rimington, and J. Moan, Formation of metal complexes of tumorlocalizing porphyrins, FEBS Lett. 172:267 (1984).

7. S. Sommer, J. Moan, and C. Rimington, Separation of different fractions of hematoporphyrin derivative by a two-phase extraction system, This volume.

8. J. Moan, C. Rimington, and A. Western, The binding of dihematoporphyrin ether (photofrin II) to human serum albumin, Clin. Chim. Acta 145:227 (1985).

9. A. G. Swincer, A. D. Ward, and G. J. Howlett, The molecular weight of haematoporphyrin derivative, its gel column fractions and some of its components in aqueous solutions, Photochem. Photobiol. 41:47 (1985).

10. J. Moan and S. Sommer, Uptake of the components of hematoporphyrin derivative by cells and tumors, Cancer Lett. 21:167 (1983).

11. S. V. Konev, "Fluorescence and Phosphorescence of Proteins and Nucleic Acids", Plenum Press, New York (1967).

12. U. Müller-Eberhard and W. T. Morgan, Porphyrin-binding proteins in serum, Ann N.Y. Acad. Sci. 244:624 (1973).

13. G. H. Beaven, S. H. Chen, A. D'Albis, and W. B. Gratzer, A spectroscopic study of the haemin-human-serum albumin system, Eur. J. Biochem. 41:539 (1974).

14. P. O. Behrens, A. M. Spiekerman, and J. R. Brown, Structure of human serum albumin, Fed. Proc. 34:591 (1975).
15. E. Reddi, F. Ricchelli, and G. Jori, Interactions of human serum albumin with hematoporphyrin and its Zn^{2+}-and Fe^{3+}-derivatives, Int. J. Peptide Res. 18:402 (1981).
16. J. E. Brown, Structure of bovine serum albumin, Fed. Proc. 34: 591 (1975).
17. G. E. Dobretsov, M. M. Spirin, O. V. Chekrygin, I. M. Karmansky, V. M. Dimitriev, and Yu. A. Vladimirov, A flourescence study of apoliprotein localization in relation to lipids in serum low densitiy lipoproteins, Biochem. Biophys. Acta, 710:172 (1982).
18. K. Felgenhauer, Quantitative and specific detection methods after disc elecrophoresis of serum proteins, Clin. Chim. Acta 27:305 (1970).
19. C. Edelstein, D. Pfaffinger, and A. M. Scanu, Advantages and limitations of density gradient ultracentrifugation in the fractionation of human serum lipoproteins: role of salts and sucrose. J. Lipid Res. 25:630 (1984).

SEPARATION OF DIFFERENT FRACTIONS OF HEMATOPORPHYRIN

DERIVATIVE BY A TWO-PHASE EXTRACTION SYSTEM

Stein Sommer, Johan Moan and Claude Rimington

Norsk Hydro's Institute for Cancer Research
The Norwegian Radium Hospital
Montebello, 0310 Oslo 3, Norway

INTRODUCTION

Hematoporphyrin derivative (Hpd), prepared as described by Lipson et al.[1] is the most widely used sensitizer in photodynamic cancer therapy (PDT). Hpd contains a number of porphyrins, some of which contribute minimally to its photodynamic action[2]. A somewhat purified version of Hpd, so-called photofrin II (PII), is also used in animal- and in clinical experiments[3,4]. PII is probably mainly composed of porphyrin dimers and/or oligomers linked together with ether- or ester-bonds, together with a variable proportion of hematoporphyrin (Hp), hydroxyethyl-vinyl-deuteroporphyrin (Hvd) and protoporphyrin (Pp). Porphyrins with low activity may be removed from Hpd by means of gel chromatography[5], dialysis or ultra-centrifugation. Below is presented a simple extraction procedure that leads to a product with properties similar to those of PII.

MATERIALS AND METHODS

Hpd was prepared from the product derived from the acetic acid/ sulphuric acid treatment of hematoporphyrin[1] (Porphyrin Products, Logan, UT) by dissolving the material in 0.1 N NaOH and adjusting the solution to neutrality one hour later by addition of HCl. A comparison of the HPLC chromatograms of Hpd made from different hematoporphyrin samples indicates that they exhibit practically no difference. The same is true when different Hpd batches, prepared in our laboratory, are compared with respect to their sensitizing ability in a cellular system. The most important factor to be kept under control is the porphyrin concentration during the preparation procedure. The stock solution of Hpd used in the present work was 5 mg/ml porphyrin in 0.15 M NaCl at pH 7.2. Photofrin II (PII) was obtained from Photofrin Inc., Cheekto-waga, N.Y.

Solutions of these materials should be stored in the frozen state; slow change may take place with increase in the Hp content.

The fractionation system to be described consisted of the use of fresh diethyl ether (stabilized by 2% ethanol) and water to which differing small amounts of acetic acid were added. The composition of the product obtained was dependent on the amount of acetic acid employed.

HPLC was carried out by means of a 0.46x15 cm LC-18 reverse phase column (Supelco Inc., Bellefonte, Penn.).

A linear gradient elution system was applied: solvent A was methanol/water in volume ratio 60/40, buffered at pH=7.0, solvent B was methanol/water in volume ratio 90/10, buffered at pH=7.5, and the gradient time (0-100%B) was 40 min. The pH was kept stable by means of 1mM phosphate buffer in both solvents. 200 μl of the porphyrins, diluted 1:400 in solvent A, were injected. The flow rate was 1 ml/min. The porphyrins eluted from the column were monitored both by absorbance (λ=392 nm) and fluorescence (λexc=330-400 nm; emitted light with wavelength lower than 410 nm was removed by a cut-off filter). Absorption spectra were recorded by means of a Cary 118 spectrophotometer.

Cellular photosensitization was studied using NHIK 3025 cells and an assay system for cell survival described earlier[6]. The cells were incubated with 12.5 μg/ml porphyrin in E2aR medium containing 3% serum for 18 hours at 37°C before irradiation. After the irradiation, the cells were cultivated for 24 hours in medium without porphyrins before fixation and staining with methylene blue.

All reagents were of the highest analytical quality.

RESULTS

A 2 mg/ml Hpd solution in water (10 ml) was shaken with 60 ml of diethyl ether in the presence of 5 μl(B) or 50 ul(A) of acetic acid. The phases were allowed to separate. A precipitate was formed between the two phases. The amount and nature of porphyrins present in this precipitate was dependent on the amount of acetic acid added.

The interphase was collected by centrifugation and washed by a small amount of ether.

The ether phase was washed with water and analyzed by means of HPLC after removal of the solvent. It was found to contain mainly Hp and Hvd. This ether phase was practically free from the higher aggregates which are known to give rise to a broad and ill-defined peak in the chromatogram of Hpd. The small amounts of well-defined peaks eluting after Hvd in the Hpd chromatogram were also strongly reduced.

The precipitated porphyrins, collected by centrifugation, were dissolved in 0.1 N NaOH, neutralized and analyzed by HPLC (Fig. 1, b and c). This fraction contained porphyrins similar to those present in PII. These porphyrins absorb maximally in the wavelength region 360-370 mm (Fig. 1, f). Fig. 1 also indicates that by properly adjusting the amount of acetic acid in the separation system, a precipitate can be obtained, the chromatogram of which (B) is practically identical with that of PII. When using very low concentrations of acetic acid, a precipitate was obtained that contained much of the higher porphyrin aggregates. These were found, however, to have a lower photosensitizing efficiency in our cellular system than did Hpd and PII.

The photosensitizing effect of the porphyrin preparations in our cellular system was studied, using the solutions whose absorption spectra are shown in Fig. 1, f. The results are shown in Fig. 1 e.

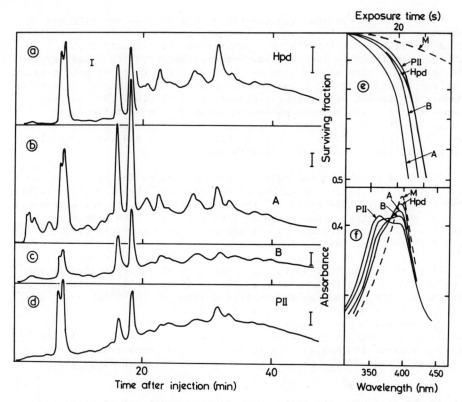

Fig. 1.
a) HPLC chromatogram of Hpd; b) and c) Chromatograms of the
precipitates in the presence of 50 µl and 5 µl of acetic acid,
respectively; d) Chromatogram of PII; e) Survival curves for the
porphyrin preparations shown in a, b, c, d and that of the monomers extracted into the ether phase (M). The absorbance data of
the solutions with which the cells were incubated are shown in f).
The bars correspond to A=0.005 at 392 nm.

DISCUSSION

    The present work shows that it is possible to obtain from Hpd a
product with chromatographic and sensitizing properties similar to those
of PII by means of a simple extraction procedure. Furthermore, it is
shown that the main sensitizing effect of these porphyrin preparations
resides neither in the monomers present nor in the high aggregates but
in some minor components with well-defined peaks in the chromatogram.
Recent studies show that uptake by cells and effect increase with HPLC
retention, i.e. with decreasing polarity of the porphyrins.

    The typical porphyrin structure is that of a flat planar molecule
possessing aromatic character. In those porphyrins devoid of free
carboxyl groups, polarity is only introduced by the pyrrolic ring
nitrogens and this will not be large unless pH conditions favour their
dissociation. Carboxylic side chains, however, impart polarity to the
molecule which increases as the extent of their ionization and electric
charge increases (i.e. with decreasing pH) in the supporting medium.

Porphyrins, therefore, can display marked surface activity and adsorption phenomena; they tend to form rather stable films at liquid-liquid and liquid-air interfaces. Studies by the Langmuir trough technique comfirm this expectation. Protoporphyrin, with 2 carboxylic side chains, located on one side of the flat disc, forms at an air-water interface condensed solid films which collapse rapidly under pressure in a manner consistent with vertical side-to-side packing. The carboxyl groups remain in the aqueous phase, cohesion of the aromatic discs through Van der Waals forces being greater than adhesion to water. Such stacks are typical of the dimer or aggregated state and can similarly form at suitable liquid-liquid interphases.

The hematoporphyrin molecule affords an interesting exception to this general behaviour. By virtue of the hydroxy-ethyl functions in its 2 and 4 positions, this region of the molecule also manifests a tendency to water solubility consequent on further polarity. The planar molecules of hematoporphyrin do not therefore form vertical stacks at an air-water interface but lie flat on the water surface.

It is thus seen that various porphyrins will be orientated at surfaces depending upon factors such as charge distribution, their structural tendency for H-bonding, the surface tension at the interphase, etc.

Species which are already highly aggregated or polymerized may present only limited superficial regions capable of reacting physically with the surrounding media. Highly carboxylated porphyrins such as uroporphyrin, and $TPPS_4$, on the other hand, do not even dimerize in solution.

The presence of detergents in an aqueous phase may drastically alter the surface activity of a porphyrin. For example, Lowe and Phillips[8] found that incorporation of Cu into protoporphyrin is greatly speeded up by adding sodium dodecylsulphate, the N atoms being, as it were, thereby exposed to the aqueous phase so that the Cu-ion collisions could be more probably effective. Another example is the effect of detergent upon the partition between an aqueous phase and octanol.

Applying the above considerations to the fractionation of Hpd described in this work, one may conclude that at the slightly acid pH conferred by the acetic acid, the polar monomers of Hp, and the Hvd well pass into the ether phase whilst the more weakly polar constitutients will tend to form less soluble stacks at the interphase and thus precipitate. Di-porphyrin ether or ester structures would be expected to be sensitive to the above conditions.

Should the acetic acid concentration fall below a critical level, large polymerized aggregates will not be resolved but will precipitate also.

ACKNOWLEDGEMENTS

The present work was supported by the Norwegian Cancer Society (Landsforeningen mot Kreft) and by the Association for International Cancer Research.

REFERENCES

1. R. Lipson, The use of a derivative of hematoporphyrin in tumor

    detection, J. Natl. Cancer Inst., 26:1 (1961).
2. J. Moan, T. Christensen and S. Sommer, The main photosensi-
    tizing components of hematoporphyrin derivative, Cancer
    Lett., 15:161 (1982).
3. V.O. Nseyo, T.J. Dougherty, D. Boyle, W. Potter,
    L.S. Englander, R.P. Huben, and J.E. Pontes, Experimental
    photodynamic treatment of canine bladder, J. Urol., 133:311
    (1985).
4. J.A.S. Carruth and A.L. McKenzie, Preliminary report of a pilot
    study of photoradiation therapy for the treatment of super-
    ficial malignancies of the skin, head and neck, Eur. J.
    Oncol., 11: in press (1985).
5. S. Sommer, J. Moan, T. Christensen and J.F. Evensen, A chroma-
    tographic study of hematoporphyrin derivatives, in: "Por-
    phyrins in Tumor Phototherapy", A. Andreoni and R. Cubeddu,
    eds., Plenum Press, New York and London (1984).
6. J. Moan, B. Høvik and S. Sommer, A device to determine fluence-
    response curves for photoinactivation of cells in vitro.
    Photobiochem. Photobiophys., 8:11 (1984).
7. J. Moan and S. Sommer, Uptake of the components of hematopor-
    phyrin derivative by cells and tumors. Cancer Lett., 21:167
    (1983).
8. M.B. Lowe, J.N. Phillips, Metalloporphyrin formation: A
    possible enzymic model, Nature, 190:262 (1961).

CHEMICAL, BIOLOGIC AND BIOPHYSICAL STUDIES ON

'HEMATOPORPHYRIN DERIVATIVE'

David Kessel†, C.K. Chang‡ and Brian Musselman¶

†Departments of Pharmacology and Medicine
Wayne State University School of Medicine
Detroit, MI 48201

‡Departments of Chemistry and ¶Biochemistry
Michigan State University
East Lansing, MI 48824

INTRODUCTION

This article will describe procedures for fractionation, analysis, and chemical, biological and biophysical characterization of the tumor-localizing product HPD. Since the latter is a mixture of different porphyrins, many of the early studies on HPD represent composite results. Such studies could initially be justified, since HPD was being used in the clinic without purification. As products with fewer components are provided for clinical work, biological and biophysical investigations on an impure product become less relevant. Results are presented which bear on the structure of the tumor-localizing component and on some of its properties related to tumor-localization.

MATERIALS AND METHODS

Hematoporphyrin (HP), hydroxyethylvinyl-deuteroporphyrin (HVD), protoporphyrin (PP) and deuteroporphyrin (DP) were obtained from Porphyrin Products, Logan, UT. t-Butylammonium phosphate (BAP) was purchased from Aldrich Chemical Co., Milwaukee, WI. HPD was prepared (1,2) by dissolving HP (100 mg) in 1.9 ml of glacial acetic acid containing 0.1 ml of H_2SO_4,. After 60 min of stirring at 20°, the solution was poured slowly into 15 ml of 3% sodium acetate solution. The precipitate of HP acetates was collected by centrifugation, washed with water until the pH of the eluate was approx. 6, lyophilized and stored at -20°. HPD was prepared from this material by dissolving 20 mg in 1 ml of 0.1 M NaOH with stirring. After 60 min, the solution was brought to pH 7.5 with 0.1 M HCl and the volume adjusted to 2 ml. This preparation was stored at -20°. In some preparations, we doubled the concentration of HP acetates in the hydrolysis step, to increase the yield of the tumor-localizing fraction. Sterility of the product was tested on Trypticase Soy Agar plates containing 5% sheep blood.

Analytical HPLC was carried out with a C-8 column and CN guard column in a Waters Z-Module system (3). The column was eluted with a 30

min gradient of 70-100% methanol (remainder aqueous 5 mM BAP, pH 3.5).
The flow rate was 1 ml/min. After 30 min, the flow rate was increased to
1.5 ml/min, and the column eluted with methanol for an additional 15 min.
The SP8700 solvent delivery system, and SP4270 plotter-integrator were
purchased from Spectra Physics. Absorbance was measured, usually at 400
nm, with a Spectraflow 730 variable wavelength detector. Fluorescence
was monitored with a Perkin-Elmer 650-10S fluorometer (excitation wavelength = 400 nm, emission = 620-630 nm). In some studies, we mixed the
eluate with THF-water (1:1), using an additional low-pressure pump and
micro-vortex mixer, just before the fluorescence detector. This solvent
was mixed with the methanolic eluate to enhance fluorescence emission by
some porphyrin components of HPD.

Preparative fractionation of HPD was carried out on columns of
Sephadex LH-20 (Pharmacia, Piscataway, NJ) which were eluted with tetrahydrofuran (THF)-methanol-5 mM aqueous phosphate buffer pH 7 (2:1:1). A
5 x 30 cm column can fractionate 100 mg of HPD applied in 40 ml of solvent (3). Smaller columns were used for fractionation of lesser amounts
of HPD or other porphyrin mixtures, as described below. The solvent was
applied with an FMI pump model RP-SYX. Pump, columns and teflon fittings
were provided by Ace Glass Co., Vineland, NJ. Absorbance of LH-20 column
eluates was monitored at 550 nm, with an ISCO model UA5 detector and
chart recorder. The initial examination of fractions was carried out
with reverse phase TLC (4). RP-18 plates were developed in 65:35
methanol/aqueous t-BAP (3 mM, pH 3.5), with appropriate chromatographic
standards (HP, HVD, PP). The following fractions were pooled: [A] those
containing mainly material which remained at the origin on the TLC
plates; [B] those which contained both material which remained at the
origin and which migrated before HVD; [C] those which contained only PP,
HVD and HP. These fractions were evaporated under reduced pressure and
analyzed via HPLC.

The molecular weight of components present in the oligomer fraction
was determined via fast atom bombardment (FAB) mass spectrometry (MS);
the FAB matrix was dithioerythritol:dithiothreitol (1:1) which contained
0.01 M trifluoracetic acid.

Stability of the tumor-localizing fraction was measured by HPLC
assay after storage. One series of experiments involved storage of 5 mg/
ml solutions at pH 7 at -70°, -20°, 4°, 37° or 100° for varying intervals. This study provided information on the relative stability of the
tumor-localizing HPD fraction. In another series of experiments, we examined the stability of the tumor-localizing fraction at 37° in 1 M HCl
or NaOH (solvent = 50% THF).

Octanol:water partitioning of HPD was carried out at 22°. A 10 μg/
ml solution of drug in 5 mM phosphate buffer pH 7 was equilibrated with
an equal volume of water-saturated l-octanol for 24 hr, the phases
separated by centrifugation, and samples of the lower and upper phases
analyzed by HPLC. A fine precipitate which collected at the interface
was dissolved in methanol for HPLC analysis.

Adducts of HP and other porphyrins were prepared using a modification of the procedure described above. Ten mg of HP acetate was dissolved in 1 ml of 0.1 M NaOH containing 10 mg of another dissolved porphyrin or drug analog as described below. After 60 min, the reaction
mixture was neutralized and analyzed by reverse-phase HPLC. A 0.8 x 40
cm LH-20 column was then used to separate the oligomer fraction as shown
in Fig. 3.

Biologic studies were carried out using the L1210 murine leukemia cell line grown in Fisher's medium containing 10% horse serum. Studies carried out in vivo involved C57/BL6 mice bearing the Lewis Lung tumor transplanted to sub-cutaneous sites. L1210 cells were incubated with varying porphyrin levels in growth medium for 10 min to 24 hr. To mimic tumor-localizing conditions, the readily-diffusible porphyrin fraction was washed out by a second incubation for 30 min at 37°, in fresh medium. Tumor-bearing animals were treated with porphyrins via the intraperitoneal route; the drug dose was usually 7.5 mg/kg.

Drug localization was assessed in L1210 cells, incubated as described above, via fluorescence microscopy; a Nikon phase-contrast instrument was fitted with a UFX-II photomicrography accessory for determination of exposure time. High-speed daylight Ektachrome film was used; the light source was a mercury lamp filtered to transmit blue light at 360-420 nm. A second filter was used so that only red light (>600 nm) reached the film.

For an estimate of porphyrin concentration by fluorescence (3), tissue (approx 150 mg) was homogenized, using a Polytron PCU-2 disrupter, in 1 ml of 50 mM HEPES buffer pH 7.0. Tissues difficult to disperse with the Polytron were first frozen at -70° and then crushed to powder. The homogenate was brought to 10 mM CTAB, shaken with 3.5 volumes of 1:1 methanol/chloroform, and the phases separated by centrifugation at 22°. The integrated corrected fluorescence emission (600-750 nm), upon excitation at 400 nm, provides an estimate of the total porphyrin content. Plasma samples (250 μl) were diluted with CTAB-HEPES buffer and extracted as described above. Tissue blanks were run to correct for endogenous porphyrin levels. For HPLC determinations, the CTAB was omitted, and the initial homogenate brought to pH 5 with 1M H_3PO_4 before addition of organic solvents. The lower phase was evaporated under nitrogen, taken up in methanol, and particulate material removed by centrifugation. Use of fluorescence for estimation of tissue levels of porphyrins is a sensitive procedure, but the low fluorescence yield of the tumor-localizing fraction complicates interpretation of results (5). Hydrolysis in 1 M HCl (10 min, 100°), followed by neutralization and addition of CTAB (to 10 mM), results in the quantitative conversion of porphyrin oligomers to monomers, permitting estimation of porphyrin content of tissues without correction for variation in fluorescence yield. Correction for losses during extraction can be determined by addition of standards not present in samples being examined, e.g., deuteroporphyrin or mesoporphyrin, which are not components of HPD.

Photosensitization of cells in culture was measured after irradiation with light of specified wave-length from a quartz-halogen lamp and a series of sharp band-pass filters (±10 nm). The total fluence required for a given level of photodamage is a function of the porphyrin and wavelength; with HPD, at 630 nm, we use 0.6kJ/m². A calibrated thermopile was used for determination of light flux.

Plasma lipoprotein fractions were separated by gradient centrifugation as described in Ref. 6. We employed samples of plasma removed from mice 4 and 24 hr after injection of HPD (7.5 mg/kg) or of the tumor-localizing fraction (5 mg/kg). The resulting patterns were photographed (Ektachrome daylight film) under UV light, then analyzed using an ISCO Model 640 Density Gradient Fractionator with absorbance monitored at 280 (protein) and 400 nm (porphyrin).

Photodamage to L1210 cells was assessed by determining survival of irradiated populations. For this purpose, experiments were carried out under sterile conditions. The effect of porphyrins + light was estimated by a soft-agar assay which provides an estimate of the number of viable cells. Tubes containing 100 - 10,000 cells were prepared with Fischer's medium containing 0.3% agar, incubated in a humidified CO_2 incubator for 7 days, and the number of cell colonies counted. The cloning efficiency of untreated cells is in the 60-70% range, so that the effect of photo-dynamic damage on viability could readily be determined.

To provide another assessment of cell damage, we also examined the effects of porphyrins + light on concentrative (active) transport of the non-metabolized amino acid cycloleucine. After exposure to porphyrins and irradiated in vitro, as described above, cell pellets (7 mg/ml) were suspended in growth medium and incubated with radioactive cycloleucine (50,000 counts/min/ml) for 5 min at 37°. The cells were then collected by centrifugation (200 x g, 30 sec), washed once with cold isotonic NaCl, and intracellular cycloleucine measured by liquid scintillation counting.

RESULTS

The HPLC profile of HPD is shown in Fig. 1, along with a preparation involving alkaline hydrolysis of a 40 mg/ml solution of HP acetates (twice the usual concentration), showing the resulting improved yield of the tumor-localizing fraction. Elution times (min): HP, 11.5; HVD, 17.7 and 18.5, PP, 24.9, tumor-localizing fraction, 22-35.

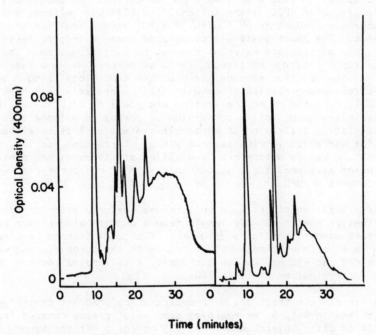

Figure 1. Reverse-phase HPLC analysis of HPD on an C_8 column. Absorbance at 400 nm was monitored. These tracings compare HPD prepared with twice the usual level of HP acetates (left) with HPD prepared in the usual manner (right).

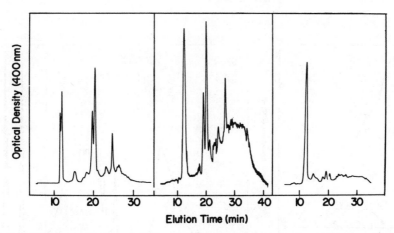

Figure 2. HPLC analysis of HPD partitioned between 1-octanol and 5 mM phosphate buffer pH 7. Left: upper (organic) phase; center: interface (dissolved in methanol); right: lower (aqueous) phase.

Sterility testing of preparations of HPD indicated presence of <5 microorganism/ml, although no precautions to insure sterility were taken. In contrast, sterility of our deionized water supply indicated the presence of 250-500 organisms/ml. The synthesis of HPD calls for reagents normally free from microorganisms: conc. acetic and sulfuric acids, 0.1 M NaOH and HCl. Furthermore, exposure of HPD solutions to low levels of illumination may have contributed to elimination of microbial contaminants (7).

HPLC analysis of HPD partitioned between 1-octanol and aqueous pH 7 buffer (Fig. 2) showed HP, HVD and PP in the upper (organic) phase, HP in the lower phase, and HP, HVD, PP + the tumor-localizing fraction at the interface.

Use of the LH-20 column for isolation of the tumor-localizing fraction of HPD is illustrated in Fig. 3. The elution profile of HPD from the column is shown, along with the HPLC analysis of the three major fractions. The first fraction (A) contained >80% of the tumor-localizing material (3). The second fraction (B) had a more complex HPLC pattern, with HP, HVD and PP present along with other porphyrins. The final fraction (C) contained only the three porphyrin monomers (HP, HVD and PP). Fraction A was used for further studies described below.

The stability of fraction A to storage at different temperatures was determined. Storage at -20° and -70° resulted in no detectable changes in the HPLC analysis after 6 months (data not shown), while storage for 3 months at 4°, or at 100° for 10 min resulted in substantial conversion to HP + HVD (Fig. 4). We also examined the relative stability of fraction A to 1 M HCl or NaOH in 50% THF. In this solvent mixture, hydrolysis to HP was demonstrated under acid or alkaline conditions (Fig. 5).

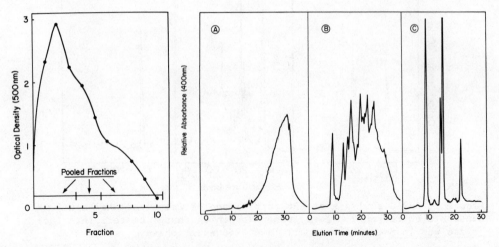

Figure 3. Left: absorbance profile (550 nm) of HPD eluted through an LH-20 column as described in the text. Right: HPLC analysis of 3 major LH-20 fractions (A-C).

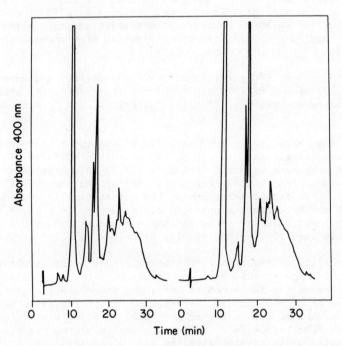

Figure 4. Stability of the tumor-localizing fraction of HPD (Fig. 3, fraction A) to storage at 100° for 10 min (left) or at 4° for 100 days (right).

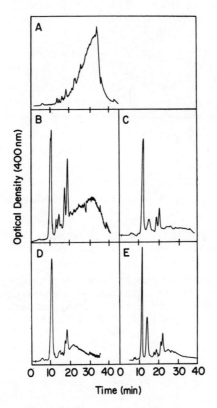

Figure. 5. <u>A</u>, HPLC elution profile of the tumor-localizing fraction of
HPD isolated by non-aqueous gel exclusion chromatography (5). <u>B</u> - <u>E</u>,
hydrolysis of this fraction in 50% tetrahydrofuran at 37°: <u>B</u>, 1M HCl for
1 hr; <u>C</u>, 1 M HCl for 24 hr; <u>D</u>, 1 M NaOH for 1 hr; <u>E</u>, 1M NaOH for 24 hr.

FAB studies of Fraction A (Fig. 6, top) showed ion clusters repre-
sentative of mono, di and tri porphyrins in the mass spectrum (top). FAB
analysis of porphyrins using an acidified matrix typically generates ions
whose mass corresponds to that of the protonated porphyrin molecule. We
detected intense ion species at m/z 1179, 1161 and 1143. The mass of two
HP molecules minus one water molecule = 1178. Protonation of this dimer
yields m/z 1179. Ions at m/z 1161 and 1143 can result from successive
loss of 1 or 2 water molecules from this dimer as <u>sec</u>-OH substituents are
dehydrated. A second, less intense series of ion clusters was present at
m/z 1201, 1183 and 1165. These are attributed to the presence of a
sodium salt in the solvent, yielding the sodium adduct of each species
described above. The appearance of ion clusters at masses corresponding
to protonated porphyrin trimers was also evident. These ions correspond
to the mol. wt. of protonated HP trimer (1795 daltons) minus 2, 3 and 4
water molecules, respectively. Sodium cation-containing species were
also present at m/z 1745 and 1763.

The data described above suggested that the tumor-localizing fraction of HPD contained porphyrin dimers and oligomers formed via elimination of water from 2 or more HP molecules. Either an ether or an ester

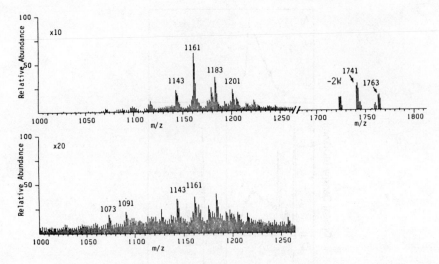

Figure 6. Partial FAB mass spectrum of (top) the tumor-localizing HPD fraction (Fig. 3, fraction A), and (bottom) HPD prepared in the presence of deuteroporphyrin. Mass regions corresponding to porphyrin dimers (m/z 1050-1250) and trimers (m/z = 1700-1850) are illustrated.

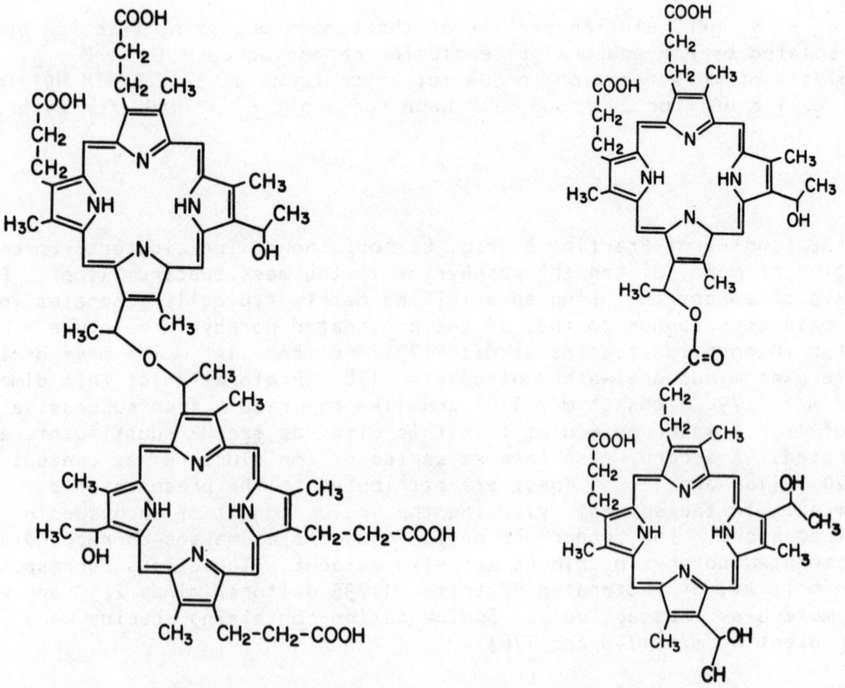

Figure 7. Dihematoporphyrin ether (left) and ester (right).

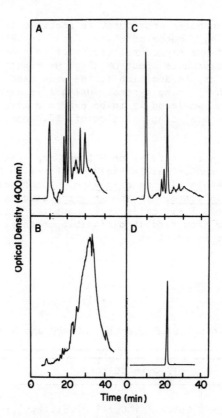

Figure 8. Reverse-phase HPLC analyses: A- HPD prepared in the presence of deuteroporphyrin as described in the text. B- Monomer-free fraction of A isolated on an LH-20 column. C- Hydrolysis products of B after treatment with 1 M NaOH in 50% THF, showing formation of hematoporphyrin and deuteroporphyrin. D. Deuteroporphyrin standard.

(Fig. 7) would fulfill these structural requirements. The instability of the oligomer fraction of HPD toward both acid and base (Fig. 5) suggested the ester. We therefore examined the reaction products formed when HP acetates were added to an alkaline solution of deuteroporphyrn (DP). The latter contains no sec-OH groups, and cannot participate in ether formation. HPLC analysis showed the usual HPD components, + unreacted DP (Fig. 8A). Elution through a 0.8 x 40 cm LH-20 column resulted in separation of porphyrin oligomers from DP and other porphyrin monomers (Fig. 8B). This fraction contains covalently-bound DP: alkaline hydrolysis in 50% tetrahydrofuran led to formation of hematoporphyrin and deuteroporphyrin (Fig. 8C). This hydrolysis pattern is consistent with the presence of an ester structure with a carboxyl group of deuteroporphyrin esterified to a sec-OH group of hematoporphyrin. FAB-MS analysis of this product is shown in Fig. 6, bottom. In addition to the ion species previously identified for HP dimer (m/z = 1179, 1161 and 1143), ions corresponding to the mass of protonated DP-HP dimer and a dehydration product are present at m/z = 1091 and 1073, respectively.

With the same synthetic procedure, we prepared an ester of HP and the naturally-occurring reduced porphyrin (chlorin) Bonellin (8). The proposed structure of one structural isomer of this product is shown in Fig. 9, left. The absorbance spectrum (Fig. 9, right) indicates a substantial band at 640 nm, in addition to the weak band at 630, characteristic of the porphyrin ring system. Using this ester for cell photosensitization studies, we found it to be 6-8 fold more lethal to L1210 cells than the tumor-localizing fraction of HPD, upon irradiation with light at 620-650 nm.

Results of biological studies on HPD and the fractions separated on the LH-20 column are summarized in Table 1. Fraction B was the superior fluorescence localizer _in vitro_ and _in vivo_. But fraction A was the better photosensitizer _in vitro_. In the _in vitro_ studies, 'localizing conditions' were employed, i.e., cells were incubated with porphyrins for 24 hr., then washed free from readily-diffusible porphyrin (30 min at 37° in growth medium) before irradiation.

Table 1.   Biological Properties of HPD and HPD Fractions

Fraction	L1210 in culture				Lewis Lung in vivo	
	Fluor[1]	Uptake[2]	CL[3]	Viability[4]	Fluor[1]	Uptake[2]
HPD	100	155	45	17	100	2.2
Fraction A	109	230	45	13	130	3.8
Fraction B	220	165	27	15	190	6.9
Fraction C	20	16	85	80	14	0.3
HP	3	5	99	100	4	0.3

All studies involved treatment of L1210 cells with 10 μg/ml of porphyrin for 24 hr, followed by washing in fresh medium for 30 min at 37°. Animals bearing the Lewis-Lung tumor were given 7.5 mg/kg of porphyrin 48 hr before measurements were taken.

1. Relative integrated fluorescence (600-750 nm) of cell homogenates in isotonic NaCl. These values are normalized so that the HPD result for each cell line = 100.

2. Porphyrin uptake measured via fluorescence of acid-hydrolyzed cell extracts. Units = μg porphyrin/gm cells.

3. Accumulation of labeled cycloleucine by L1210 cells after treatment with porphyrin + light. The control value (100) represents a 2.5-fold concentrative uptake of this non-metabolized amino acid.

4. Viability (% control) of cells after phototherapy (630 nm, 0.6 kJ/m^2) determined by a clonogenic assay.

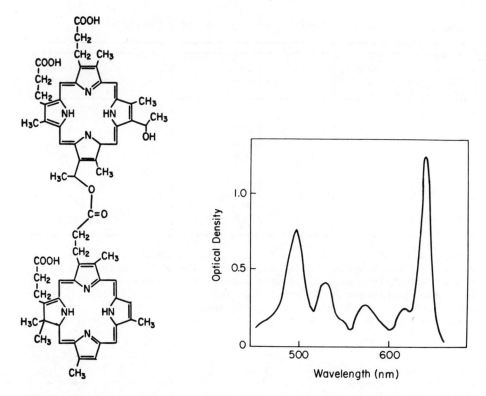

Figure 9. Left: structure of the hematoporphyrin-bonellin ester (one isomer shown). Right: absorbance spectrum of the ester.

The fluorescence yield of the tumor-localizing fraction was inherently less than that of HP regardless of solvent conditions (Table 2). To obtain a fluorescence yield equivalent to that of HP, this fraction was first hydrolyzed in 0.5 M HCl (100°, 10 min) and then neutralized. After addition of CTAB to 10 mM, fluorescence could be used to quantitate the drug concentration.

Examination of L1210 cells under the fluorescence microscope revealed a diffuse peripheral red fluorescence after 10 min incubations with HPD, characteristic of drug accumulation at membrane loci. After 24 hr incubations with either HPD or the fraction A (Fig. 3), followed by washing in fresh medium at 37° for 30 min to remove readily-diffusible porphyrin, discrete spots of cytoplasmic fluorescence were observed. These results are shown in color photographs at the front of this book.

A study of the persistence of HPD components in circulating blood of the mouse after injection of 7.5 mg/ml revealed a 2-phase exodus (Fig. 10, left). An initial phase (half-life = 2.4 hr) was followed by a slower second-phase of drug loss (half-life = 24.5 hr). HPLC analysis (fluorescence detection) showed that a complex pattern of porphyrins was responsible for tumor and plasma fluorescence 48 hr after drug administration (Fig. 10, right). Ultracentrifugation studies showed that porphyrin was initially bound to low-density (LDL) and high-density (HDL) plasma lipoprotein; >90% of the total porphyrin was bound to the HDL fraction after 48 hr.

Table 2. Fluorescence yields of porphyrins

Porphyrin	Solvent	Fluorescence yield
HP	MeOH	3400
	TMW	6400
	CTAB	4695
DHE	MeOH	625
	TMW	1850
	CTAB	1900

Solvents: MeOH, methanol; TMW, 2:1:1 tetrahydrofuran/methanol, 5 mM aqueous sodium phosphate buffer pH 7; CTAB, 10 mM CTAB detergent in 50 mM phosphate buffer pH 7.0. Fluorescence yield = ratio of (integrated corrected fluorescence)/(absorption at 400 nm). For all studies, absorbance (400 nm) = 0.1. HP was chromatographically pure; DHE (dihematoporphyrin ester) was Fraction A, Fig. 3.

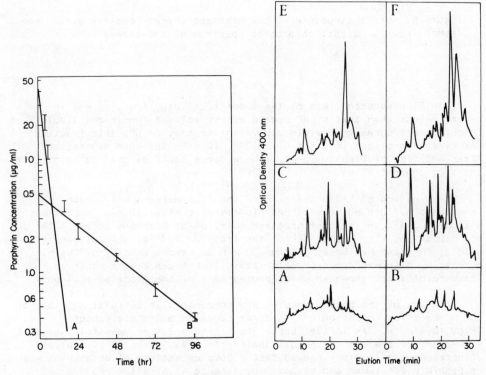

Figure 10. Left, plasma fluorescence as a function of time after injection of HPD (7.5 mg/ml) in the mouse. Right, HPLC analysis (fluorescence detector) of: A control plasma; B control tumor; C (plasma) and D (tumor) 48 hr after HPD injection; E (plasma) and F (tumor) 48 hr after injection of the tumor-localizing component of HPD (Fig 3A, 4 mg/kg).

DISCUSSION

Use of the C_8 column for reverse-phase HPLC analysis of HPD resulted in >99% recovery of all material applied to the column. One column was sectioned after 100 determinations and found to contain no discoloration from porphyrin retention. In contrast, we could not elute the tumor-localizing fraction of HPD from a C_{18} column at pH <7.5. Use of the CN guard column resulted in no porphyrin hold-up, while protecting the main column from particulate contamination. As suggested by the data shown in Table 2, the tumor-localizing fraction of HPD could not be readily detected by fluorescence measurements of HPLC eluates (5). Addition of THF and water via a post-column pump promoted the fluorescence yield of this fraction, as suggested by data shown in Table 2.

Since formation of the tumor-localizing fraction of HPD is a condensation reaction, increasing the level of HP acetates during alkaline hydrolysis results in an increased yield of product (Fig. 1).

Partition of HPD between octanol and water gave an unexpected result. The tumor-localizing fraction was found at the interface, along with some other HPD components. HP remained in the aqueous phase, while the HVD and PP components partitioned into the octanol. The octanol/water model is sometimes used as a model system for the extracellular:membrane interface, but in this case, the model is apparently inappropriate.

The HPLC analysis of HPD indicates that preparative reverse-phase fractionation cannot yield a pure preparation of the tumor-localizing fraction which elutes as a broad band from 22-35 min. The appearance of PP in the middle of this band would complicate such attempts at purification. Use of an organic solvent and the LH-20 column resulted in separation of HPD into three major fractions. The first contained >80% of the tumor-localizing component. The final fraction was composed of porphyrin monomers (HP, HVD and PP). An intermediate fraction had a complex pattern, perhaps reflecting the presence of HPD components better separated by the reverse-phase system than is the bulk of the tumor-localizing material. Moan had described, in a report on HPLC analysis of the tumor-localizing fraction of HPD, the likely presence of a broad poorly-defined fraction together with a series of fractions better separated by HPLC (8); the data shown in Fig. 5 confirms this prediction.

Photosensitization studies in culture indicate that the first LH-20 fraction is the major sensitizer, while fraction B is associated with fluorescence localization (Table 1). The low degree of lethality associated with fraction B may be a function of the site(s) of intracellular binding of this material.

Using an organic solvent and an LH-20 column (Fig. 3), we isolated the tumor-localizing fraction of HPD. While this product was readily converted to HP + HVD at 100°, slow hydrolysis was observed even at 4° (Fig. 4). This result indicates that solutions of HPD need to be stored below 0° if degradation is to be avoided. The hydrolysis of the tumor-localizing component of HPD by aqueous acid, but not base suggested an ether structure (10). But in 50% THF, this fraction was more stable to acid than to alkaline hydrolysis (Fig. 5). This result appears to remove a major argument for an ether linkage. Since mass spectrometry had suggested a structure via HP molecules covalently joined with loss of the elements of water (Fig. 6), the alternative ester structure was considered. Our evidence for the formation of an HP-DP structure, described above, suggests that the ester linkage (Fig. 7) is the more likely conformation. The mechanism of formation of a base-catalyzed ester linkage presumably involves the nucleophilic substitution reaction illustrated in the following scheme:

Previous studies have documented photosensitization at membrane loci when irradiation closely followed incubation of cells in HPD (11,12). In view of the complex nature of HPD, this result may not reflect the situation which occurs in vivo, where irradiation usually occurs 24-48 hr after HPD administration. Since HPD contains porphyrins which are known to photosensitize at membrane loci (13,14), the results obtained with short-term incubations are not surprising. But when irradiation occurred after long-term exposure of cells to HPD in vivo, intracellular damage was observed (15). This latter result corresponds to our results using fluorescence microscopy which show that the tumor-localizing fraction of HPD localizes at small vesicles in the cytoplasm.

By dissolving the HP acetate mixture in a solution of 0.1 M NaOH containing the natural product bonellin, we prepared the product shown in Fig. 9 (left). This product has the strong absorption band at 640 nm (Fig. 9, right) characteristic of a chlorin structure (8). We found this product to be 5-fold more effective than the HP analog (Fig. 7, right) for photodynamic killing of L1210 cells under 620-650 nm light. This result suggests a program of drug development, with a view toward preparation of more effective photosensitizers. It may also be feasible to prepare dimeric or polymeric photosensitizers with less affinity for skin, therefore minimizing the major adverse reaction to photodynamic therapy: photosensitization of the skin.

Drug distribution studies indicate both a fast and slow phase of HPD loss from (Fig. 10), the latter representing porphyrin bound to HDL. Other studies (16,17) have indicated binding of HP and PP to plasma lipoprotein. It is tempting to speculate that an initial binding of porphyrin to LDL might explain the tumor-localizing phenomenon. The relative affinity of the tumor-localizing component of HPD for different tissues (2) generally reflects the level of LDL receptors in different tissues (18). Only the phenomenon of long-term retention of porphyrin at tumor loci remains to be explained.

ACKNOWLEDGMENTS

This work was supported in part by grant CA 23378 from the National Cancer Institute, NIH, DHHS. Mass spectrometry studies were carried out at Michigan State University. The mass spectrometry instrumentation was supported by the Biotechnology Resources Program of the Division of Research Resources of the NIH, RR 00480-16. Additional mass spectra were obtained by Lubo Baczynskj, Upjohn Co, Inc., Kalamazoo, MI. HPLC analyses were carried out by May-Ling Cheng; photosensitization studies by Deborah Khalil.

REFERENCES

1. Lipson, R., E. Baldes and A. Olsen (1961) The use of a derivative of hematoporphyrin in tumor detection. J. Natl. Cancer Inst. 26, 1-8.
2. Gomer, C.J. and T.J. Dougherty (1979) Determination of [³H]-and

226

[¹⁴C]hematoporphyrin derivative distribution in malignant and normal tissue. Cancer Res. 39, 146-151.

3. Kessel, D. and M-L Cheng. (1985) Biologic and biophysical proper-
    ties of the tumor-localizing component of 'hematoporphyrin deriva-
    tive'. Cancer Res, In Press.
4. Kessel, D. and T.C. Chou (1983) Tumor-localizing components of the
    porphyrin preparation hematoporphyrin derivative. Cancer Res. 43,
    1994-1999.
5. Kessel D., and M-L. Cheng (1985) On the preparation and properties of
    the tumor-localizing component of HPD. Photochem. Photobiol. 41,
    277-282.
6. Chapman, M.J., S. Goldstein, D. Lagrange and P.M. Laplaud (1981) A
    density gradient ultracentrifugal procedure for the isolation of
    the major lipoprotein classes from human serum. J. Lipid Res. 22,
    339-358.
7. Bertoloni, G., M. Dall'Aqua, M. Vazzoler, B. Salvato and G. Jori
    (1982) Photosensitizing action of hematoporphyrin on some bac-
    terial strains. Med. Biol. Environ. 10, 239-242.
8. Kessel, D. and C.J. Dutton (1984) Photodynamic effects: porphyrin vs.
    chlorin. Photochem. Photobiol. 40, 403-406.
9. Moan, J. and S. Sommer (1983) Uptake of the components of hematopor-
    phyrin derivative by cells and tumours. Cancer Lett. 21, 167-174.
10. Dougherty, T.J. (1984) The structure of the active component of hema-
    toporphyrin derivative. In Porphyrin Localization and Treatment of
    Tumors. (Edited by D.R. Doiron and C.J. Gomer). Alan R. Liss, New
    York, pp. 301-314.
11. Kessel, D. (1976) Effects of photoactivated porphyrins at the cell
    surface of leukemia L1210 cells. Biochem. 16, 3443-3449.
12. Dubbelman, T.M.A.R. and J. Van Steveninck (1984) Photodynamic effects
    of hematoporphyrin-derivative on transmembrane transport systems
    of murine L929 fibroblasts. Biochim. Biophys. Acta 771, 201-207.
13. Girotti, A. (1979) Protoporphyrin-sensitized photodamage in isolated
    membranes of human erythrocytes. Biochem. 18, 4403-4411.
14. Dubbelman, T.M.A.R., A.F.P.M. De Goeij and J. Van Steveninck (1978)
    Protoporphyrin-sensitized photodynamic modification of proteins in
    isolated human red blood cell membranes. Photochem. Photobiol.
    28, 197-204.
15. Christensen, T., T. Sandquist, K. Feren, H. Waksvik and J. Moan
    (1983) Retention and photodynamic effects of haematoporphyrin
    derivative in cells after prolonged cultivation in the presence of
    porphyrin. Br. J. Cancer 48, 35-43.
16. Reyftmann, J.P., P. Morliere, S. Goldstein, R. Santus, L. Dubertret,
    D. and Lagrange (1984) Interactions of human serum low density
    lipoproteins with porphyrins: a spectroscopic and photochemical
    study. Photochem. Photobiol. 40, 721-729.
17. Jori, G., M. Beltramini, E. Reddi, B. Salvato, A. Pagnan, L. Ziron,
    L. Tomio, and T. Tsanov (1984) Evidence for a major role of plasma
    lipoproteins as hematoporphyrin carriers in vivo. Cancer
    Lett. 24, 291-297.
18. Brown, M.S., P.T. Jovanen and J.L. Goldstein (1980) Evolution of the
    LDL receptor concept-from cultured animal cells to intact animals.
    Proc. N.Y. Acad. Sci. 348, 48-68.

SOME PREPARATIONS AND PROPERTIES OF PORPHYRINS

Tilak P. Wijesekera and David Dolphin

Department of Chemistry
University of British Columbia
Vancouver, B.C., V6T 1Y6, Canada

INTRODUCTION

Porphyrins and the related tetrapyrrolic macrocyclic compounds, chlorins, bacteriochlorins and corrins are prosthetic groups of a large number of biological molecules which serve diversified roles in nature. These macrocycles, coordinated to a central metal ion, perform functions such as oxygen transport and storage (hemoglobin and myoglobin), electron and energy transfer (cytochromes and chlorophylls) and biocatalysis (coenzyme B_{12}, cytochrome P-450). The metal-free ("free-base") porphyrins on the other hand are generally present in organisms as precursors of metalloporphyrins and are accumulated and/or excreted in certain physiological disorders such as porphyrias (1). The selective biodistribution of free-base porphyrins was reported by Policard (2) as early as 1924, based on the observed orange-red fluorescence from neoplastic tissue. The preferential localization of certain parenterally administered porphyrins in tumor cells, coupled with the property of intense fluorescence, have since been used for convenient detection of such tissue *in situ*. In addition, the ability of porphyrins to act as photosensitizers, especially in the production of highly reactive singlet oxygen which in turn exerts cytotoxic action, has been used extensively in the novel treatment of malignant tissue, a technique referred to as photoradiation therapy (PRT).

Attention has been focussed primarily on a porphyrin preparation known as hematoporphyrin derivative (HPD) whose tumor localizing properties were first described by Lipson et al. (3). HPD, prepared (4) by the acetylation of hematoporphyrin (HP) in 5% H_2SO_4 in CH_3CO_2H and subsequent treatment with aqueous NaOH (0.1 M), is a mixture of porphyrins whose chemical composition, tumor localization and photosensitization properties have been the subjects of intense investigation for the past decade (5-7). In order to avoid ambiguity of results due to the possible variations of composition (8) in different preparations of HPD, most studies have been carried out using a commercial preparation known to as Photofrin I. HPD has been partially purified using gel-filtration chromatography (see under chromatography) and the more active "aggregated"* fraction is also available commercially as Photofrin II. Although considerable progress

---

* In the literature the term aggregation is used to refer to both the physical intermolecular association and intramolecular covalent oligomerization.

has been made in the studies related to HPD therapy, numerous questions are still left unanswered. An in-depth description of the chemical and physical properties of porphyrins may provide a better understanding of the structure-activity relationship of this material, and could also help in the design and synthesis of related new drugs with improved localization, photosensitization and tumoricidal properties.

STRUCTURE AND NOMENCLATURE OF PORPHYRINS

The porphyrin nucleus (Figure 1) with its conjugated 18 π electron system (conforming to Huckel's 4n+2 rule for aromaticity) constitutes a highly stable macrocyclic system. It is stable towards concentrated sulfuric acid and trifluoroacetic acid, both of which are often used to remove coordinated metals. Perchloric and hydroiodic acids as well as permanganate are known to destroy the porphyrin nucleus. Chlorin (the parent macrocycle of a variety of chlorophylls) with one peripheral double bond reduced and bacteriochlorin with two double bonds reduced, both retain the aromaticity, the conjugated 18 π electron system is maintained, since only the cross-conjugated double bonds are reduced. The corrin nucleus (the parent macrocycle of B_{12} coenzyme) is neither aromatic nor planar but is highly colored due to the extended conjugation and cobalt coordination.

(A) FISCHER

(B) PORPHYRIN 1–24

PORPHYRIN NUCLEUS–ALTERNATIVE NUMBERING SCHEMES

CHLORIN

BACTERIOCHLORIN

CORRIN

Figure 1. Nomenclature and numbering of tetrapyrrolic macrocycles

The nomenclature commonly used in porphyrin chemistry was developed by Hans Fischer and is based on the numbering scheme shown in Figure 1A, where the pyrrolic β-positions are numbered from 1-8 and the meso positions $\alpha, \beta, \gamma, \delta$. Fischer nomenclature involves a very large number of trivial names which do not convey any structural information. In addition, it involves a "type-isomer" system to distinguish between the different positional isomers. Two different substituents, one each on each pyrrole ring produces four type-isomers (e.g. coproporphyrins and uroporphyrins). Three different substituents A, B and C, with one

230

substituent A on each pyrrole ring, one substituent B on each of two pyrrole rings and one substituent C on each of the other two pyrrole rings produce 15 possible isomers. Protoporphyrin and hematoporphyrin fall into this group with the natural isomer being designated IX. This isomer number is not used now to describe the natural porphyrins (Table I).

Table I
Trivial names for some common porphyrins

Trivial name	Substituents[a] and Locants								
	1	2	3	4	5	6		7	8[b]
	2	3	7	8	12	13	15	17	18[c]
Etioporphyrin I	Me	Et	Me	Et	Me	Et	H	Me	Et
Coproporphyrin II	Me	Cet	Cet	Me	Me	Cet	H	Cet	Me
Uroporphyrin III	Cm	Cet	Cm	Cet	Cm	Cet	H	Cet	Cm
Protoporphyrin[d]	Me	Vn	Me	Vn	Me	Cet	H	Cet	Me
Hematoporphyrin[d]	Me	CH(OH)CH$_3$	Me	CH(OH)CH$_3$	Me	Cet	H	Cet	Me
Deuteroporphyrin[d]	Me	H	Me	H	Me	Cet	H	Cet	Me
Mesoporphyrin[d]	Me	Et	Me	Et	Me	Cet	H	Cet	Me
Rhodoporphyrin[e]	Me	Et	Me	Et	Me	CO$_2$H	H	Cet	Me
Phylloporphyrin[e]	Me	Et	Me	Et	Me	H	Me	Cet	Me
Pyrroporphyrin[e]	Me	Et	Me	Et	Me	H	H	Cet	Me
Chlorocruoroporphyrin (Spirographisporphyrin)	Me	CHO	Me	Vn	Me	Cet	H	Cet	Me

[a] The following abbreviations are used: Me for CH$_3$; Et for CH$_2$CH$_3$; Vn for CH=CH$_2$; Cm for CH$_2$CO$_2$H (carboxymethyl); Cet for CH$_2$CH$_2$CO$_2$H (carboxyethyl).
[b] Fischer numeration.
[c] Porphyrin 1-24 numeration.
[d] Natural isomer - formerly type IX.
[e] Natural isomer - formerly type XV.

It has been long recognized that the Fischer nomenclature was inadequate to name the large number of synthetic and newly isolated porphyrins. The International Union of Pure and Applied Chemistry (IUPAC) and the International Union of Biochemistry (IUB) through a joint commission on biochemical nomenclature have issued recommendations (9) for naming porphyrins, based on the 1-24 numbering system shown in Figure 1B. Systematic names of substituted porphyrins are formed by the application of the rules of systematic organic nomenclature. Etioporphyrin I is accordingly named as 2,7,12,17-tetraethyl-3,8,13,18-tetramethylporphyrin.

Considering the fact that porphyrin chemists working with biological systems still prefer to use the Fischer nomenclature, eleven well established trivial names have been retained in the new recommendations. Moreover, the designations of the positional isomers in the etio-, copro- and uroporphyrin series introduced by Fischer have also been accepted. For more complex cases, the use of type isomers is not recommended. As an alternative to a systematic name, compounds closely related to the accepted trivially named porphyrins have been named semisystematically as derivatives of such porphyrins. Thus, chlorocruoroporphyrin (Table I) is semisystematically named as 3-formyl-8-vinyldeuteroporphyrin whereas the systematic name would be 8-formyl-3,7,12,17-tetramethyl-13-vinyl-porphyrin-2,18-dipropanoic acid. The porphyrins whose trivial names have been retained, are ranked according to (i) the number of component rings

(ii) the number of carbon atoms (iii) molecular weight, and the porphyrin of higher rank number is preferred to one of lower number in the selection of the parent for a semisystematic name. A condensed version of the recommendations, adapted for the use of those of biochemical interest has appeared (10) in the literature. The semisystematic nomenclature based on the 1-24 numeration will be used throughout this discussion.

ABSORPTION SPECTRA

Porphyrins exhibit characteristic absorption and fluorescence properties in the visible region which make them useful as photo-sensitizers. The metal-free porphyrin has an intense absorption (λ_{max} ~10^5) around 400 nm, known as the Soret band and four bands, designated I, II, III and IV between 450 and 700 nm. The intensity and the exact peak positions are dependent on the solvent as well as the concentration. More importantly, correlations have been shown to exist between the nature of the porphyrin side chains and the positions and the relative inten-sities of the absorption bands (11).

Four basic types of spectra have been identified on the basis of the relative intensities of the four visible bands (Figure 2). Gouterman (11) has proposed an interpretation of these spectral differences based on the perturbations of the π-electron levels, by the peripheral substituents.

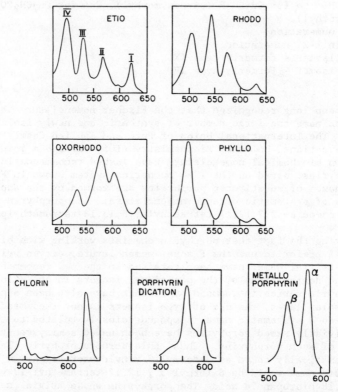

Figure 2. Typical visible absorption spectra of tetrapyrrolic macrocycles

(a) Etio-type:

This type of spectrum is characterized by a IV > III > II > I order of band intensities (Figure 2). Porphyrins in which six or more β-positions have alkyl substituents, with the other two being unsubstituted, will exhibit an etio type spectrum, irrespective of the relative orientations of the substituents. In addition to etioporphyrin isomers (from which the name is derived), most naturally occurring porphyrins such as copro-, uro-, hemato-, proto- and deuteroporphyrins exhibit this type of spectrum (Table II).

Table II
Visible spectral data for some natural porphyrins

Porphyrin	Solvent	λ nm ε mM	Soret	Band IV	Band III	Band II	Band I	Ref
Uroporphyrin I or III octamethyl ester	CHCl$_3$	λ	406	502	536	572	627	13
		ε	215	15.8	9.35	6.85	4.18	
Coproporphyrin I or III tetramethyl ester	CHCl$_3$	λ	400	498	532	566	621	14
		ε	180	14.34	9.92	7.13	5.0	
Protoporphyrin IX dimethyl ester	CHCl$_3$	λ	407	505	541	575	630	14
		ε	171	14.15	11.6	7.44	5.38	
Hematoporphyrin dimethyl ester	Pyridine	λ	402	499.5	532	569.2	623	15
		ε	175.5	14.7	9.04	6.57	4.35	
Deuteroporphyrin dimethyl ester	CHCl$_3$	λ	399.5	497	530	566	621	14
		ε	175	13.36	10.1	8.21	4.95	
3,8-diformyldeutero porphyrin dimethyl ester	CHCl$_3$	λ	435	526	562.5	595	651	14
		ε	137.5	12.6	7.70	6.48	3.48	
3-Formyl-8-vinyl- deuteroporphyrin dimethyl ester	CHCl$_3$	λ	420	518.5	559	584	642	16
		ε	163	10.6	15.0	9.48	2.00	

HPD in saline solution (as commercially supplied) exhibits an etio type spectrum. The spectrum shows averaging from the different components and Figure 3 gives the absorption spectrum of the most hydrophobic component obtained by Hanzlik et al. (12). The Soret band which appears between 360-400 nm has been shown to be sensitive to the aggregation state of the porphyrin (see under aggregation).

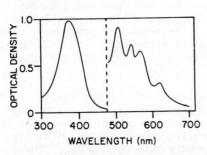

Figure 3. Absorption spectrum of the hydrophobic fraction of HPD (Ref. 12)

(b) Rhodo-type:

One strongly electron-withdrawing group (e.g. formyl, acetyl or carboxyl) conjugated with the porphyrin ring causes band III to be more intense than band IV resulting in the rhodo type spectrum (III > IV > II > I) shown in Figure 2 (named after rhodoporphyrin - Table I). In addition, this produces a bathochromic shift of all bands in the spectrum. The change from an etio to a rhodo spectrum (rhodofying effect) is minimal with vinyl substituents. The visible spectrum of a porphyrin carboxylic acid has been shown (17) to change from a rhodo type (in pyridine) to etio type (in methanolic sodium methoxide). An etio type spectrum is also produced when two adjacent pyrrole units carry electron-withdrawing substituents, e.g. 3,8-diformyldeuteroporphyrin. Although the rhodofying effect of one group is cancelled by the other, the effect upon the red-shift of absorption bands, is additive (Table II).

(c) Oxorhodo-type:

This spectral pattern is characteristic of porphyrins having two electron-withdrawing groups on diagonally opposite pyrrole rings. This can be viewed as a further enhancement of the rhodofying effect. Oxo-rhodoporphyrin (CH_3CO substituted for C_2H_5 at position 3 of rhodo-porphyrin - Table I), a degradation product of chlorophyll, has given its name to this spectroscopic class in which the intensities of absorption maxima follow the order III > II > IV > I (Figure 2). The effect of the carbonyl group on the porphyrin spectrum has been shown to be reversed by oxime formation (18) e.g. the oxorhodo spectrum of oxorhodoporphyrin changes to a rhodo type and the rhodo spectrum of rhodoporphyrin changes to an etio type. Another interesting observation that has been made (17) is that when the electron-withdrawing group is a β-keto ester ($\omega = COCH_2CO_2R$), a rhodo spectrum is observed as expected, but when it is a ketomalonate [$\omega = COCH(CO_2R)_2$], an oxorhodo spectrum is observed. Both spectra change to etio type in methanolic sodium methoxide due to isomerization to the enolate anion.

(d) Phyllo-type:

This spectral pattern (IV > II > III > I; Figure 2), named after phylloporphyrin (Table I) is distinguished from the etio type by less intense bands III and I. Two substitution patterns on the periphery produce the phyllo spectrum: (i) a single meso-alkyl substitution and (ii) four or more unsubstituted β-positions.

In addition to external substitution whose effects on optical spectra are not pronounced, changes in the conjugation path affect the porphyrin spectra significantly. These include tetrabenzporphyrins (benzo rings fused to the four pyrroles), tetraazaporphyrins (meso carbons replaced by nitrogens), and of more biological importance, the chlorins and the bacteriochlorins (Figure 1). The reduction of the $^{17}\Delta$ exo double bond, although not affecting the aromaticity of the molecule, produces a visible spectrum characterized by a major long wavelength peak at 650-680 nm with less intense peaks between 450-650 nm (Figure 2). This gives a green color to these compounds.

Metalation of the porphyrin (the dianion formed by the removal of the NH protons) which acts as a tetradentate ligand often changes the four band spectrum to one with two bands, designated α and β, between 500 and 600 nm while retaining the "Soret" absorption around 400 nm. This is due to the change in the conjugated ring symmetry from D_{2h} to D_{4h}. The relative intensities and the absorption maxima of the α and β bands depend on the metal as well as on the nature of the porphyrin ligand. In acid medium, porphyrins produce dicationic species which exhibit two major bands in the visible region with weaker bands appearing as shoulders (Figure 2). The spectral simplification is a result of the approach towards D_{4h} symmetry. In a recent review, Gouterman (19) discusses the optical spectra and the electronic structure of porphyrins and related molecules.

FLUORESCENCE SPECTRA

Fluorescence is the emission of energy from the lowest excited singlet state and the energy of the fluorescence photon will be less (hence higher wavelength) than that of the absorbed photon. For porphyrins, fluorescence provides a very sensitive method for their detection and in the case of HPD related treatment, fluorescence, coupled with its ability to reach and maintain a higher concentration in malignant tissue than in non-malignant tissue, have formed the basis of tumor diagnosis. Fluorescence has also been used to provide information on the concentration of HPD as a function of dosage and time of injection (20).

The fluorescence excitation maxima for free-base porphyrins are, in general, close to the absorption maxima, both in the Soret and the longer wavelength regions. Usually, the fluorescence emission spectrum shows a distinct major peak at 615 nm and another at approximately 675 nm. Figure 4 shows the fluorescence excitation and emission spectra for the most hydrophobic fraction of HPD (12). The fluorescence from porphyrins is thus in the red region (600-800 nm) of the spectrum and is relatively free from interferences due to emissions from other compounds present in biological specimens.

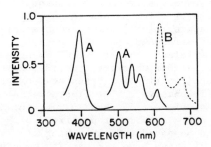

Figure 4.   Fluorescence excitation (A) and emission (B) spectra of the hydrophobic fraction of HPD (Ref. 12)

The fluorescence spectra of HPD upon interaction with biological material could be somewhat different. Berns et al. (21) in their studies with HPD treated mouse cells incubated at 37°, observed in addition to the distinct major peaks, a minor peak at 590 nm. The existence of the 590 nm peak has previously been reported in solid mouse tumors treated with HPD (in vivo) (22) and also in certain buffer solutions of hematoporphyrin (23). Berns et al. (21) report that the 590 nm peak increases in intensity at the expense of the other two bands and in 20 hr, becomes the major peak. The spectrum is unchanged even when the cells are centrifuged, the supernatant discarded and the cells resuspended in fresh saline, indicating that the fluorescent material is bound to the cells. Andreoni and Cubeddu (24) made similar observations regarding the change, with time, of the fluorescence emission spectra in aqueous solution (Figure 5). This effect was pronounced with the hydrophobic gel filtration fraction of HPD and the commercially available Photofrin II and has been suggested to be due to a new molecular species originating from a binding of "monomers" to polymeric porphyrins.

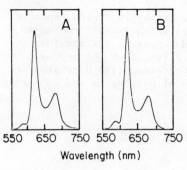 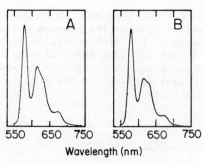

Wavelength (nm)   Wavelength (nm)

(a)   Initial   (b)   After 48 hours

Figure 5.   Variation of fluorescence emission spectra with time (Ref. 24)
            A. Gel-filtration fraction of HPD   B. Photofrin II

Chlorins, in cell-free environments have not received the same
systematic study as porphyrins. Differences between the fluorescence of
porphyrins and chlorins relate to the changes observed in the absorption
spectra. The lowest energy absorption band of chlorins has an ε_{max} of 5-10
x 10^4 while in porphyrins, the value is approximately an order of magnitude less. Since the natural radiation fluorescence rate is roughly
proportional to the molar extinction coefficient (25), the fluorescence
quantum yield should be substantially higher for chlorins, provided that
the radiationless decay rates from the excited singlet states are comparable. This is a feature that is generally observed. The fluorescence
spectrum of chlorophyll in ethyl ether solutions (26) shows a very intense
band at 670 nm and a weaker band at 720 nm, a spectrum red shifted with
respect to that of porphyrins. It should be noted that quenching of
fluorescence (27) as well as appearance of a new band (28) at 755 nm have
been observed with chlorophyll under conditions which produce aggregation.

Metal ion chelation affects porphyrin fluorescence dramatically. The
presence of heavy metal ions, whether actually complexed with the ground
state organic molecule or not, is known to quench fluorescence. The heavy
metals increase the radiationless decay rate for the intersystem crossing
to the excited triplet state resulting in a decrease in the fluorescence
quantum yield. Phosphorescence, the emission from the triplet state, is
expected to increase. Iron and most paramagnetic transition metals lead
to a marked quenching of porphyrin fluorescence, due not only to the heavy
atom effect, but also due to the unpaired electrons inducing fast triplet
state formation. Becker and Allison (29) have investigated a wide range of
metalloporphyrins and have observed reasonably strong fluorescence from
complexes of closed-shell diamagnetic metals such as Mg(II) and Zn(II).

An important aspect that should be considered in the use of the
absorption and fluorescence properties in tumor detection and photoradiation therapy is light penetration in tissues. Most tissues are non-
homogeneous in nature and tend to exhibit strong absorption and scattering
for visible light. *In vitro* and *in vivo* studies (30,31) of the attenuation
coefficients and penetration depths for tissues have shown that red light,
although poorly absorbed by porphyrins, is the best for photoradiation
therapy due to higher tissue transparency. Thus the weakest band of HPD
(ca. 630 nm) is commonly used for phototherapy, but maximum tissue

penetration has been observed in the near infrared from 700-850 nm and
1000-1100 nm. It is reasonable therefore to expect chlorins, with
stronger absorption bands between 650-700 nm, to be more efficient photosensitizers than the corresponding porphyrins for excitation in this range
of the spectrum. In fact, Kessel and Dutton (32), using Sarcoma-180 tumor
cells in culture, observed a 10 fold greater efficiency of photodynamic
inactivation for the chlorin, bonnellin, compared with the structurally
related mesoporphyrin.

PHOTOPROCESSES OF EXCITED PORPHYRINS

(a) Photodynamic Effect

Several studies have established that biological systems can be
damaged by the simultaneous exposure to light, oxygen and a photosensitizer, a phenomenon referred to as the photodynamic effect. The basic
processes involved are:

(i) Photoexcitation of the ground state sensitizer molecule (S) to
the first excited singlet state (^1S)

$$S \xrightarrow{\ h\nu\ } {}^1S$$

(ii) Spin inversion of the excited singlet to the triplet state (^3S)
which has a lower energy but a longer lifetime

$$^1S \longrightarrow {}^3S$$

(iii) The excited triplet ^3S undergoing one of the following two
types of reactions:

Type I

The excited triplet sensitizer interacts directly with the substrate
molecule, either by hydrogen atom abstraction or by electron transfer to
produce a reactive radical species. These radicals may interact with
oxygen to produce oxidized products via reactive species such as O_2^-, $OH^\bullet$
or H_2O_2.

Type II

The triplet sensitizer ^3S reacts with the ground state triplet oxygen
producing the excited singlet oxygen

$$^3S + {}^3O_2 \longrightarrow S + {}^1O_2$$

Singlet oxygen is a highly reactive species which selectively attacks
electron rich substrates to give peroxides or other oxidized species.
In biological systems, the substrates could be amino acid residues of
proteins, bases in nucleic acids or lipids in membranes. The overall
effect could be disturbances in metabolism, mutations or membrane defects
leading to changes in permeability. This destructive influence of the
photodynamic effect has been used to destroy tumor tissue in which certain
porphyrins preferentially accumulate. The competition between type I and
type II reaction pathways is controlled by the relative concentrations of
the substrate and oxygen as well as the rates of reaction of the triplet
sensitizer (porphyrin) with substrate and oxygen. The reaction rates in

turn depend on the nature of the reaction medium and the aggregation state of the porphyrin.

A feature of porphyrins that makes them particularly useful as donors of triplet excitation is the relatively small energy gap between the lowest singlet and triplet states and the corresponding high intersystem crossing efficiencies. The exact mode of porphyrin mediated photodynamic cytotoxicity has been under investigation by several research groups and there appears to be some disagreement in the literature concerning the pathway of photosensitized cell destruction. In homogeneous cell-free systems, the results of Reddi et al. (33) strongly indicated a singlet oxygen mediated type II mechanism for the hematoporphyrin sensitized photodegradation of tryptophan. This is confirmed by the observations of Moan (34) which suggest that type II processes play a dominant role in tryptophan degradation in aqueous solution sensitized by hematoporphyrin as well as the commercially available Photofrin I and Photofrin II. A competing type I process has been observed by Cannistraro et al. (35) and Jori et al. (36) in homogeneous aqueous solutions as well as in aqueous dispersion of ionic and neutral micelles. Grossweiner et al. (37) investigating the lysis of phosphatidylcholine (PC) liposomes (model membranes), sensitized to visible light by hematoporphyrin observed that the lytic mechanism changed from a type II (single oxygen mediated) pathway at low porphyrin concentrations to a type I (anoxic) pathway at high concentrations. This observation has been used to suggest a possible shift of mechanism from type II in the non-aggregated state of porphyrin (low concentrations) to type I in the aggregated state (high concentrations). Using the "aggregated fraction" of HPD as the photosensitizer, Grossweiner (38) recently presented results that provide unambiguous proof that the damage in PC liposomes was singlet oxygen mediated.

Studies of Dougherty (39) and Gomer et al. (40) with tumor tissue subjected to HPD photosensitization indicate that, limit of blood flow to such tissues prevents photodynamic cytotoxicity. This strongly suggests that a singlet oxygen mediated type II mechanism in solely responsible for *in vivo* phototoxicity.

(b) Alterations of the Macrocycle and/or Side Chains

(i) The extreme light sensitivity of protoporphyrin (1) in solution is known to be due to its self-sensitized photooxidation (Scheme 1). The

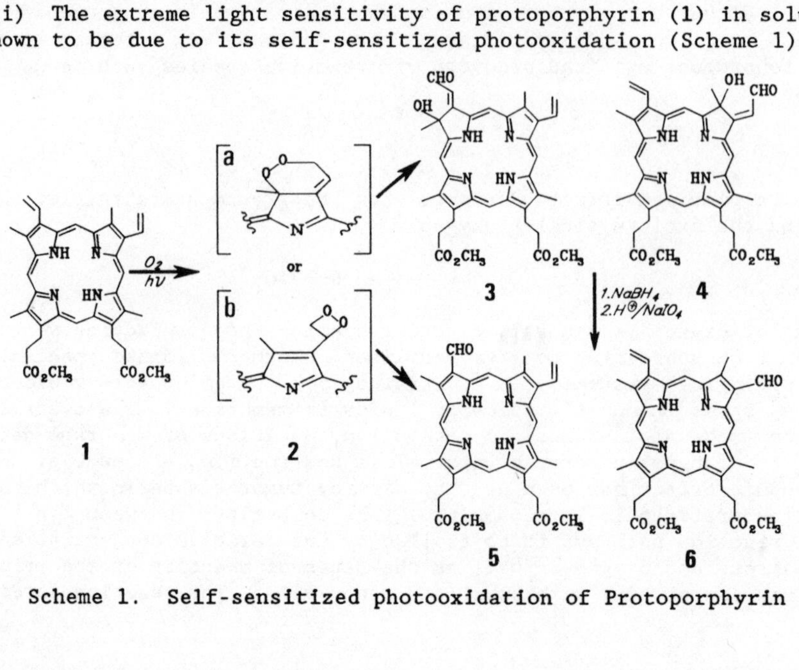

Scheme 1. Self-sensitized photooxidation of Protoporphyrin

major products of this reaction have been characterized as the isomeric hydroxy-formylethylidene porphyrins 3 and 4 commonly referred to as photoprotoporphyrins. The change in the peripheral conjugation from a porphyrin type to a chlorin type gives these products a strong visible absorption at ca. 670 nm (41) which results in a green colored solution. The synthetic value of these products was first explored by Inhoffen et al. (41, 42) who converted the chromatographically purified isomers 3 and 4 separately to 3-formyl-8-vinyldeuteroporphyrin (5) and its 3,8 isomer 6 respectively (Scheme 1). The isomer 5, commonly known as chlorocruoro-porphyrin, as its ferric complex, is the prosthetic group of chloro-cruorin, the oxygen carrying pigment of certain polychete worms. The porphyrins 5 and 6 together with the 3,8-diformyldeuteroporphyrin have been observed as minor photooxidation products of 1 in organic solvents but, have been reported as the dominant products when 1 is irradiated in aqueous micelles or phospholipid vesicles (43).

A singlet oxygen mediated mechanism can explain the formation of both types of products. A Diels-Alder type reaction between $^1O_2$ and the diene unit formed from the vinyl and the endocyclic double bond of either ring A or ring B of 1 could lead to an endoperoxide 2a which rearranges to give 3 or 4. Alternatively, a 1,2 addition of $^1O_2$ to a vinyl double bond could produce a dioxetane 2b which would cleave lead-ing to the formyl products 5 or 6 (44). A type I electron transfer mechanism from the excited porphyrin (producing a cation radical) to oxygen (producing a superoxide ion) has also been suggested (43) for the photodegradation of protoporphyrin. Cox et al. (45), using trapping reagents for singlet oxygen and superoxide ion, were able to demonstrate that the major pathway was via singlet oxygen. Further, these workers were able to show that the products of photooxidation (3, 4, 5, 6) themselves are good sensitizers of singlet oxygen. But the lower reaction rates of the monoformylmonovinyldeuteroporphyrins attributable at least in part to the increased oxidation potential (suggested to be due to the electron-withdrawing formyl group), make these compounds effective agents for promoting photodynamic action.

(ii) Photooxidation has a synthetic usefulness in porphyrin chemistry. Mauzerall and Granick (46) have reported the photooxidation of uroporphyrinogen (7) to the corresponding porphyrin 10 in solution containing oxygen (Scheme 2). Visible spectral evidence suggested that

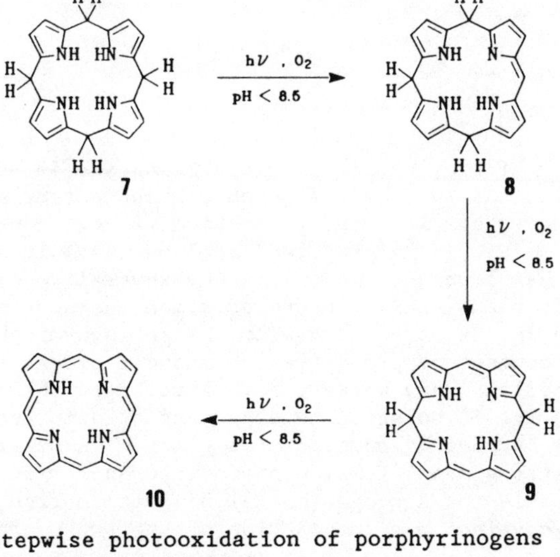

Scheme 2. Stepwise photooxidation of porphyrinogens (substituents omitted for clarity)

the oxidation proceeds via a porphomethene (8) and a porphodimethene (9) intermediate. The long induction period exhibited by this reaction is significantly shortened by the addition of a small quantity of uroporphyrin (10), which probably acts as a photosensitizer. A similar photooxidation has been used by Paine and Dolphin (47) in their improved synthesis of octaethylporphyrin, a useful model substance in porphyrin chemistry. By effecting the synthesis of the macrocycle from the monopyrrole precursor without aeration, it has been possible to crystallize the corresponding porphyrinogen in high yield. This is subsequently photooxidized in acetone solutions over a period of 1-2 weeks.

(iii) Photoreductions of porphyrins under a variety of conditions are also known. Mauzerall (48,49) has reported a study of the photochemical reduction of uroporphyrin in the presence of reducing agents such as ascorbic acid, EDTA and ethylacetoacetate. The products have been shown to be the di-, tetra- and hexahydroporphyrins, 11, 12 and 7 respectively (Scheme 3).

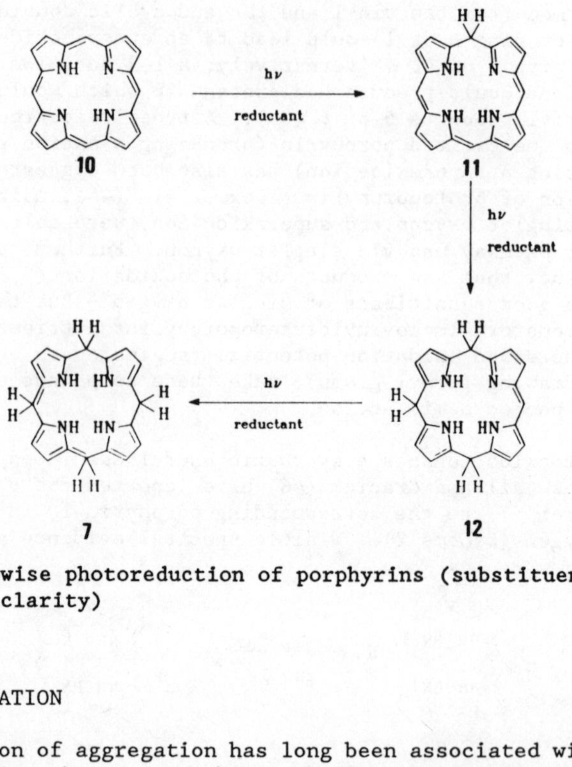

Scheme 3.   Stepwise photoreduction of porphyrins (substituents omitted for clarity)

PORPHYRIN AGGREGATION

The phenomenon of aggregation has long been associated with changes in spectral properties observed as the porphyrin concentration in solution is increased. Dougherty et al. (50) were the first to report that the most active component of HPD is "aggregated" (polymerized) in aqueous solution. This observation has since been confirmed by several research groups. The importance of this phenomenon in photoradiation therapy comes from the fact that aggregation is associated with change in hydrophobicity, absorption and fluorescence properties, transport across membranes and localization in cells. Many workers have attributed the observed property changes of porphyrins in solution to a process of dimerization based on a "face-to-face" or "sandwich" model of two units. In metalloporphyrins and chlorophylls, a strong metal to side-chain interaction is known to be primarily responsible for aggregation (51,52) but in free base porphyrins only a weak, non-covalent π-π interaction of the two macrocyclic units holds them together.

Aggregation is the only phenomenon which has been consistently related to deviation in Beer's law (in aqueous solution) indicating the presence of more than one absorbing species in solution. Associated with aggregation, there is a distinct hypsochromic shift in the Soret absorption maximum (accompanied by a broadening of the band) and a smaller bathochromic shift in the bands between 500-700 nm (Figure 6). In addition, a decrease in the extinction is also observed. The "monomer" is characterized by a Soret absorption ca. 400 nm whereas the "dimer" shows a Soret band ca. 365-370 nm. The observation of an isosbestic point in dilution studies where the number of chromophore molecules in the light path (concentration x pathlength) is kept constant, is considered as good evidence for a monomer-dimer equilibrium versus extensive aggregation.

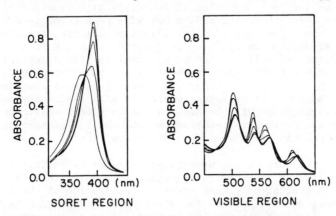

Figure 6. Effect of concentration on hematoporphyrin absorption spectra. (concentration x path length kept constant)

Visible spectroscopy has not been of much use in the study of aggregation of porphyrins in non-aqueous media. At concentrations generally used in visible spectroscopy, no Beer's law deviations related to aggregation have been observed, probably due to the lower dielectric constants of such solvents. However, doubly-linked synthetic cofacial porphyrin dimers exhibit (53,54) the same shifts (as compared with the monomers) in their visible spectra as those observed for dimerization in aqueous solution. Most of the information on aggregation of porphyrins in organic solvents comes from proton magnetic resonance spectroscopy, where the concentrations used are higher. Natural porphyrins have generally been studied as their methyl esters in deuterochloroform solution. The shifts of various resonances with dilution have been examined in terms of distances between the two porphyrins in a dimer and their orientations with respect to each other (52,55). For protoporphyrin and related porphyrins, Caughey et al. (56) have pointed out that in the absence of steric factors, electron withdrawing groups at the 3,8-positions enhance dimerization. A dimer model, with the 13,17-propionate chains of the two rings pointing in opposite directions (Figure 7) has been suggested, with the possibility of a dipole-dipole

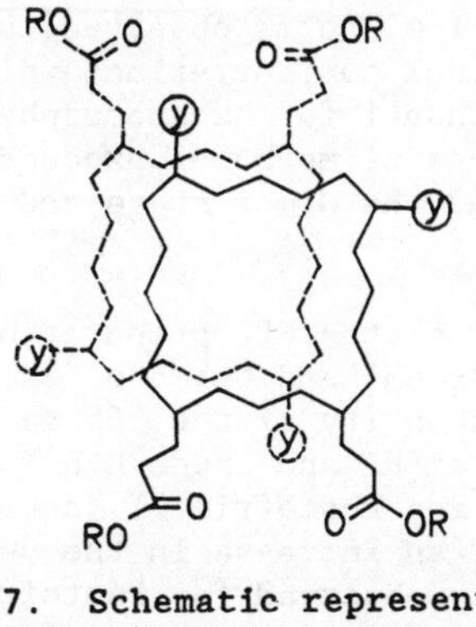

Figure 7. Schematic representation of a porphyrin dimer in solution

type interaction (57). NMR results as well as the possible steric effects of side chains indicate that the two rings would be slightly displaced. Porphyrins do not show any dimerization in strongly acidic media since the charge on the dications will oppose any interaction.

Water soluble porphyrins are characterized by ionizable substituents. The two major types of porphyrins in this group are (i) the synthetic meso-aryl porphyrins with carboxy or sulfonato groups and (ii) the natural porphyrins with acetic and propionic acid side chains on the pyrrolic β-position. The extent of dimerization of these porphyrins is strongly dependent on the nature of the aqueous medium. Several studies (58-60) have shown that added alcohol decreases the formation of the dimer, suggested to be due to a reduction of the dielectric constant of the medium. Ionic strength of the medium has also been found to affect dimerization. Some porphyrins do not dimerize at ionic strengths near zero whereas extensive dimerization is observed in 0.1 M ionic strength. It has been suggested (61) that added electrolyte aids dimerization by providing a large concentration of counterions, thus screening the repulsive like charges of the two porphyrins. Aggregation is also dependent on the pH of the medium. At pH values where the side chains are not neutralized, aggregation will be minimal due to electrostatic repulsion. The effect of an increase in temperature would be to increase the amount of monomer in solution in any monomer-dimer equilibrium.

Of the natural porphyrins, uroporphyrin is unique in that in aqueous solution, it shows no tendency to aggregate (62, 63). This is probably due to the fact that at pH values where neutralization of the 8 carboxylic acid groups does not take place, the electrostatic repulsion is high. Brown et al. (64) investigated the aggregation of protoporphyrin, deuteroporphyrin, hematoporphyrin and coproporphyrin III in aqueous solution and observed that at a concentration of 4 μM, the aggregation of protoporphyrin was much greater than for the other three porphyrins, suggesting a state higher than a dimer. For the latter three porphyrins, within the concentration range 80 nM to 4 μM, monomer-dimer equilibria were observed as indicated by single isosbestic points. At high concentrations, a superimposition of "micellization" (large variations of spectra with no single isosbestic points) was observed. Using the magnitude of the extinction coefficients and the sharpness of the Soret peaks as indicators, the extent of dimerization at any given concentration (proto- > deutero- > hemato- > copro-) was shown to be related to the hydrophobicity of the side chain.

The importance of the phenomenon of aggregation in porphyrin mediated photoradiation therapy comes primarily from the observation that it alters the normal photosensitization and fluorescence behaviour of porphyrin monomers. Kessel and Rossi (65) in their studies of porphyrin sensitized photooxidation of tryptophan in water-methanol mixtures observed that the optimal rates of photooxidation were at methanol concentrations which produced the porphyrin dimer spectra; 30% methanol for hematoporphyrin and 60% methanol for protoporphyrin. Lower amounts of methanol produced spectral changes reflecting aggregation beyond the dimer stage and also impaired photooxidations.

The absorption and fluorescence behaviour of the HPD preparations have been studied by several research groups. Andreoni and Cubeddu (24) observed a marked increase in the relative intensity of the 365 nm band compared with the band ca. 400 nm in going from HP and crude HPD to the aggregated (gel-filtration) fractions of HPD and Photofrin II (aqueous solutions; same concentrations) indicative of an increase in the degree of aggregation. The lower fluorescence intensity observed for Photofrin II when compared with HP (by a factor of approximately 9) accounts for the

242

large amounts of non-fluorescent aggregated species present in the former. Studies of Moan and co-workers (66,67) have shown that the hydrophobic fraction of HPD is less effective in the photooxidation of tryptophan and also produce a low quantum yield of singlet oxygen.

All of the identified components of HPD, i.e. HP, isomeric hydroxy-ethylvinyldeuteroporphyrins (HVD) and Pp are known to aggregate and the ability to aggregate in aqueous environments is therefore not sufficient to ensure tumor localization. Although Pp and HVD's are known to be taken up by tissue cultures, only the "aggregate" fraction of HPD is selectively retained. Furthermore, once within the cell, any interaction with cellular components may change the aggregation state as well as the sensitization properties of the porphyrin. A preliminary approach to investigating this type of factors has been the study of aggregation and photosensitization in model systems which simulate in very simple ways, some properties of living cells. The two systems commonly studied are the detergent miscelles (SDS or CTAB), which act as very simple membranous structures, and liposomes, the bilayered phospholipid vesicles which simulate cell membranes. Poletti et al. (68) in their *in vitro* studies with Photofrin II observed that the addition of SDS or CTAB miscelles or Human Serum Albumin (HSA; known to bind porphyrins), induced spectral changes that suggest disaggregation and also an enhancement of fluores-cence quantum yields. Similar observations have been made by Hisazumi et al. (69), not only in miscelles but also in cells derived from a human bladder carcinoma. Grossweiner and co-workers (38,70) have reported that the aggregated fraction of HPD diffuse easily into PC liposomes with a Soret shift from 364 nm to 398 nm and a 4-fold increase in its fluores-cence, suggesting disaggregation or monomerization in the lipid regions. Emiliani and Delmelle (71) who studied the photodamage induced on cholesterol embedded in egg lecithin liposomes observed that the solubilization of the sensitizer in the lipid bilayer is a prerequisite for its photosensitizing activity at the membrane level.

CHROMATOGRAPHIC PURIFICATION OF PORPHYRINS

The relatively high molecular weight and the low volatility make separations of porphyrins by classical methods, rather difficult. However, due to the inherent color and fluorescence properties of porphyrins, chromatographic methods are best suited for their purification. It should be emphasized that since porphyrins can undergo self-sensitized photo-degradation, chromatographic experiments should be performed in the dark or under subdued light conditions, at least until the photosensitivity of the sample is known.

The chromatographic purification methods available for porphyrins have been reviewed (72,73). The diverse nature of the peripheral substituents determines the best chromatographic method suitable for the purification of any porphyrin. The first consideration should be thin-layer and conventional column chromatography which have been commonly used with cellulose, silica or alumina as adsorbents. Paper chromatography is now mainly of historical interest. Preparative tlc with approximately 10-times greater capacity than paper is often very useful particularly since preliminary qualitative tlc may be scaled up with greater reproduci-bility. For large scale separations and to effect a gross separation of a desired porphyrin from non-porphyrin contaminating material, conventional column chromatography is still found to be very useful. The use of alumina is preferred for porphyrins that contain structural features which increase basicity since they tend to be less mobile on silica, probably due to protonation. Alumina, especially the basic grade should be avoided if the porphyrin contains carbonyl functional groups.

The presence of carboxylic acid groups in many of the naturally occurring porphyrins imposes certain restrictions on the chromatographic omethod of choice. Firstly, the acid groups make the porphyrin extremely polar, so that mobility on normal phase adsorbents can be achieved only by the use of solvent mixtures containing polar components such as alcohols, bases (e.g. pyridine, lutidine) and carboxylic acids (e.g. acetic, formic) and water. Secondly, most natural porphyrins differ by the number of the acid groups and some, only by their arrangement (e.g. uroporphyrins I and III) which make their adsorption behaviour quite similar, and hence chromatographic separation extremely difficult. The isolation of such porphyrins from other water soluble metabolites is best achieved by esterification of the carboxyl groups, preferably as methyl esters. Esterification can easily be accomplished by the treatment of the carboxylic acid in THF-Et$_2$O solution (acidic pH) with diazomethane. Diazomethane is generated as an ether solution from the commercially available Diazald (N-methyl-N-nitroso-p-toluenesulfonamide) by the reaction of an alcoholic solution of KOH. A microscale technique has also been reported (74). Another common method for esterification is to stir the porphyrin carboxylic acid in 5% H$_2$SO$_4$ in methanol, overnight in the dark. Esterification has also been carried out with boron trifluoride-methanol (75) and also with trimethyl orthoformate-methanol-sulfuric acid. If deesterification is required after purification, it may be achieved by acid or base hydrolysis (76). The most common method is simply to dissolve the porphyrin ester in 25% (w/v) HCl and allow to stand at room temperature in the dark to effect complete hydrolysis (10-48 hr). The acid is then recovered by standing the container over KOH in a vacuum desiccator or by adjusting the pH and extracting into ether. Alternatively, the porphyrin ester, dissolved in tetrahydrofuran is stirred overnight (in the dark) with an equal volume of 2N aqueous KOH. After separation of the colorless organic phase, the aqueous phase is acidified to collect the porphyrin acid.

Numerous methods are available for the separation of porphyrin esters but high performance liquid chromatography (hplc) is now the technique of choice. The analysis of porphyrins from body tissues and fluids accounts for the majority of work in this area. Using a Porasil T (Waters Assoc.) column and isocratic elution with light petroleum-dichloromethane (5:4), Carlson and Dolphin (77) achieved base line separations of porphyrin methyl esters containing 2 to 8 carboxylic acid groups. Similar results have been reported (78) for a silica column with gradient elution - hexane to ethyl acetate or with other eluents such as benzene- ethyl acetate-chloroform (7:1:2) (79).

Quantitation in analytical hplc is generally achieved by coupling an absorption or fluorimetric detector to the system. But a complicating factor has always been the variation of the intensity of absorption or fluorescence of different porphyrins at the wavelength of detection and the efficiency of porphyrin extraction from natural sources. In order to estimate the quantity of each porphyrin in routine clinical analysis, Carlson et al. (80) have developed a useful internal standard, 8-hydroxy-methyl-3-vinyldeuteroporphyrin 13. Figure 8 shows the hplc resolution of porphyrins with 2 to 8 carboxylate ester groups in the presence of the internal standard (IS). The internal standard 13 and its 3,8-isomer (ISI) were prepared by the NaBH$_4$ reduction of the isomeric mixture of compounds 5 and 6 (Scheme 1) followed by chromatography. This internal standard has been successfully used in the hplc analysis of porphyric urine samples, both as free acids and as methyl esters. Battersby et al. (81) using isocratic elution with acetonitrile-water (7:3) from a reversed phase (μ-Porasil C$_{18}$) column readily separated the ethyl esters of copro-porphyrin isomers I and II from III and IV. The more difficult separation of the III and IV isomers was achieved by recycling the esters ten times

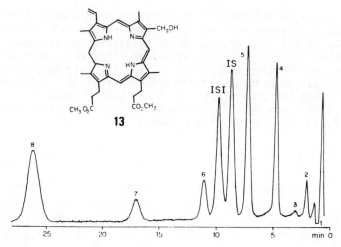

Figure 8. Hplc separation of porphyrins in the presence of an internal standard

with ether-n-heptane (2:3), 90% saturated with water. Wright et al. (82) have described a rapid and effective method for the simultaneous separation of all four coproporphyrin isomers as the free acids. They used an ODS-Hypersil column with 26% (v/v) acetonitrile in 1M-ammonium acetate solution as the mobile phase (pH adjusted to 5.15 with 1 M-acetic acid).

The separation of the type isomers uroporphyrin I and III, a hitherto difficult chromatographic problem has been achieved (83) as their methyl esters on a μ-Porasil (Waters Assoc.) column by recycling with n-heptane-acetic acid-acetone-water (1800:1200:600:1). The resolution of the isomers was found to be greater than 90% after 5 cycles. Jackson et al. (84) have recently separated the type I and III isomers of uroporphyrin octamethyl ester on a Hypersil column using hexane-ethyl acetate (1:1) as eluting solvent.

For the correct diagnosis of certain diseases such as porphyrias, it is sometimes necessary to analyze biological samples directly without prior isolation or derivatization of the porphyrins. The analysis of urinary porphyrin carboxylic acids by direct injection of the urine or acidified urine has been achieved on reversed phase hplc columns, with filtration or centrifugation as the only specimen pretreatment. Bonnett et al. (85) separated urinary porphyrins using μ-Bondapak C_{18} (Waters Assoc.) columns and aqueous methanol solvent system with tetrabutyl ammonium as counter ion added to the eluent. Johansson and Niklasson (86) obtained good separation using a reversed phase column coated with tributylphosphate (TBP) and applying a pH gradient (pH 4.4-6.5) in the eluent using phosphate buffers containing methanol (9:1). TBP has strong H-bonding properties and thus the columns give high retention of carboxylic acids.

The first attempt of the chromatographic analysis of HPD was made by Clezy et al. (87). Their method involved the esterification of the crude acetylated reaction mixture and separation by preparative tlc. The commercially obtained starting material (HP) was purified by esterifica-

245

tion, chromatography and subsequent hydrolysis. The major product in the complex reaction mixture has been identified as the 3,8-diacetoxyethyl-deuteroporphyrin (15; Figure 9). Minor quantities of protoporphyrin (1), monoacetoxyethyl monovinyldeuteroporphyrins (16, 17), isomeric mono-hydroxyethyl monovinyldeuteroporphyrins (18, 19) and the isomeric monoacetyxyethyl monohydroxyethyldeuteroporphyrins (20, 21) have also been identified. When the pure porphyrin 15 was base treated and introduced into tumor tissue, stronger fluorescence was observed (88) compared to that observed for the surrounding tissues.

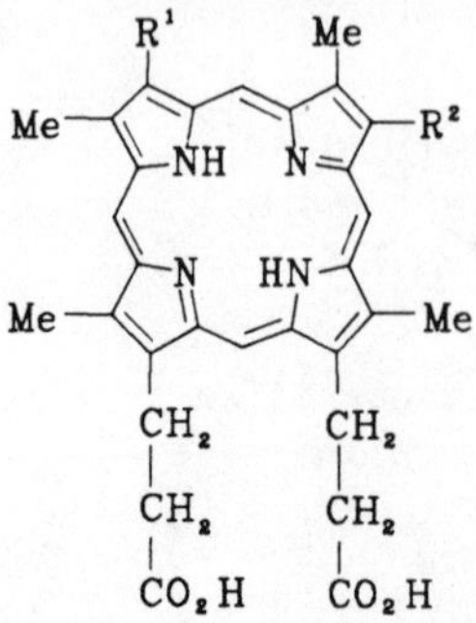

		R_1	R_2
	1	$CH=CH_2$	$CH=CH_2$
	14	$CH(OH)CH_3$	$CH(OH)CH_3$
	15	$CH(OAc)CH_3$	$CH(OAc)CH_3$
	16	$CH=CH_2$	$CH(OAc)CH_3$
	17	$CH(OAc)CH_3$	$CH=CH_2$
	18	$CH=CH_2$	$CH(OH)CH_3$
	19	$CH(OH)CH_3$	$CH=CH_2$
	20	$CH(OAc)CH_3$	$CH(OH)CH_3$
	21	$CH(OH)CH_3$	$CH(OAc)CH_3$

Figure 9.   Structures of some HPD components

Bonnett et al. (89) using their previously reported (85) reversed phase hplc technique for the analysis of porphyrin carboxylic acids, carried out a similar study on the acetylated product of hematoporphyrin, without subjecting to esterification. Their results were in agreement with those of Clezy et al. (87) in that the product was a mixture of porphyrins 15-21. When the individual fractions, dissolved in DMSO-PBS solution, were treated with base and tested for biological activity (90, 91), the monoacetates 20 and 21 appeared to be the most active although in general, all acetates were found to be active. When the acetylated mixture was base treated and subjected to hplc analysis, HP (14), HVD (18, 19) and Pp (1), all biologically inactive, were the only products eluting out of the column with aqueous methanol (1:3) containing 3% glacial acetic acid. When the spent column was eluted with a powerful eluent THF-H₂O-DMSO, the material eluted was found to have the highest activity. Since this material was retained when the polar and non-polar monomeric porphyrins were eluted, Bonnett and Berenbaum (90) suggested that the active component is probably a covalently bonded dimer or oligomer, with an ether (Figure 10a) or an ester (Figure 10b) linkage.

Similar elution patterns and activity profiles have been observed by several other research groups (50,67,92) in their analytical reversed phase hplc studies. Dougherty et al. (50) estimated approximately 42% HP and 34% HVD isomers (18, 19) as major components, both being biologically inactive. They also investigated the separation of the HPD components by gel-filtration, using aqueous conditions under which non-covalent polymers are expected to be more stable. Using a Bio Gel P-10 column with a nominal exclusion limit of 20,000, three distinct regions were observed. The most rapidly moving dark brown fraction A comprised of 40-50% of the mixture with a red-brown fraction B constituting 25% and a final dark red band comprising approximately 28%. All biological activity was found to reside in fraction A.

By using reversed-phase hplc with H₂O-THF-CH₃OH (1:1:1) at pH 5.7- 5.8 as the eluting solvent, Dougherty et al. (93) observed that only the known

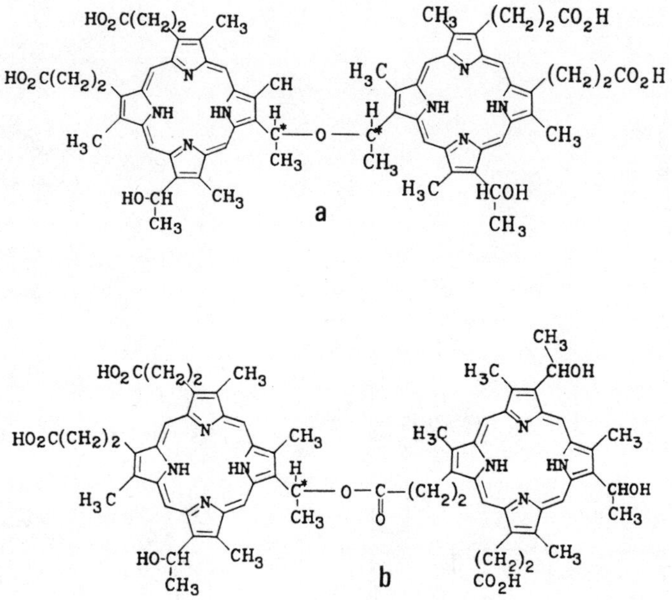

Figure 10. Proposed structures for HPD active component (Ref 90)

porphyrins HP, HVD isomers and Pp eluted from the column. On changing the solvent system to THF-H$_2$O (9:1), a fraction was observed which corresponded to fraction A isolated by gel filtration. This fraction A was largely aggregated (Soret $\lambda \sim$ 365 nm) in H$_2$O-THF-CH$_3$OH (1:1:1), a solvent system in which no known porphyrin is aggregated. On the other hand, in THF-H$_2$O (9:1), this fraction was primarily disaggregated (Soret $\lambda \sim$ 400 nm), further strengthening Bonnett and Berenbaum's (90) suggestion that the non-polar material is not a physical association (aggregation) of known porphyrins. Fraction A was found to be stable to base (1 N aqueous NaOH) but was decomposed by acid (1 N aqueous HCl), producing HP and HVD as the main products. This led Dougherty and co-workers to propose (93) the ether linked dimer, dihematoporphyrin ether (DHE; Figure 10a) as the active component of HPD. They have provided mass spectrometric (FAB) and ^{13}C NMR evidence in support of this structure.

Hplc studies of Ward and co-workers (94) have shown the possibility of elution of the more hydrophobic components of HPD by the increase in pH of the eluting solvent. Figure 11A shows the hplc trace of a HPD analysis using aqueous methanol (1:9) as eluent with a pH gradient from 4 to 8. In addition to the major peaks of HP and the two isomers of HVD, at least 10 other peaks (one of which is Pp) were evident. At pH=4, the same solvent system did not elute the more hydrophobic components. When these components were eluted using THF-H$_2$O (9:1) and reanalyzed by hplc under the pH gradient elution mentioned above, traces of HP, HVD, Pp and a number of poorly resolved strongly retained material were observed. It should be noted that the fraction isolated by Dougherty et al. (93) from hplc using this same solvent system THF-H$_2$O (9:1), was found to correspond to the gel-filtration fraction A and was identified as DHE.

Ward and co-workers (94) analyzed the composition of HPD by gel-filtration chromatography as well. Figure 11B upper trace shows the two main fractions (fast moving "aggregate" and slow moving "monomer") eluted

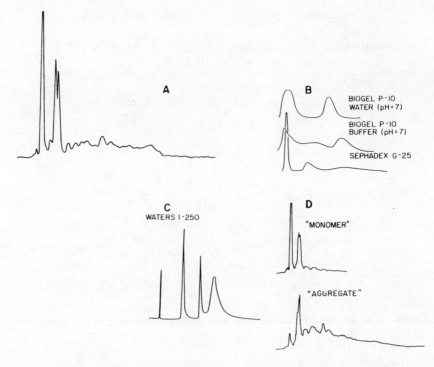

Figure 11.   Chromatographic analysis of HPD (Ref. 94)

from a Bio Gel P-10 column using water, the fast moving fraction being the
biologically active component. When the column was eluted with water
buffered at pH 7, three poorly resolved bands were observed (Figure 11B,
middle trace). The total separation of the fast moving band was observed
when a Sephadex G-25 column was used (Figure 11B, lower trace). The three
bands termed "aggregate", "dimer" and "monomer" (based on gel-filtration
behaviour and not on a molecular weight basis) exhibited Soret band maxima
(in aqueous solution) at 364 nm, 372 nm and 404 nm respectively.   In
ethanol - 0.1 N sodium hydroxide (1:1), all fractions exhibited sharp
Soret bands with the absorption maximum for the "aggregate" band being at
393 nm and that for the "monomer" being at 396.   Although this is
apparently a result of the change in the aggregation state, when HPD was
chromatographed on a Sephadex G-25 column using the same solvent system
(ethanol - 0.1 N sodium hydroxide 1:1), they observed that the amount of
the "aggregate" fraction remained essentially unchanged.

Figure 11C shows the gel filtration behaviour observed by these
workers on Waters I-250 hplc columns, the bands corresponding to
aggregate, dimer and monomer. When the base treated hematoporphyrin
diacetate 15 was analyzed under the same conditions, significantly greater
amounts of the aggregated fraction were observed. The "aggregate" fraction from the Sephadex G-25 gel-filtration column (Figure 11B, lower
trace) when analyzed on reversed-phase hplc under pH gradient elution
described for crude HPD, was shown to be composed of very small amounts of
HP, larger quantities of HVD's and a majority of the less polar material
(Figure 11D).   The "monomer" band (from BioGel P-10; pH = 7 buffer)
contained mainly HP and HVD under similar conditions.

One important feature of the gel-filtration fraction A (or Photofrin
II) of Dougherty et al. (90) (characterized as DHE) that is yet to be

firmly established in that it is one single chemical compound, probably in varying states of aggregation and not a mixture of chemical entities, of which DHE is just one. Chromatographic analysis of Ward and co-workers (94) described above strongly suggest it to be a mixture. The fact that this material exists in a "dimer" form in solvents that produce monomers in all known porphyrins and that its visible spectrum changes from a "dimer" type to a "monomer" type not on dilution but on changing to certain solvents, suggest that the interactions that hold the dimer in a "face-to-face" orientation are extremely strong and intramolecular. In studies on covalently linked etioporphyrin I dimers (singly linked), Selensky et al. (95) observed that the spectra and the quantum fluorescence yields were not markedly different from those of the monomer whereas Moan (34) has reported that the hydrophobic fration had the lowest quantum fluorescence yield of all components of HPD.

CPK models of both the ether linked and ester linked dimers indicate that, in addition to the π-π interactions of the two macrocycles, H-bonding interactions between the non-bonding β-substituents strongly favor "face-to-face" dimers although the 3-atom ether linkage would impose more of a steric strain in attaining this orientation than the 5 atom ester linkage. The ester linked dimer (Figure 10b) had been ruled out due to the base stability of the aggregate fraction (93). A CPK model of such a dimer shows that, in order to attain the tetrahedral transition state at the carbonyl carbon, required for base hydrolysis, considerable lateral movement of the two rings is required. This may not be favored by the strong interactions already present in the dimer in aqueous solution. However, in solvents such as THF, which strongly favor disaggregation (as indicated by visible spectroscopy, Ref. 93), base hydrolysis has been achieved by Kessel and co-workers. These results are described elsewhere in this volume.

SYNTHETIC ASPECTS OF PORPHYRINS

In the laboratory, as well as in nature, porphyrins are synthesized from pyrroles which in turn are readily obtained from acyclic precursors. The substitution pattern at the porphyrin periphery depends primarily on the β-substitution of the starting pyrroles. The porphyrin macrocycle can be constructed in three fundamentally different ways. They are:

(a) The single step condensation of monopyrroles
(b) The coupling of two dipyrrolic intermediates, commonly referred to as a "2+2 synthesis".
(c) The stepwise condensation of individual pyrroles leading to a linear tetrapyrrole and a final head-to-tail cyclization.

Method (a) has very limited synthetic value, primarily due to the symmetric nature of the porphyrins produced and also due to the limit on the type of substituents that can be introduced. Methods (b) and (c), together form the major route to porphyrins as well as other related macrocycles since it is now possible to synthesize systems with diverse peripheral substituents in the required relative orientations. Nevertheless, the experience required in the preparation and handling of the appropriately substituted monopyrrolic and polypyrrolic intermediates has limited the use of such methods to a few research groups. For many, whose interests lie in porphyrins of biological importance, nature provides a good source of the porphyrins with a particular substitution pattern and simple chemistry performed on the side chains, provide the required porphyrin. A detailed account of the methods for the construction of the macrocycle is beyond the scope of this discussion. A brief overview will be provided here and the readers directed to recent reviews in the appropriate areas.

A. Porphyrins Directly from Monopyrroles

Single step porphyrin synthesis has been widely used in the preparation of meso-tetraarylporphyrins. Adler et al. (96,97) obtained a yield of 20% for the synthesis of 5,10,15,20-tetraphenylporphyrin (TPP) from the reaction of equimolar amounts of benzaldehyde and pyrrole in refluxing propionic acid. The product crystallized out leaving most of the poly-pyrrolic by-products in solution, facilitating the purification of the product. Higher yields (35-40%) have been obtained in acetic acid but the high solubility of the porphyrin in this solvent made purification more difficult.

By using appropriately substituted benzaldehyde starting materials, this method has been extended to the synthesis of porphyrins with the desired substituents on the phenyl group (Scheme 4). Water-soluble porphyrins for biological studies 22, 23 and 24 have been synthesized by this method. Winkelman (98) observed that compound 22 was preferentially localized in tumors. This observation has subsequently been confirmed (99), using ^{3}H, ^{35}S and ^{14}C labelled material. However, Carrano et al. (100) showed, by chemical extraction from tissue digests, that compound 22 could be recovered in greater amounts from the kidneys than from tumors. Thaller et al. (101) investigated the potential use of 22, 23 and 24, coordinated to γ-emitting nuclides for tumor scanning, and observed that the ferric complexes of 23 and 24 had high initial uptake in tissue culture systems but tumor uptake in the *in vivo* model was poor.

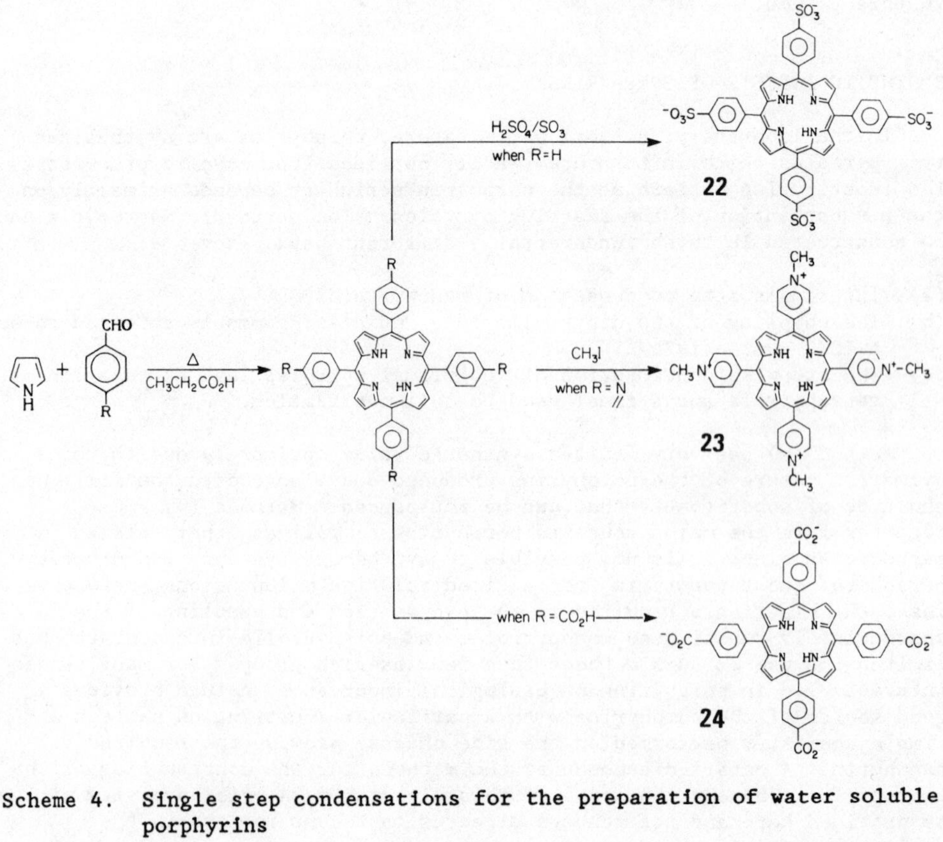

Scheme 4. Single step condensations for the preparation of water soluble porphyrins

250

The single step condensation to prepare porphyrins has been used extensively in the synthesis of porphyrins used as models for dioxygen carriers (102-104). An interesting feature in the preparation of the ortho-substituted tetraarylporphyrins is the possibility of the existence of atropisomers. This occurs as a result of the perpendicular orientation of the phenyl groups with respect to the porphyrin macrocycle. These isomers have been separated in the cases of the ortho-hydroxy (105) and ortho-amino (102) compounds.

β-Octaalkylporphyrins can also be obtained by single step condensation of monopyrroles. Triebs and Haberele (106) obtained 77% yield of octa-methylporphyrin (OMP) by reacting 2,3-dimethylpyrrole with formaldehyde in acetic acid and pyridine. The monopyrroles with a potential carbonium ion at one α-position (-CH$_2$OH or -CH$_2$NR$_2$) and unsubstituted at the other α-position, could be made to undergo self-condensation to produce octa-alkyl porphyrins. Octaethylporphyrin (OEP), a widely used model compound in porphyrin chemistry has been synthesized by this method (107, 108). The mechanistic aspects of the single step porphyrin synthesis have been discussed by Kim et al. (109).

B.   Systematic Stepwise Synthesis of Porphyrins

The strategy involved here is the initial synthesis of the mono-pyrrole with the required β-substituents and the subsequent manipulation of the α-substituents to give the reactive functional groups used in the stepwise coupling to produce dipyrrolic, tripyrrolic and tetrapyrrolic intermediates. The synthesis of pyrroles and dipyrrolic intermediates and the subsequent 2+2 coupling of the latter to give porphyrins, have been discussed in detail by Paine (110).

A synthetic route of great value is that developed by MacDonald (111) (Scheme 5). This procedure involves the one-step condensation of a dipyrromethane dialdehyde 25 with a di-unsubstituted pyrromethane 26 or its synthetic analogue, the dicarboxy derivative, in the presence of hydroiodic acid in acetic acid solution. The reaction which proceeds via bilene-b 29 and porphodimethene 30 intermediates produces the porphyrin 31 on neutralization of the acid (sodium acetate) and aeration in the dark. In order to obtain a single porphyrin product, one of the two components 25 or 26 must be symmetrical in the β-substitution. This restriction allows the method to be used for isomer types II, III and IV but not I. With monoaldehydes 27 and/or 28 a single product is produced only when the dipyrromethane is allowed to self-condense. This produces types I and II isomers. If two different pyrromethanes (e.g. 32, 33) are used, a mixture of porphyrins (34, 35, 36) result but with the advent of hplc methods these can be separated if the synthesis can be so designed as to produce porphyrins with different number of ester groups. A large number of porphyrins of biological significance (e.g. protoporphyrin, uroporphyrin III, coproporphyrin III) have been synthesized by this method.

Dipyrromethene intermediates (a methene linking two pyrrole units - 39 Scheme 6) have not been used extensively in the single step 2+2 synthesis of porphyrins. Coproporphyrin I and etioporphyrin I isomers have been synthesized by the self-condensation of 5-methyl-5'-bromodipyrromethenes (with appropriate β-substituents) in refluxing formic acid with one equivalent of bromine (112). On the other hand, porphyrins with lower symmetry than type I have been synthesized (113) by refluxing 5,5'-dimethyldipyrromethenes with 5,5'-dibromidopyrromethene in formic acid and bromine. It should be noted that in this type of a 2+2 condensation, one unit has to be symmetric in order to obtain a single porphyrin product.

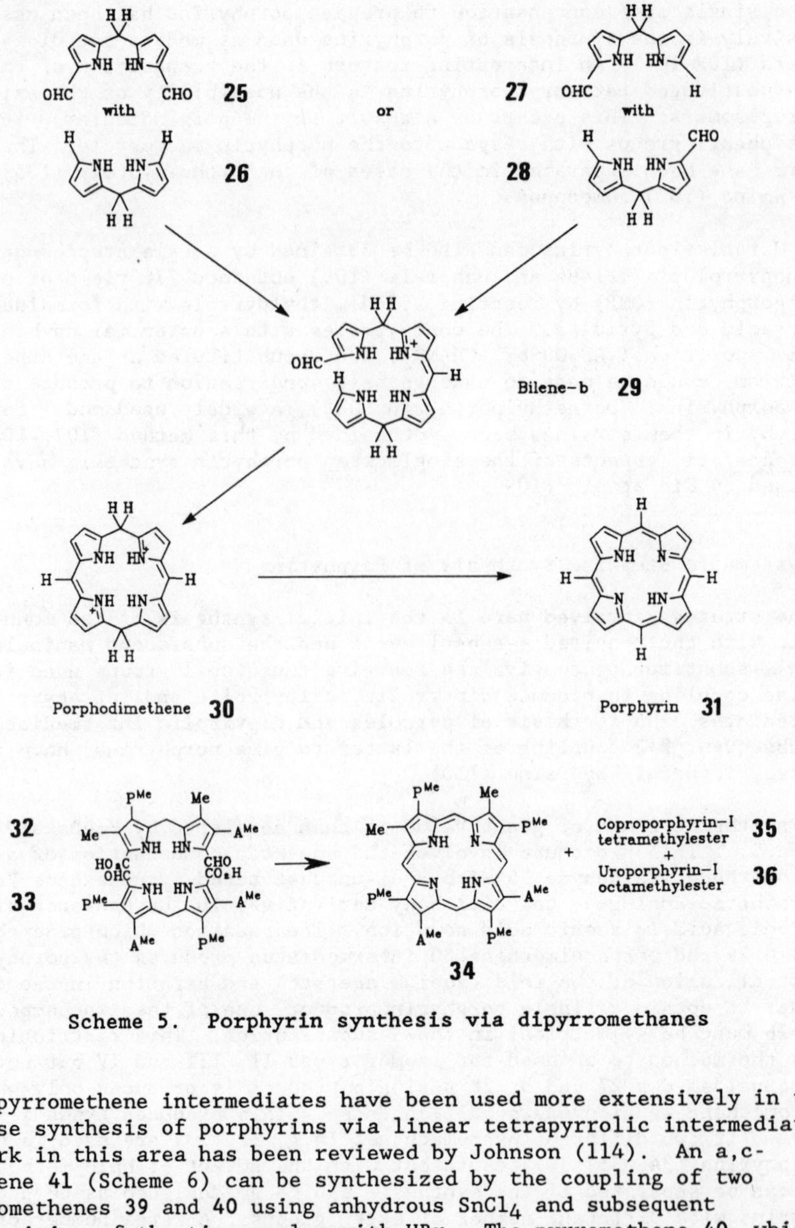

OHC   CHO   **25**

with   or   **27**   OHC   H

26   **28**   H   CHO

OHC   Bilene-b   **29**

Porphodimethene   **30**   Porphyrin   **31**

32   **33**   **34**

Coproporphyrin-I tetramethylester   **35**
+
Uroporphyrin-I octamethylester   **36**

Scheme 5. Porphyrin synthesis via dipyrromethanes

Dipyrromethene intermediates have been used more extensively in the stepwise synthesis of porphyrins via linear tetrapyrrolic intermediates. The work in this area has been reviewed by Johnson (114). An a,c-biladiene 41 (Scheme 6) can be synthesized by the coupling of two dipyrromethenes 39 and 40 using anhydrous SnCl$_4$ and subsequent decomposition of the tin complex with HBr. The pyrromethene 40, which is readily obtained by the bromination (110) of 39, could be coupled with another 5-methyl-5'-unsubstituted pyrromethene (prepared by the acid catalyzed condensation of monopyrroles 37 and 38) carrying different β-substituents. The biladiene 41 is cyclized to the porphyrin (114) with pyridine in dimethylsulfoxide over several days.

Two dipyrromethanes of the types 43 and 44 (Scheme 7) can be condensed under acidic conditions to give another type of linear tetrapyrrole, b-bilene 45, which is cyclized (115) using Cu(II) salts. One terminal methyl group completes the macrocycle while the other is eliminated. A truly stepwise procedure (116) in which the four pyrrole units are linked one by one is also shown in Scheme 7. By using a differentially protected dipyrromethane diester 47, it has been possible to produce a tripyrrene 49

252

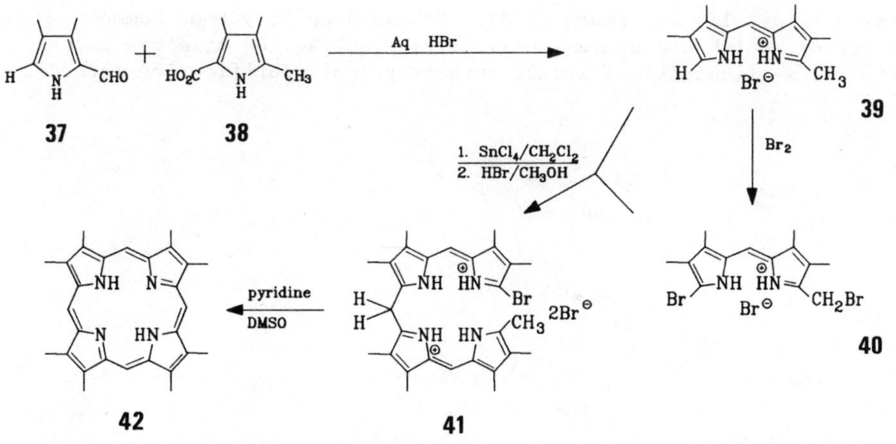

Scheme 6. Porphyrin synthesis via dipyrromethenes and a,c-biladienes

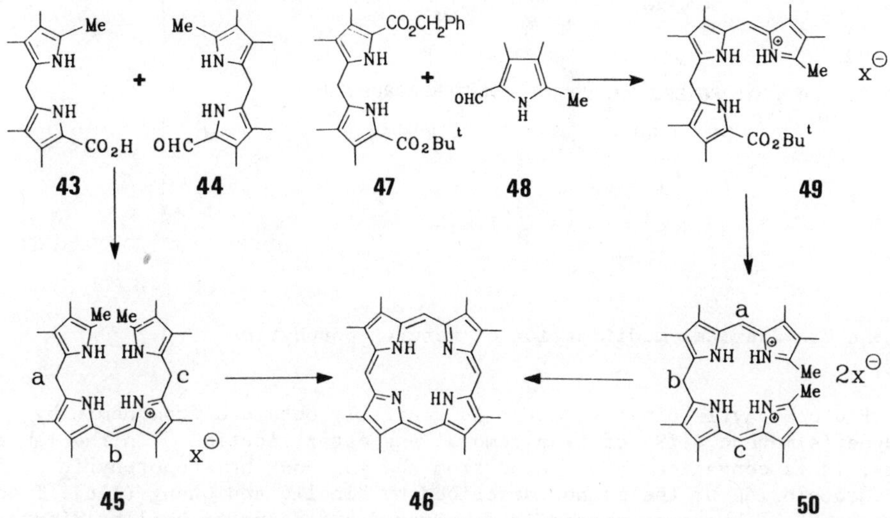

Scheme 7. Porphyrin synthesis via b-bilenes, tripyrrenes and
a,c-biladines

(hydrogenolysis of the benzyl ester with subsequent acid catalyzed
condensation with a formylpyrrole 48), which can then be condensed with a
different formylpyrrole after the removal of the t-butyl ester under
acidic conditions. The a,c-biladiene 50 so obtained can be cyclized using
Cu(II) salts. A recent review (117), discusses in detail, the literature
on the total synthesis of natural pyrrole pigments.

C. Modification of Natural Porphyrins

Although this method is limited to the synthesis of porphyrins having
the basic substitution pattern of natural porphyrins (e.g. protoporphyrin,
hematoporphyrin), the relatively low cost of these starting materials and
the availability of a number of simple chemical transformations have made
this method the most suitable for most investigators. The methods
available for the isolation and modification of porphyrins have been

reviewed by Dinello and Chang (118). Scheme 8 depicts some common conversions, which are conveniently carried out as the dimethyl esters, due to the availability of facile chromatographic purification methods.

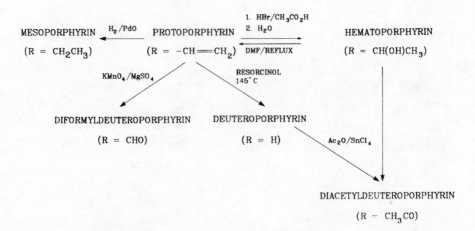

Scheme 8. Chemical modification of natural porphyrins

Protoporphyrin dimethyl ester is generally obtained from hemin by Caughey's method (119) of iron removal and esterification. In the laboratory, it is conveniently prepared from the low cost hematoporphyrin dihydrochloride by the method described by Dinello and Chang (118). Twenty five grams of hematoporphyrin is dehydrated by adding to boiling dimethyl-formamide (2 L) in a 4 L erlenmeyer. After approximately 30 sec, the flask is cooled, the solvent removed using a rotary evaporator (oil pump), the residue dissolved in minimum formic acid (90%) and precipitated by adding diethyl ether. The product is then filtered, washed thoroughly and dried in a vacuum dessicator before esterification with 5% H_2SO_4 - CH_3OH (v/v) overnight in the dark. Protoporphyrin dimethyl ester is purified by column chromatography on silica (Woelm, activity I, 70-150 mesh), using 0.5% CH_3OH in CH_2Cl_2 as eluent.

In the laboratory, hematoporphyrin can be prepared by the treatment of hemin with HBr in acetic acid followed by the decomposition of the HBr adduct with water. Decomposition with methanol gives hematoporphyrin dimethyl ether. Hematoporphyrin is one of the most labile of the natural porphyrins and the hydration dehydration equilibrium between hemato-porphyrin and protoporphyrin has been discussed by Falk (14). The presence of labile hydroxyethyl groups makes esterification by diazo-methane (119), the method of choice for hematoporphyrin.

Mesoporphyrin is prepared by the catalytic hydrogenation of proto-porphyrin or its dimethyl ester at elevated temperature. Caughey et al. (119) observed that high yields (>80%) are obtained when hydrogen is bubbled through a formic acid solution of hemin over PdO at 94-98°.

Diformyldeuteroporphyrin is readily obtained by the $KMnO_4$ oxidation of the vinyl side chains of protoporphyrin. Caughey et al. (119) have reported that addition of $MgSO_4$ to the reaction mixture in acetone, suppresses further oxidation to carboxyl groups. In practice, $KMnO_4$ oxidation produces, in addition to the diformyldeuteroporphyrin, the isomeric 3(8)-formyl-8(3)-vinyldeuteroporphyrins and has been separated (119) from the latter by column chromatography on alumina using 2,3-dichloroethane-chloroform (2:1) as eluent. The isomeric mixture of formylvinyldeuteroporphyrins in turn can be separated (16) on silica gel thick layer plates using chloroform as the developing solvent although they can be readily obtained from the purified photoprotoporphyrin isomers 3 and 4 (Scheme 1).

Deuteroporphyrin is prepared from deuterohemin which in turn is obtained from hemin by a process of devinylation in a resorcinol melt (118, 119). Hemin is mixed with approximately 3 times its weight of resorcinol and heated in an oil bath at 140-150° for 45 minutes to effect the devinylation. The crude deuterohemin is obtained by triturating the reaction mixture with ether. Deuteroporphyrin dimethyl ester is prepared by iron removal, esterification and chromatography (119).

The crude deuterohemin could be acetylated using acetic anhydride and anhydrous stannic chloride at 0° (118). After quenching the reaction mixture with 0.1 N HCl, the crude diacetyldeuterohemin is isolated and purified by chromatography. The diacetyldeuteroporphyrin dimethyl ester is obtained by treating the purified hemin with anhydrous HCl in anhydrous methanol which demetalates and esterifies.

Clezy et al. (87) prepared the 3,8- differentially substituted deuteroporphyrins by partially oxidizing hematoporphyrin with MnO_2 in refluxing benzene. The isomeric 3(8)-acetyl-8-(3)-hydroxyethyldeutero-porphyrins were carefully separated, initially by silica gel column chromatography using benzene-chloroform (2:3) as eluting solvent, and subsequently on preparative tlc. The remaining hydroxymethyl group can then be dehydrated to a vinyl group by heating with dimethylformamide and benzoyl chloride on a steam bath for 1 hr. The pure 3(8)-monohydroxy-ethyl-8(3)-monovinyldeuteroporphyrin isomers 18 and 19 (Figure 9) have been prepared (89) directly from hematoporphyrin by partial dehydration in dimethylformamide at 65° for 1 hr followed by separation on preparative hplc.

In order to investigate the aggregation, photosensitization, permeability across membranes and tumor localizing properties of porphyrins carrying a permanent charge, we (120) have undertaken the synthesis of several nitrogen analogues of hematoporphyrin (Scheme 9). The reaction of protoporphyrin dimethyl ester with anhydrous hydrogen bromide in dry dichloromethane gave the HBr adduct which, when treated with ammonia, ethyl amine or diethyl amine, produced the corresponding 1-aminoethyl derivative. By controlling the reaction conditions, it has been possible to prepare the 3,8-bisaminoethyldeuteroporphyrins 52 as well as the 3(8)-monoaminoethyl-8(3)-monovinyldeuteroporphyrins 51. Column chromatography on alumina and preparative tlc on silica have been used to purify the products. The syntheses of the amine linked dimer 54 as well as an amide linked dimer are in progress.

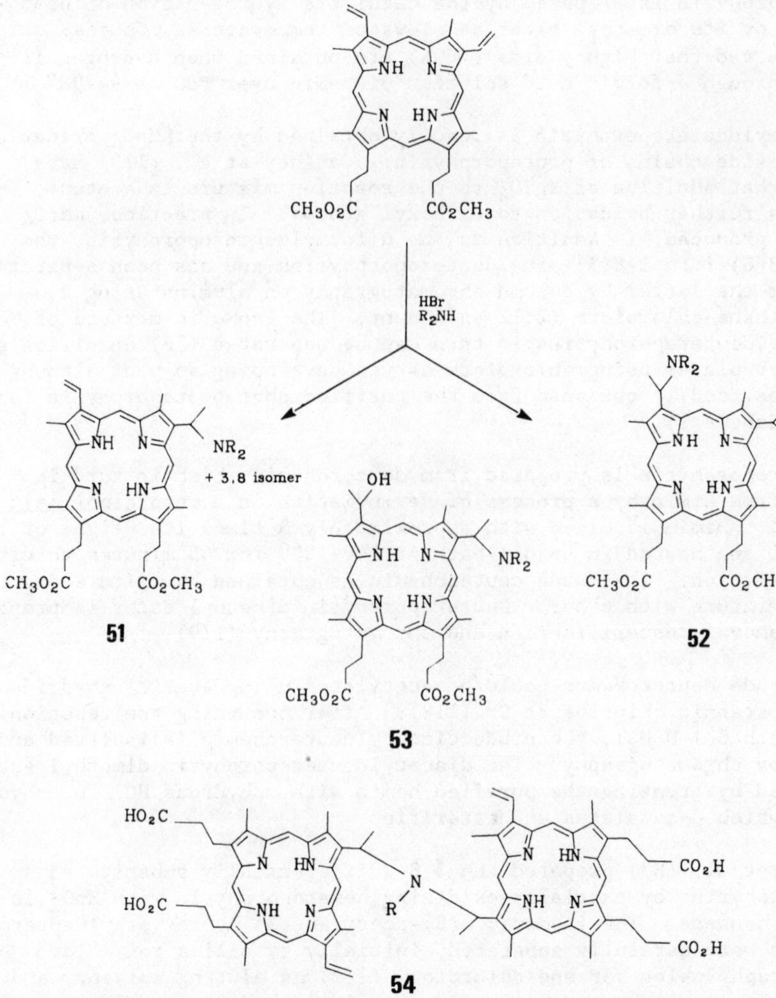

Scheme 9.   Synthesis of some nitrogen analogues of HPD

METALATION AND DEMETALATION

A simple metalation process (Eq. 1) involves the reaction of a porphyrin (H_2P) with a metal carrier (MX_2) to produce a metalloporphyrin (MP) with the displacement of the two inner NH protons.

$$H_2P \; + \; MX_2 \; \rightleftharpoons \; MP \; + \; 2HX \qquad (1)$$

The tetradentate dinegative porphinato ligand (P^{2-}) is very versatile in its coordination chemistry and almost all metals and some metalloids have been combined with it. Metals associated with porphinato or related macrocyclic ligands in nature include Mg (in chlorophylls), Fe (in the hemins) and Co (in coenzyme B_{12}), with Cu, Ni, V, Mn and Zn being observed to a lesser extent. The use of metalloporphyrins in photoradiation therapy is minimal due to their low fluorescence efficiencies, but porphyrins coordinated to radioactive metals have been investigated for tumor scanning (101) as well as for other clinical applications (121-123).

Five essential stages can be identified (124,125) in the synthesis of a metalloporphyrin:

(a) Protonation/deprotonation equilibria

$$H_4P^{2+} \rightleftharpoons H_2P^+ \rightleftharpoons H_2P \rightleftharpoons HP^- \rightleftharpoons P^{2-} \qquad (2)$$

Deprotonation of H_2P occurs so as to produce P^{2-} which coordinates to the metal ion. The presence of strong acids therefore retard metalation reactions (Eq. 2).

(b) Release of metal ion from the "metal carrier"

The metal carrier discociates in the reaction medium producing a coordinatively unsaturated species which can combine with P^{2-}. Usually, the choice of the carrier depends on its activity, availability and also on its solubility in organic solvents which are used to dissolve the porphyrin.

(c) Formation of the equatorial MN_4 system

$$Mn^{n+} + P^{2-} \longrightarrow MP^{(n-2)+} \qquad (3)$$

The initial metalation reaction involves the addition of the 4 pyrrole nitrogen atoms to the metal. For divalent metal ions which prefer a square planar arrangement (e.g. Ni^{2+}), the above reaction produces a stable neutral metalloporphyrin.

(d) Charge neutralization

For metal ions with a charge greater than 2, step (c) (Eq. 3) produces a charged metalloporphyrin. In order to attain electroneutrality, axial coordination is built up with anionic ligands. For Fe(III) and Cl$^-$, the square pyramidal Fe(P)Cl (hemins) that is formed, has the metal centre coordinatively saturated, thus producing a stable system.

(e) Completion of the coordination sphere

For metal ions that prefer octahedral geometry, the coordination sphere is completed by further complexation with neutral donor ligands such as water, pyridine etc.

No single metal insertion procedure is applicable to all metalation reactions or to all porphyrins . Metalating systems classified by the solvent as well as the metal carrier are discussed in detail by Buchler (124,125). The metal insertion into the porphyrin may be monitored by visible spectrophotometry. Special precautions should be taken with porphyrins carrying labile side-chains; e.g. vinyls in protoporphyrins. Some common metalating methods are briefly mentioned below.

1. Acetate method (126-128).

The metal salt, usually the acetate, is heated with the porphyrin in acetic acid at 100°. The product is crystallized either directly by cooling, or by the addition of water or methanol. This method is applicable to the synthesis of metalloporphyrins of most divalent metal ions, except for species that are unstable in acetic acid. For Mn(II) and Fe(II), autoxidation under aerobic conditions produce higher oxidation states.

2. Pyridine method (129).

For metalloporphyrins labile towards acetic acid, this method is useful since pyridine is a good solvent for porphyrins as well as metal salts. The metalloporphyrins are usually isolated as the pyridinates. Pyridine forms complexes with metals of high charge and thereby retard the metalation process of the porphyrin.

3. Metal carbonyl method (130,131).

This method has proved to be especially useful for the preparation of porphyrins with metals of groups VI to VIII. The metal carbonyls or the carbonyl chloride are heated with the porphyrin in inert solvents such as benzene, toluene or decalin.

4. Acetylacetonate method (132,133).

The ready availability, solubility in organic solvents and weakness of the acid liberated make metal acetylacetonates very useful "metal carriers". With metal ions of high charge and small size, a weakly acidic solvent e.g. phenol, is required to liberate the "active" metalating species. This method has been used with metals of groups IIIa and IIIb.

5. Dimethylformamide method

Adler et al. (134) have investigated the complexation reaction of a series of metals (as anhydrous metal chlorides) using refluxing dimethyl-formamide as the solvent. The HCl produced in the reaction escapes at reflux temperature of the solvent and the crude product is crystallized by diluting with water and cooling.

6. N-substituted porphyrin method (135-137).

The chemistry of N-substituted porphyrins (56, Scheme 10) differs significantly from that of the non-N-substituted analogues 55 presumably due to distortion from planarity of the aromatic ring system. As a result, N-substituted porphyrins are known (135) to form complexes (57) with metal ions more rapidly. Nucleophilic displacement of the N-substituent from 57 through a carbocationic intermediate produces the metallo-porphyrin. Scheme 10 gives the basic steps involved in this novel and convenient method for the metalation of porphyrins. The dealkylation is usually carried out using di-n-butylamine as the nucleophile, but the solvent dimethylsulfoxide or dimethylformamide itself acted as the nucleophile when the metal was Pd (121,137). The rate of dealkylation is not affected greatly by the nature of the porphyrin ring (138) and therefore the results deduced from synthetic porphyrins can be used with naturally occurring porphyrins. The nature of the N-substituent (R) has been shown (139) to have a profound effect on the dealkylation step.

The process of demetalation or removal of the metal ion is the reverse of equation (1). Metalloporphyrin stability is therefore often defined in terms of the degree of resistance towards the displacement of the metal atom by acids. Five stability classes have been identified (14) for metalloporphyrins:

a	100% H_2SO_4	Not completely demetalated
b	100% H_2SO_4	Completely demetalated
c	$HCl/H_2O - CH_2Cl_2$	Demetalated
d	100% CH_3CO_2H	Demetalated
e	$H_2O - CH_2Cl_2$	Demetalated

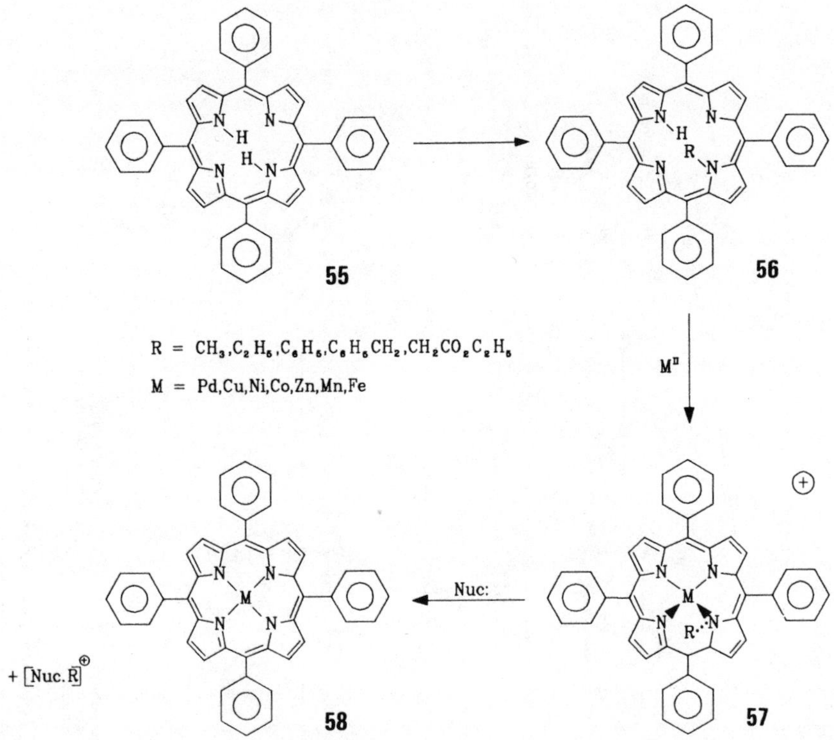

R = CH₃,C₂H₅,C₆H₅,C₆H₅CH₂,CH₂CO₂C₂H₅

M = Pd,Cu,Ni,Co,Zn,Mn,Fe

Scheme 10. Preparation of Metalloporphyrins via N-alkyl porphyrins

Divalent metal ions may also be removed from metalloporphyrins of class b, with trifluoroacetic acid.

BIOSYNTHESIS

In order to conserve energy and materials, nature uses a relatively small number of building blocks to synthesize many complex molecules. Uroporphyrinogen III (59; Scheme 11) has been identified as the key intermediate in the biosynthesis of all of the natural tetrapyrrolic macrocycles. The transformation of 59 to the iron and magnesium containing prosthetic groups, and the isolation of intermediates and enzymes involved have been discussed in detail elsewhere (140, 141).

Enzymic decarboxylation of the four acetic acid side chains of 59 affords coproporphyrinogen III (60). Subsequently, the 3,8-propionate side chains of 60 are oxidatively decarboxylated to vinyl groups, to produce protoporphyrinogen (61) which undergoes aromatization and gives proto-porphyrin (62). Although the autoxidation of a porphyrinogen to a porphyrin occurs readily, the enzymes responsible for the transformations 59 to 60 and 60 to 61 do not accept the corresponding porphyrins as the substrates. Insertion of Fe^{2+} into 62 (via ferrochelatase) produces protoheme (63) which itself is the prosthetic group of hemoglobin, myoglobin and the cytochromes-b, while some peripheral modifications lead to other types of hemes.

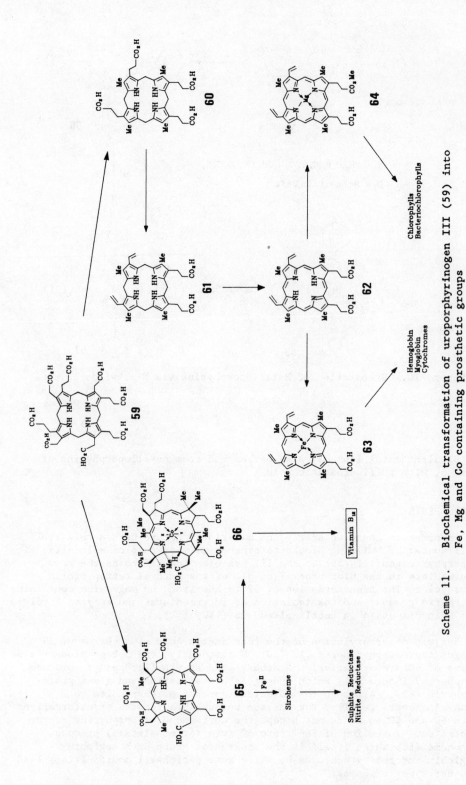

Scheme 11. Biochemical transformation of uroporphyrinogen III (59) into Fe, Mg and Co containing prosthetic groups

Protoporphyrin (62) serves as the precursor to chlorophylls as well as bacteriochlorophylls. Insertion of Mg^{2+} and the esterification of the 13-propionic acid group produces 64 which then undergoes an oxidative cyclization involving the 13-propionate side chain and the 15-meso position to produce a carbocyclic ring. Reduction of ring D gives rise to the "chlorin" system which is then transformed into the various chlorophylls by side chain modifications (140,142).

The intermediacy of uroporphyrinogen III (59) in the biosynthesis of the highly reduced corrin system (Figure 1) was confirmed only after the observation of the intact incorporation of specifically labelled 59 into cobyrinic acid 66. The conversion of 59 to 66 requires several distinct changes, namely, methylation at C-1,2,5,7,12,15 and 17, decarboxylation of C-12 actetate, loss of C-20 with the direct C-1, C-19 bond formation, introduction of Cobalt and reduction of the macrocycle. Extensive labelling studies and degradation analysis used in understanding these reactions have been discussed recently (143). It is interesting to note that sirohydrochlorin (65) which as its Fe(II) complex, is the prosthetic group of the sulfite and nitrite reductase enzymes, serves as a precursor to cobyrinic acid 66 and is itself generated from uroporphyrinogen III.

The origin of the key biosynthetic intermediate, uroporphyrinogen III from acyclic precursors, is outlined in Scheme 12. Pioneering work (140) of Shemin, Granick, Bogorad, Neuberger and Rimmington had shown that two

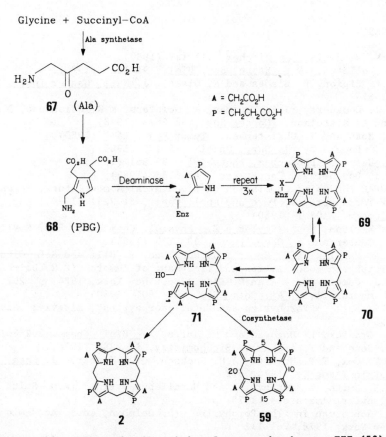

Scheme 12. Biosynthetic origin of uroporphyrinogen III (59)

enzymes are required to catalyze the conversion of four molecules of the monopyrrole porphobilinogen (PBG, 68) into uroporphyrinogen III (uro'gen III) (59). PBG itself is produced from glycine and succinyl coenzyme A via δ-aminolevulinic acid (Ala, 67). Although the two enzymes, deaminase and cosynthetase together catalyze the conversion of PBG to uro'gen III, in the absence of cosynthetase, only the unnatural isomer uro'gen I (72) is produced. Uro'gen I, which is the expected product from simple head-to-tail condensation of four PBG units, is not converted into uro'gen III by cosynthetase or by the deaminase-cosynthetase system. The understanding of the highly specific method of rearrangement of ring D to produce the type III porphyrin system of 59 during the condensation, has involved extensive synthetic, isotopic labelling and enzymatic studies. It is now known (143,144) that deaminase catalyzes the production of an enzyme bound unrearranged linear tetrapyrrole 69, which is in equilibrium with an enzyme stabilized methylenepyrrolenine 70. Deaminase is not an enzyme for ring-closure, and its product, the unrearranged hydroxymethylbilane 71 is the substrate for cosynthetase, which produces uro'gen III via a spiro-cyclic intermediate (145). In the absence of cosynthetase, the unrearranged bilane 71 is released by deaminase, into the medium where it is cyclized nonenzymatically to uro'gen I.

ACKNOWLEDGEMENT

This work was supported by the U.S. National Institutes of Health (AM 17989) and by the Canadian Natural Sciences and Engineering Research Council.

REFERENCES

1. T.K. With, Int. J. Biochem., 11 189 (1980).
2. A. Policard, C.r. Seanc. Soc. Biol., 91 1423 (1924).
3. R.L. Lipson, E. Blades and A. Olsen, J. Natl. Cancer Inst., 26 1 (1961).
4. T.J. Dougherty, J.E. Kaufman, A. Goldfarb, K.R. Weishaupt, D.G. Boyle and A. Mittelman, Cancer Res., 38 2628 (1978).
5. J. Moan and T. Christensen, Tumor Res., 15 1 (1980).
6. T.J. Dougherty, J. Surg. Oncol., 15 209 (1980).
7. D. Kessel, Photochem. Photobiol., 39 851 (1984).
8. T.J. Dougherty, Cancer Res., 42 1188 (1982).
9. IUPAC-IUB Joint Commission on Biochemical Nomenclature, Nomenclature of Tetrapyrroles, Pure and Appl. Chem., 51 2251 (1979); Eur. J. Biochem., 108 1 (1980).
10. P. Karlson, Hoppe-Seyler's Z. Physiol. Chem., Bd. 362, S VII (1981).
11. M. Gouterman, J. Chem. Phys., 30 1139 (1959).
12. C.A. Hanzlik, W.H. Knox, T.M. Nordlund, R. Hilf and S.L. Gibson in "Porphyrin Localization and Treatment of Tumors" (D.R. Doiron and C.J. Gomer, eds.) Alan R. Liss, Inc., New York, 1984, p. 201.
13. D. Mauzerall, J. Am. Chem. Soc., 82 2601 (1960).
14. J.E. Falk, "Porphyrins and Metalloporphyrins", Elsevier, Amsterdam 1964.
15. S. Granick, L. Bogorad and H. Jaffe, J. Biol. Chem., 202 801 (1953).
16. M. Sono and T. Asakura, Biochemistry, 13 4386 (1974).
17. M.T. Cox, T.T. Howarth, A.H. Jackson, G.W. Kenner, J. Chem. Soc. Perkin Trans I., 1974, 512.
18. K.M. Smith in "Porphyrins and Metalloporphyrins" (K.M. Smith, ed.), Elsevier, Amsterdam, 1975, p 23.
19. M. Gouterman in "The Porphyrins" (D. Dolphin, ed.), Academic Press, New York, 1978, Vol III, p 1.

20. A.E. Profio and J. Sarnaik in "Porphyrin Localization and Treatment of Tumors", (D.R. Doiron and C.J. Gomer, eds.), Alan R. Liss, Inc., New York, 1984, p. 163.

21. M.W. Berns, M. Wilson, P. Rentzepis, R. Burns and A. Wile, _Lasers Surg. Med._, 2 261 (1983).

22. A.J.M. van der Putten and M.J.C. van Gemert in "Proceedings of Laser '81 Opto-Electronik, Munchen, West Germany (1981).

23. A. Pasqua, A. Poletti and S.M. Murgia, _Med. Biol. Environ._, 10 287 (1982).

24. A. Andreoni and R. Cubeddu in "Porphyrins in Tumor Phototherapy" (A. Andreoni and R. Cubeddu, eds.), Plenum Press, New York, 1984, p. 11.

25. P.G. Seybold and M. Gouterman, _J. Mol. Spectrosc._, 31 1 (1969).

26. F.P. Zscheile and D.G. Harris, _J. Phys. Chem._, 47 623 (1943).

27. R. Livingston, W.F. Watson and J. McArdle, _J. Am. Chem. Soc._, 71 1542 (1949).

28. J. Fernandez and R.S. Becker, _J. Chem. Phys._, 31 467 (1959).

29. a) R.S. Becker and J. Allison, _J. Phys. Chem._, 67 2662 (1963).
    b) R.S. Becker and J. Allison, ibid., 67 2669 (1963).
    c) J. Allison and R.S. Becker, ibid., 67 2675 (1963).

30. D.R. Doiron, L.S. Svaasand, A.E. Profio in "Porphyrin Photosensitization" (D. Kessel and T.J. Dougherty, eds.), Plenum Press, New York, 1983, p. 63.

31. D.R. Doiron in "Porphyrin Localization and Treatment of Tumors" (D.R. Doiron and C.J. Gomer, eds.), Alan R. Liss, Inc., 1984, p. 41.

32. D. Kessel and C.J. Dutton, _Photochem. Photobiol._, 40 403 (1984).

33. E. Reddi, G. Jori and M.A.J. Rodgers, _Studia Biophys._, 94 13 (1983).

34. J. Moan, _Photochem. Photobiol._, 39 445 (1984).

35. S. Cannistraro, G. Jori, A. van der Vorst, _Photobiochem. Photobiophys._, 3 353 (1982).

36. G. Jori, E. Reddi, L. Tomio and F. Calzavara in "Porphyrin Photosensitization" (D. Kessel and T.J. Dougherty, eds.), Plenum Press, New York 1983 p. 193.

37. L.I. Grossweiner, A.S. Patel and J.B. Grossweiner, _Photochem. Photobiol._, 36 159 (1982).

38. L. Grossweiner in "Porphyrin Localization and Treatment of Tumors", (D.R. Doiron and C.J. Gomer, eds.), Alan R. Liss, Inc., New York, 1984, p. 391.

39. T.J. Dougherty, CRC Critical Reviews in Oncology/Hematology, 2 83 (1984).

40. C.J. Gomer, D.R. Doiron, S. Dunn, N. Rucker, N. Nazum and S. Fountain, _Photochem. Photobiol._, 37S S91 (1983).

41. H.H. Inhoffen, K. Bliesener and H. Brockmann, _Tetrahedron Lett._, 1966, 3779.

42. H.H. Inhoffen, H. Brockmann and K.M. Bliesener, _Justus Liebigs Ann. Chem._, 730 173 (1969).

43. G.S. Cox and D.G. Whitten, _J. Am. Chem. Soc._, 104 516 (1982).

44. A.P. Schaap and K.A. Zaklika in "Singlet Oxygen" (H.H. Wasserman and R.W. Murray, eds.), Academic Press, New York, 1979, p. 173.

45. G.S. Cox, C. Bobillier and D.G. Whitten, _Photochem. Photobiol._, 36 401 (1982).

46. D. Mauzerall and S. Granick, _J. Biol. Chem._, 232 1141 (1958).

47. J.B. Paine and D. Dolphin, unpublished results.

48. D. Mauzerall, _J. Am. Chem. Soc._, 82 1832 (1960).

49. D. Mauzerall, _J. Am. Chem. Soc._, 84 2437 (1962).

50. T.J. Dougherty, D.G. Boyle, K.R. Weishaupt, B.A. Henderson, W.R. Potter, D.A. Bellnier and K.E. Wityk in "Porphyrin Sensitization" (D. Kessel and T.J. Dougherty, eds.), Plenum Press, 1983, p 3.

51. R.J. Abraham, F. Eivazi, H. Pearson and K.M. Smith, _J. Chem. Soc. Chem. Commun._, 1976, 698, 699.

52. J.J. Katz, L.L. Shipman, T.U. Cotton and T.R. Janson in "The Porphyrins" (D. Dolphin, ed.) Academic Press, New York, 1978, Vol V, p. 401.

53. J. Hiom, J.B. Paine III, U. Zapf and D. Dolphin, Canad. J. Chem., 61 2220 (1983).

54. J. P. Collman, C.S. Bencosme, C.E. Barnes and B.D. Miller, J. Am. Chem. Soc., 105 2704 (1983).

55. T.R. Janson and J.J. Katz in "The Porphyrins" (D. Dolphin, ed.) Academic Press, New York, 1979, Vol. IV p. 1.

56. W.S. Caughey, H. Eberspaecher, W.H. Fuchsman, S. McCoy and J.D. Alben, Ann. N.Y. Acad. Sci., 153 722 (1969).

57. T.R. Janson and J.J. Katz, J. Magn. Reson., 6 209 (1972).

58. W.A. Gallagher and W.B. Elliot, Ann. N.Y. Acad. Sci., 206 463 (1973).

59. R.F. Pasternack, P.R. Huber, P. Boyd, G. Eugasser, L. Francesconi, E. Gibbs, P. Fasella, G.C. Venturo and L. de C. Hinds, J. Am. Chem. Soc., 94 4511 (1972).

60. W.I. White and R.A. Plane, Bioinorg. Chem., 4 21 (1974).

61. W.I. White in "The Porphyrins" (D. Dolphin, ed.), Academic Press, New York, 1978, Vol. V, p. 303.

62. D. Mauzerall, Biochemistry, 4, 1801 (1965).

63. E. Reddi, E. Rossi and G. Jori, Med. Biol. Environ., 9 337 (1981).

64. S.B. Brown, M. Shillcock and P. Jones, Biochem. J., 153 279 (1976).

65. D. Kessel and E. Rossi, Photochem. Photobiol., 35 37 (1982).

66. J. Moan and S. Sommer, Photobiochem. Photobiophys., 3 93 (1981).

67. J. Moan, S. Sandberg, T. Christensen and S. Elander in "Porphyrin Photosensitization" (D. Kessel and T.J. Dougherty, eds.), Plenum Press, New York, 1983, p. 165.

68. A. Poletti, S.M. Murgia, A. Pasqua, E. Reddi and G. Jori, in "Porphyrins in Tumor Phototherapy" (A. Andreoni and R. Cubeddu, eds.), Plenum Press, New York, 1984, p. 37.

69. H. Hisazumi, N. Miyoshi, O. Ueki and K. Nakajima in "Porphyrin Localization and Treatment of Tumors" (D.R. Doiron and C.J. Gomer, eds.), Alan R. Liss, Inc., New York, 1984, p. 443.

70. G.C. Goyal, A. Blum and L.I. Grossweiner, Cancer Res., 43 5826 (1983).

71. C. Emiliani and M. Delmelle, Photochem. Photobiol., 37 487 (1983).

72. D. Dolphin in J. Chrom. Library (E. Heftmann, ed.), Elsevier Scientific Publishing Company, 1983 Vol. 22B p. B 403.

73. W.I. White, R.C. Bachmann and B.F. Burnham in "The Porphyrins", (D. Dolphin, ed.), Academic Press, New York, 1978, Vol I p. 553.

74. D.P. Schwartz and R.S. Bright, Anal. Biochem., 61 271 (1974).

75. R.E. Carlson and D. Dolphin in "High Pressure Liquid Chromatography in Clinical Chemistry" (P.F. Dixon, C.H. Gray, C.K. Lim and M.S. Stoll, eds.) Academic Press, London, 1976 p. 87.

76. J.-H. Fuhrhop and K.M. Smith in "Laboratory Methods in Porphyrin and Metalloporphyrin Research", Elsevier Scientific Publishing Company, Amsterdam (1975).

77. R.E. Carlson and D. Dolphin in "Porphyrins in Human Diseases", (M. Doss, ed.), Karger, Bassel 1976 p. 465.

78. N. Evans, D.E. Games, A.H. Jackson and S.A. Matlin, J. Chromatogr., 115 325 (1975).

79. C.H. Gray, C.K. Lim and D.C. Nicholson in "High Pressure Liquid Chromatography in Clinical Chemistry (P.F. Dixon, C.H. Gray, C.K. Lim and M.S. Stoll, eds.), Academic Press, London 1976, p. 79.

80. R.E. Carlson, R. Sivasothy, D. Dolphin, M. Bernstein and A. Shavi, Anal. Biochem., 140 360 (1984).

81. A.R. Battersby, D.G. Buckley, G.L. Hodgson, R.E. Markwell and E. McDonald in "High Pressure Liquid Chromatography in Clinical Chemistry (P.F. Dixon, C.H. Gray, C.K. Lim and M.S. Stoll, eds.), Academic Press, London, p. 63.

82. D.J. Wright, J.M. Rideout and C.K. Lim, Biochem. J., 209 553 (1983).

83. J.C. Bommer, B.F. Burnham, R.E. Carlson and D. Dolphin, Anal. Biochem., 95 444 (1979).

84. A.H. Jackson, K.R.N. Rao and S.G. Smith, _Biochem J._, **203** 515 (1982).

85. R. Bonnett, A.A. Charalambides, K. Jones, I.A. Magnus and R.J. Ridge, _Biochem. J._, **173** 693 (1978).

86. I.M. Johansson, F.A. Niklasson, _J. Chromatogr._, **275** 51 (1983).

87. P.S. Clezy, T.T. Hai, R.W. Henderson and L. van Thuc, _Aust. J. Chem._, **33** 585 (1980).

88. R.W. Henderson, G.S. Cristie, P.S. Clezy and J. Lineham, _Br. J. Exp. Pathol._, **61** 345 (1980).

89. R. Bonnett, R.J. Ridge, P.A. Scourides and M.C. Berenbaum, _J. Chem. Soc., Perkin Trans. I_, 1981, 3135.

90. R. Bonnett and M.C. Berenbaum in "Porphyrin Photosensitization", (D. Kessel and T.J. Dougherty, eds.), Plenum Press, New York, 1983, p. 24).

91. M.C. Berenbaum, R. Bonnett and P.A. Scourides, _Br. J. Cancer_, 45 571 (1982).

92. D. Kessel and T. Chou, _Cancer Res._, 43 1994 (1983).

93. T.J. Dougherty, W.R. Potter, K.R. Weishaupt in "Porphyrin Localization and Treatment of Tumors" (D.R. Doiron and C.J. Gomer, eds.), Alan R. Liss, Inc., New York, 1984, p. 301.

94. A.G. Swincer, V.C. Trenerry and A.D. Ward in "Porphyrin Localization and Treatment of Tumors" (D.R. Doiron and C.J. Gomer, eds.), Alan R. Liss, Inc., New York, 1984, p. 285.

95. R. Selensky, D. Holten, M.W. Windsor, J.B. Paine III, D. Dolphin, M. Gouterman and J.C. Thomas, _Chem. Phys._, **60** 33 (1981).

96. A.D. Adler, F.R. Longo and W. Shergalis, _J. Am. Chem. Soc._, **86** 3145 (1964).

97. A.D. Adler, F.R. Longo, J.D. Finarelli, J. Goldmacher, J. Assour and L. Korsakoff, _J. Org. Chem._, **32** 476 (1967).

98. J. Winkelman, _Cancer Res._, **22** 589 (1962).

99. J. Winkelman, G. Slater and J. Grossman, _Cancer Res._, **27** 2060 (1967).

100. C.J. Carrano, M. Tsutsui and S. McConnel, _Cancer Treat. Rep._, **61** 1297 (1977).

101. R.A. Thaller, D.M. Lyster and D. Dolphin in "Porphyrin Photosentization" (D. Kessel and T.J. Dougherty, eds.), Plenum Press, New York, 1983, p. 265.

102. J.P. Collman, R.R. Gagne, C.A. Reed, T.R. Halbert, G. Lang an W.T. Robinson, _J. Am. Chem. Soc._, **97** 1427 (1975).

103. J. Almog, J.E. Baldwin, R.L. Dyer and M. Peters, _J. Am. Chem. Soc._, **97** 226 (1975).

104. M. Momenteau, J. Mispelter, B. Looke and E. Bisagni, _J. Chem. Soc., Perkin Trans. I_ 1983, 189.

105. L.K. Gottwald and E.F. Ullman, _Tetrahedron Lett._, 1969, 3071.

106. A. Triebs and N. Heberle, _Justus Liebigs Ann. Chem._, **718** 183 (1968).

107. J.B. Paine III, W.B. Kirshner, D.W. Moskowitz and D. Dolphin, _J. Org. Chem._, **41** 3857 (1976).

108. C.-B. Wang and C.K. Chang, _Synthesis_, (1979) 548.

109. J.B. Kim, A.D. Adler and F.R. Longo in "The Porphyrins" (D. Dolphin, ed.), Academic Press, New York 1978, Vol. I, p. 85.

110. J.B. Paine III in "The Porphyrins" (D. Dolphin, ed.), Academic Press, New York, 1978, Vol. I, p. 101.

111. a) G. P. Arsenault, E. Bullock, S.F. MacDonald, _J. Am. Chem. Soc._, **82** 4384 (1960).
    b) E.J. Tarlton, S.F. MacDonald and E. Baltazzi, ibid., **82** 4389 (1960).

112. K.M. Smith, _J. Chem. Soc. Perkin Trans. I_, 1972, 1471.

113. J.B. Paine, C.K. Chang and D. Dolphin, _Heterocycles_, **7** 831 (1977).

114. A.W. Johnson in "The Porphyrins (D. Dolphin, ed.), Academic Press, New York, 1978, Vol. I, p. 235.

115. P.S. Clezy and L. van Thuc, _Aust. J. Chem._, **37** 2085 (1984).

116. a) J.A.P. Baptista d'Almeida, G.W. Kenner, J. Rimmer and K.M. Smith, Tetrahedron, 32 1793 (1976).
   b) K.M. Smith and G.W. Craig, J. Org. Chem., 48 4302 (1983).
117. A.H. Jackson and K.M. Smith in "Total Synthesis of Natural Products" (J. ApSimon, ed.), Wiley-Interscience, New York, 1984, Vol, 6, p. 237.
118. R.K. Dinello and C.K. Chang in "The Porphyrins (D. Dolphin, ed.), Academic Press, New York, 1978, Vol. I, p. 289.
119. W.S. Caughey, J.O. Alben, W.F. Fugimoto and J.L. York, J. Org. Chem., 31 2631 (1966).
120. T. P. Wijesekera and D. Dolphin, unpublished results.
121. J.D. Doi, D.K. Lavallee, S.C. Srivastava, T. Prach. P. Richards and R.A. Fawwaz, Int. J. Appl. Radiat. Isot., 32 877 (1981).
122. C. Cloutour, D. Ducassou, J.-C. Pommier and L. Vuillemin, Int. J. Radiat. Isot., 33 1311 (1982).
123. R. Vaum, N.D. Heindel, H.D. Burns, J. Emrich and N. Foster, J. Pharm. Sci., 71 1223 (1982).
124. J.W. Buchler in "Porphyrins and Metalloporphyrins" (K.M. Smith, ed.), Elsevier, Amsterdam, 1975, p. 157.
125. J.W. Buchler in "The Porphyrins (D. Dolphin, ed.), Academic Press, New York, 1978, Vol. I, p. 389.
126. J.O. Alben, W.H. Fuchsman, C.A. Beaudreau and W.S. Caughey, Biochemistry, 7 624 (1968).
127. P.J. Crook, A.H. Jackson and G.W. Kenner, Justus Liebigs Ann. Chem., 748 26 (1971).
128. J.W. Buchler and L. Puppe, Justus Liebigs Ann. Chem., 740 142 (1970).
129. J.-H. Fuhrhop and D. Mauzerall, J. Am. Chem. Soc., 91 4174 (1969).
130. M. Tsutsui, M. Ichakawa, F. Vohwinkel and K. Suzuki, J. Am. Chem. Soc., 88 854 (1966).
131. M. Tsutsui, C.P. Hrung, D. Ostfeld, T.S. Srivastava, D.L. Cullen and E.F. Meyer, Jr., J. Am. Chem. Soc., 75 3952 (1975).
132. J.W. Buchler, M. Folz, H. Habets, J. van Kaam and K. Rohbock, Chem. Ber., 109 1477 (1976).
133. J.W. Buchler, L. Puppe, K. Rohbock and H.H. Schneehage, Ann. N.Y., Acad. Sci., 206 116 (1973).
134. A.D. Adler, F.R. Longo, F. Kampas and J. Kim, J. Inorg. Nucl. Chem., 32 2443 (1970).
135. M.J. Bain-Ackerman and D.K. Lavallee, Inorg. Chem., 18 3358 (1979).
136. D.K. Lavallee and D. Kuila, Inorg. Chem., 23 3987 (1984).
137. J.D. Doi, C. Compito-Magliozzo and D.K. Lavallee, Inorg. Chem., 23 79 (1984).
138. D. Kuila and D.K. Lavallee, Inorg. Chem., 22 1095 (1983).
139. D.K. Lavallee and D. Kuila, Rev. Port. Quim., 27 251 (1985).
140. A.R. Battersby and E. McDonald in "Porphyrins and Metalloporphyrins" (K.M. Smith, ed.), Elsevier, Amsterdam, 1975, p. 61.
141. R.B. Frydman, B. Frydman and A. Valasinas in "The Porphyrins (D. Dolphin, ed.), Academic Press, New York 1979, Vol. VI, p. 1.
142. O.T. Jones in "The Porphyrins (D. Dolphin, ed.), Academic Press, New York 1979, Vol. VI, p. 179.
143. A.R. Battersby and E. MacDonald in "B$_{12}$ (D. Dolphin, ed.), John Wiley & Sons Inc., 1982, Vol. 1, p. 107.
144. A.R. Battersby, C.J.R. Fookes, G. Hart, G.W.J. Matcham and P.S. Pandey, J. Chem. Soc., Perkin Trans. I, 1983, 3041.
145. A.R. Battersby, C.J.R. Fookes, K.E. Gustafson-Potter, G.W.J. Matcham and E. McDonald, J. Chem. Soc. Chem. Commun., 1979, 1155.

THE STRUCTURE OF PROTOPORPHYRIN AND OF THE AGGREGATE

FRACTION OF HAEMATOPORPHYRIN DERIVATIVE IN SOLUTION

A.D. Ward and A.G. Swincer

Department of Organic Chemistry
University of Adelaide,
P.O. Box 498,
Adelaide, South Australia, 5001

INTRODUCTION

The complexity of haematoporphyrin derivative (HPD) either in total or as its aggregate fraction (Photofrin II) has meant that many studies of the effects of this material, both *in vitro* and *in vivo*, have been by comparison with other porphyrins and by deduction from the behaviour of these porphyrins, particularly haematoporphyrin (HP) and protoporphyrin (PP). Many of our studies on these materials, particularly those involving chemistry and solution behaviour, have indicated that HPD, HP and PP show a remarkably wide variation in their physical and chemical properties and that they are particularly different from each other in aqueous solution. Of these three porphyrins HP is the most "normal" in its properties and the one which shows the expected behaviour in aqueous solution. It is appropriate therefore to discuss it first and then to compare and contrast other species with it and with each other.

MATERIALS AND METHODS

Nuclear magnetic resonance (NMR) spectra were recorded using Bruker WP-80DS and CXP-300 spectrometers. Deuterium oxide (D_2O) solutions of porphyrins were prepared using slightly basic conditions involving addition of a little sodium deuteroxide (NaOD) or potassium carbonate (K_2CO_3) since some of the porphyrins were not sufficiently soluble in neutral solution to give the concentrations required for the NMR measurements.

Molecular weights were determined using a Knauer vapour pressure osmometer with benzil as a reference. The preparation of the porphyrins and other experimental details have been previously described.[1]

RESULTS AND DISCUSSION

Haematoporphyrin (HP)

The difficulty in obtaining pure HP has been summarised.[2] Commercial HP generally contains approximately 10% of the isomers of hydroxyethyl-vinyldeuteroporphyrin as well as much smaller amounts of other material,

as determined by HPLC.[1] These impurities, as well as the ability of HP
to change on standing in aqueous solution,[1,3,4] largely account for the
conflicting reports in the literature concerning the effectiveness of HP
as a cytotoxic agent. Pure HP (> 98% by HPLC) can be obtained by an
extraction procedure, although the method is only convenient for small
amounts of material.[1] Pure HP and commercial HP are inactive *in vitro*
and *in vivo*[1,4] indicating that the *in vivo* activity attributed to some
commercial HP samples is actually due to impurities or to the slow
formation of cytotoxic material.

HP gives normal and relatively well resolved ^{1}H and ^{13}C NMR signals
in both D_2O (Figure 1) and deuterated dimethyl sulfoxide (d_6-DMSO) solutions. The better resolution obtained from the DMSO solution would be
expected on the grounds that porphyrin molecules associate more in aqueous
solution than in organic solvents.[5] The chemical shift values, coupling
constants etc. of HP are all very similar to those of the dimethyl ester
of HP whose NMR spectra are well documented.[6]

The molecular weight of HP has been measured using ultracentrifugation methods[7] and by vapour pressure osmometry. Ultracentrifugation
studies showed that HP in aqueous solution had a low molecular weight[7]
(ca. 1000) although the technique does not enable a distinction to be
made between a monomer, dimer or slightly higher oligomer species. A
vapour pressure osmometry determination of the molecular weight of HP in
aqueous solution was not possible due to the detergent-like behaviour of
aqueous HP solutions; however in DMSO the molecular weight was that of a
monomeric species.

Thus HP shows the behaviour that would be expected of a porphyrin
with two side chains containing carboxylic acid groups. The association
between HP molecules that has been reported to occur in aqueous solution
is not sufficient to perturb the NMR spectrum in water other than to make
the signals slightly broader and more complex. The better NMR resolution
in DMSO solution would suggest that this association does not occur in
this solvent; other studies[5] suggest that this is likely to be true for
other organic solvents as well.

The Diacetate of Haematoporphyrin (HPdiAc)

This derivative of HP, obtained by conventional acetylation procedures,[8] gives normal and expected ^{1}H and ^{13}C NMR spectra in d_6-DMSO
solution; addition of D_2O to this solution does not change the signals.
Addition of sodium deuteroxide (NaOD) in D_2O to the solution provides
basic conditions similar to those used to form HPD. HPLC of the resulting
product shows a clean formation of HP, the two isomers of hydroxyethylvinyldeuteroporphyrin and PP. HPLC analysis showed that the material that
makes up HPD aggregate was absent. HP is the major product and its ^{1}H
NMR signals can be clearly observed in the basic solution.

HPdiAc is the major component produced when HP reacts with acetic
acid and sulfuric acid in the first step of the process by which HPD is
produced. The next step involves treating the product from the first
step with aqueous base. HPdiAc is insufficiently soluble in D_2O to
obtain an NMR spectrum, however addition of a small amount of NaOD in D_2O
forms a solution which does not give any ^{1}H NMR signals under the normal
conditions. As will be seen later, this is exactly the situation that
applies to a solution of HPD aggregate in water. We have shown[1] that
HPdiAc when treated with dilute aqueous base forms about 60% of the
aggregate fraction of HPD.

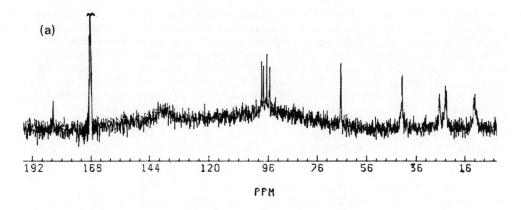

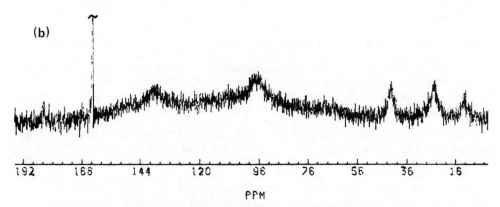

Figure 1. ^{13}C NMR spectra of (a) HP and (b) HPD in $D_2O/$ K_2CO_3 at about equal concentration (14 mg/ml). The peaks of the HP spectrum have been assigned to CO_2H (183), CO_3^{--} (168), pyrrole carbons (138), meso carbons (96-99), CHOH (66), β-CH_2 (42), β-CH_3 (26), α-CH_2 (24), pyrrole CH_3 (12).

The above results demonstrate clearly that the reaction of HPdiAc with base is very solvent dependant; in organic solvents, or mainly organic solvents, the normal hydrolysis and elimination reactions occur but in aqueous solution these are relatively minor processes with the dominant reaction being the formation of HPD aggregate.

Protoporphyrin (PP)

PP is a minor component of both HPD and of HPD aggregate (Photofrin II). It can be readily obtained[1] from HP. PP gives no observable ^1H or ^{13}C NMR signals in aqueous solution but shows the normal spectral signals for both nuclei in d_6-DMSO solution. Addition of one drop of D_2O to the DMSO solution causes a substantial broadening of all the ^1H signals. Addition of several drops of D_2O causes, after several hours standing, the complete disappearance of all ^1H signals. This loss of signal is not due to deuterium-hydrogen exchange since the recovered material shows the normal molecular weight by mass spectrometry and no signals in the ^2H NMR region. Addition of about an equal volume of deuterated methanol (CD_3OD) to a D_2O solution of PP causes the appearance of the signals due to the meso and vinyl protons of the PP structure. It is not possible to observe other signals of PP because solvent signals due to the incompletely labelled solvents obscure the remainder of the spectrum. This data shows that PP does not give detectable NMR signals in aqueous conditions but does in the organic solvents used; addition of water to DMSO solutions decreases the signals, ultimately to the stage that they cannot be observed whereas addition of the organic solvent methanol to an aqueous solution enables signals to be detected. Tetrahydrofuran (THF) is an exceptionally good solvent for many of the porphyrins associated with HPD, however it is only a strong solvent for these porphyrins if a little (ca. 10%) water is present. PP in d_8-THF containing ca. 10% D_2O shows normal ^1H and ^{13}C NMR spectra.

Using high power proton-enhanced (PE) NMR conditions an aqueous solution of PP gave a PE ^{13}C spectrum.[9] This is only possible if PP is a large molecule with a particle size larger than 30 nm in diameter.[9] Preliminary studies using photon correlation measurements showed[10] that aqueous solutions of PP contained particles as large as 60-80 nm in diameter. A measurement of the molecular weight of PP in aqueous solution was not possible using an ultracentrifuge experiment since under the conditions used PP precipitated from solution.[7] This behaviour is typical of that shown[7] by molecules with a molecular weight greater than 100,000. The molecular weight of PP determined in DMSO solution by a vapour phase osmometer experiment was that of a monomer (500 ± 100).

The above data show that PP exhibits very different properties in aqueous solution from those that it shows in some organic solvents. All of the data obtained from aqueous solution is consistent with the presence of a high molecular weight and/or a highly aggregated species. In contrast, the data from organic solvents is essentially that expected for typical dicarboxylic acid porphyrins and, in particular, is similar to that of HP in organic solvents. The behaviour of PP in water has been attributed[9] to the formation of a stable micelle structure. This structure does not form in the organic solvents described above.

HPD Aggregate

HPD aggregate gives no observable signals when the ^1H NMR spectrum is measured in D_2O under the usual conditions; in d_6-DMSO the ^1H signals are weak and poorly resolved. The ^{13}C NMR spectrum of HPD aggregate in D_2O is rather weak and considerably broader than an equivalent spectrum for HP (Figure 1). The signal due to the alpha carbons of the side chains of HPD aggregate that are derived from the hydroxyethyl groups of HP is very broad compared to the other signals. In d_6-DMSO the ^{13}C NMR spectrum of HPD aggregate shows better resolution than in aqueous solution; the alpha carbons signal is still rather broad and is noticeably weaker than other comparable signals.

Application of the high power proton-enhanced NMR conditions[9] to aqueous solutions of HP and HPD aggregate failed to yield any detectable spectral signals showing that both HP and HPD aggregate are of much smaller size in aqueous solution than is PP. However, molecular weight measurements[7] show that HPD aggregate, with an average molecular weight of 21,500 in water, is a much larger molecule than is HP which has an average molecular weight of ca. 1,000 in water. Vapour pressure osmometry experiments showed that HP is monomeric in DMSO solution while HPD aggregate has a much higher molecular weight (greater than 2,000) in DMSO whose value can not be adequately measured using this technique.

All the above data can be summarised as follows. The NMR spectra and the molecular weight determinations of the three materials HP, PP and HPD aggregate show that they are quite different in some important ways. HP behaves as a normal organic molecule and is, at most, only slightly associated in aqueous solution. PP exists as a large and strongly associated species (probably a micelle) in aqueous solution but behaves as a monomer in organic solvents. HPD aggregate behaves as a relatively large and stable species in both aqueous and organic conditions. Hence the bonding or association involved in HPD aggregate is unlikely to be the same as that occurring with PP in aqueous solvents. The NMR signals of HPD aggregate are considerably broader in aqueous solution compared to the signals from a DMSO solution indicating that the nuclei of HPD aggregate have different relaxation times in aqueous solution compared to those in organic solvents. This in turn suggests that HPD aggregate has a different structure or conformation in water compared to the situation in organic solvents. We suggest that the above results are best explained by a polymer structure for HPD aggregate. This would explain the relatively weak NMR signals for HPD aggregate and the stability in both aqueous and organic solvents. The molecular weight data[7] indicate that, on average, 30-40 porphyrin units are linked in the polymer structure of HPD aggregate.

A dimer structure has been proposed for HPD aggregate.[11,12] The evidence on which this structure was based came largely from mass spectrometry and NMR data. The published mass spectrum clearly shows the presence of peaks due to trimer species as well as those due to a dimer. Given that HPD aggregate is a reasonably labile polymer it is likely that the benzyl ether or ester linkages may cleave even under FAB mass spectrometry conditions. The most obvious peaks in the resulting spectrum would be those of lowest mass, i.e. monomer, dimer and trimer. It is also possible that during the esterification process and/or its workup to obtain the methyl esters of HPD aggregate used for these measurements the polymer may have been degraded to smaller fragments. This comment also applies to the material used for the NMR studies. For our work all NMR samples were checked by HPLC after the NMR measurements to ensure that they had not changed. The NMR data quoted is difficult to interpret completely, however it appears that much, if not all, of the data quoted for the free base, i.e. HPD aggregate, is also compatible with a polymer structure involving an ether linkage. The assignment of a dimer structure seems to have been largely because of the FAB mass spectrum data.

The Structure and Reactivity of HP, PP and HPD Aggregate in Solution

The above data does not provide any firm evidence that clearly indicates the functional groups that are involved in the polymer linkages. The ^{13}C NMR spectrum of HPD aggregate shows that the relaxation time for the alpha carbons of the groups derived from the hydroxyethyl side chains of HP is different from that of the remaining carbons. This would be expected if the alpha carbons are part of the backbone of the polymer chain. These carbons could be part of a polyether system (Figure 2a) or a polyester system (Figure 2b).[13,14] The chemical shift of these alpha carbons (61 ppm) suggests that an oxygen atom is still attached to them and is not compatible with a carbon-carbon bond (10-30 ppm) that would be obtained from the polymerisation of vinyl side chains.[13] The carboxyl signal of HPD aggregate appears, in the ^{13}C NMR spectrum, as a normal sharp singlet whose chemical shift and relative intensity are very similar to those of the carboxyl carbons in HP. This would not be expected if one of the carboxyl carbons was involved, as an ester, in the polymer chain for, in particular, the intensity of the signal should be diminished and the peak broadened as is the case for the alpha carbons signal. The mono-methyl esters of HP show separate ^{13}C signals for the ester carbon and

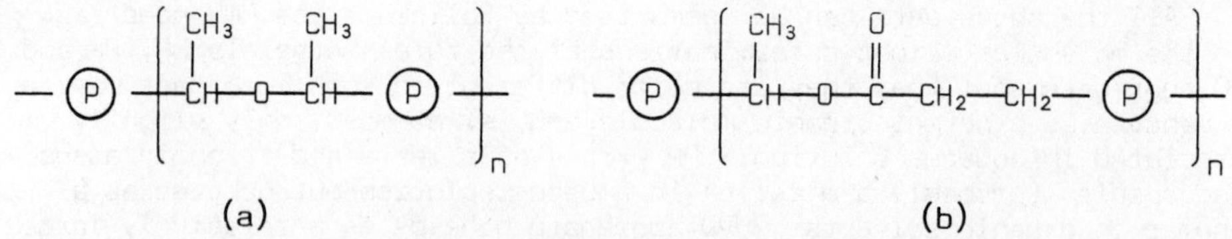

$$-\text{(P)}-\left[\begin{array}{c} \text{CH}_3 \quad\quad \text{CH}_3 \\ | \quad\quad\quad | \\ \text{CH}-\text{O}-\text{CH} \end{array}-\text{(P)}\right]_n \qquad\qquad -\text{(P)}-\left[\begin{array}{c} \text{CH}_3 \quad\quad \text{O} \\ | \quad\quad\quad || \\ \text{CH}-\text{O}-\text{C}-\text{CH}_2-\text{CH}_2 \end{array}-\text{(P)}\right]_n$$

(a)   (b)

Figure 2. Schematic representations of (a) the polyether structure and (b) the polyester structure for HPD aggregate.

(P) represents the porphyrin ring system including the side chains not shown.

the carboxylic acid carbons, suggesting that an ester polymer would have two carboxyl signals rather than one.

HPD aggregate is stable to reflux conditions in dilute sodium hydroxide for an hour but is quite rapidly hydrolysed by hot aqueous acidic conditions.[15] An ester polymer structure would be expected to be hydrolysed by basic as well as acidic conditions. The dimethyl ester of HP is fully hydrolysed by dilute aqueous sodium hydroxide at room temperature within several hours. Kessel has recently shown[16] that HPD aggregate is labile to both acid and base in aqueous tetrahydrofuran. The substantially increased reactivity of nucleophiles in less solvating conditions is well known and may explain the increased reactivity of the polymer towards hydroxide in aqueous tetrahydrofuran. The polyether system involves benzylic carbons at both ends of the ether linkage; this would certainly increase the reactivity compared to normal ethers. Besides a direct displacement reaction at the benzylic carbon, a leaving group attached at the benzylic carbon could also be substituted by an elimination-addition sequence as shown in the Scheme. Such a process should be base catalysed as shown.

The above observations do not explain why the aggregate fraction forms only when HPdiAc or the solid from which HPD is formed is treated with aqueous base. We suggest that this feature is a consequence of the aggregation or association of porphyrin molecules that occurs only in water. Porphyrin acids are known[5] to form sandwich aggregates because of the π-π interactions between the planar porphyrin units. This aggregation decreases the wavelength and the intensity of the Soret band of porphyrins in water. In organic solvents, where this aggregation does not occur,[5] a sharp Soret band is observed at higher wavelength. Thus in water HPdiAc molecules will be associated, with reactive groups on separate porphyrin systems held only a short distance from each other, enabling a polymer system to form readily. In organic solvents HPdiAc molecules are well separated by solvent molecules, enabling normal hydrolysis of the acetate groups to occur at a faster rate than the polymerisation process.

The two most likely aggregation orientations of HPdiAc molecules are shown in Figure 3. The first, Figure 3a, has the polar acetoxyl groups of two separate porphyrins associating via favourable dipole-dipole interactions. This orientation would require at least one of the carboxylate groups in each porphyrin molecule to be adjacent to another in the next porphyrin. This situation may be favourable if the two like charges of the carboxylate ions are separated and stabilised by hydrogen bonding to one, or more, water molecules as depicted in Figure 3b. The other alternative, Figure 3c, has the acetoxyl group of one porphyrin associating with the carboxylate group of the next via dipole-dipole interactions. Whichever of these two situations is the actual case should determine the

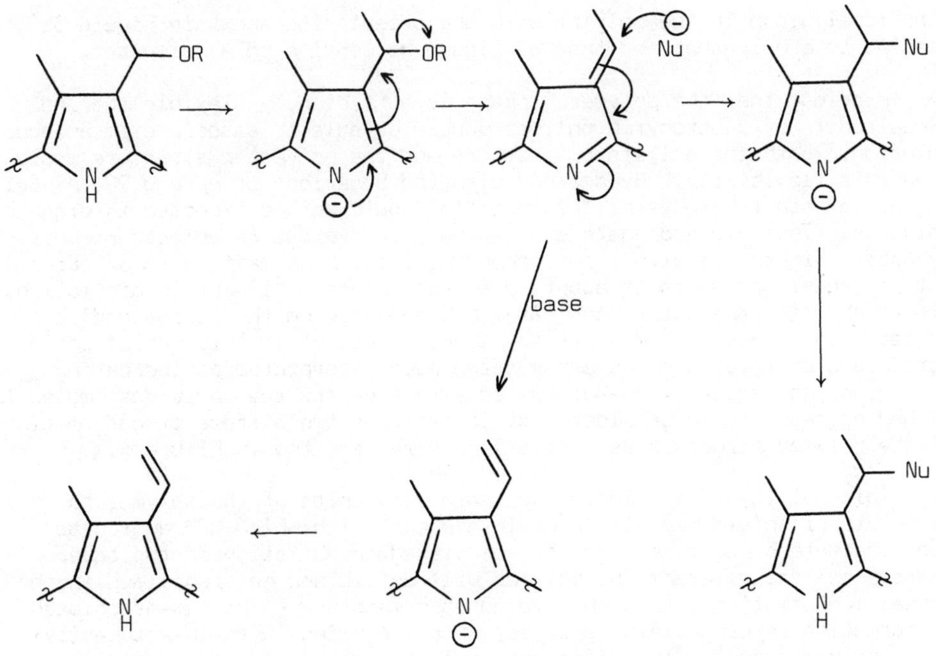

Scheme: Elimination-addition sequence for the substitution of groups at the benzylic carbon. Only one pyrrole ring of the porphyrin system is shown.

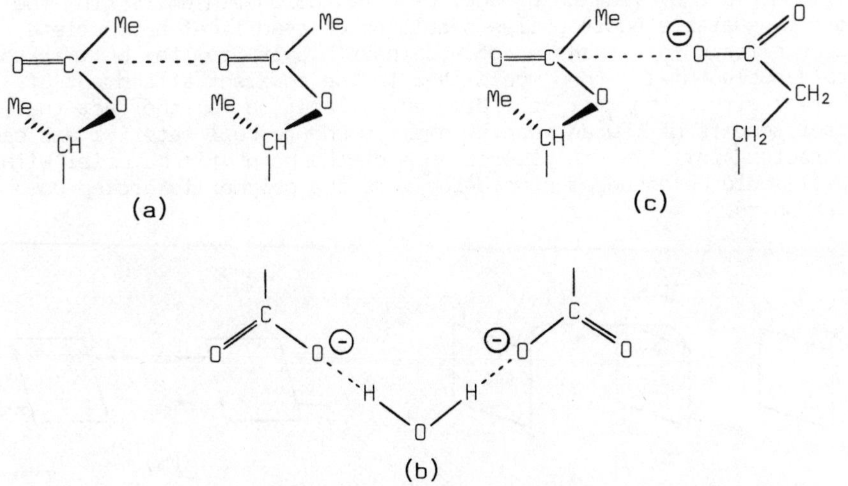

Figure 3: Favourable dipole-dipole interactions between (a) two acetoxyl groups, (b) two carboxylate groups and a water molecule, and (c) an acetoxyl group and a propionate side chain, attached to separate porphyrin ring systems held close by π-π interactions.

functional group in the polymer with the orientation shown in Figure 3a leading to a polyether and that of Figure 3c leading to a polyester.

Provided that the polymer linkage is sufficiently flexible the individual units of a porphyrin polymer should be able to associate in aqueous solution. Both the polyether structure and the polyester structure would have this flexibility. By analogy with the behaviour of HP and PP, as well as that of other porphyrins,[5] association would not be expected in organic solvents. Thus HPD aggregate can behave as a typical associated porphyrin in water, Figure 4a, with a corresponding Soret band maximum (λmax 365 nm) but in organic solvents it adopts a non-associated (linear) conformation, Figure 4b, with the Soret band (λmax 400 nm) showing the corresponding increase in intensity and shift to higher wavelength. This variation in Soret band position and shape has often been interpreted as indicating an aggregated versus a non-aggregated situation for the individual molecules of HPD aggregate; we consider that it reflects two different conformations of the polymer structure as schematically represented in Figure 4.

This ability of the individual porphyrin units of the polymer to associate or unfold may also explain the ease with which HPD aggregate can penetrate a membrane. As the associated conformation in the aqueous phase meets the membrane the polymer will unfold and be stabilised in the linear conformation within the hydrophobic membrane. The non-associated conformation is essentially a linear sheet enabling it to insert easily into the membrane.[17] The different conformations could also explain the difference in fluorescence noted for HPD aggregate. In aqueous solution the association of individual porphyrins will lead to substantial self-quenching of fluorescence; PP a strongly associated system is also essentially non-fluorescent in water at concentrations similar to those used for HPD aggregate. PP is quite strongly fluorescent in DMSO solution where it is present as a monomer. The linear conformation is not so conducive to self quenching allowing HPD to exhibit significant fluorescence in organic solvents and in hydrophobic regions, e.g. membranes.

Since the HPD polymer structure is reactive towards nucleophilic attack, particularly in a non-aqueous environment, the displacement of a benzylic group and cleavage of one, or more, porphyrin units from the polymer by suitable nucleophiles should occur readily at appropriate cellular sites. This transfer of a porphyrin unit from the polymer to a suitable nucleophilic group would lead to the covalent attachment of porphyrins within the cell providing an explanation for the fact that significant amounts of HPD aggregate remain bound to cell material and cannot be extracted from it. The fluorescence of the porphyrin material within the cell would be enhanced considerably as the polymer is broken down to smaller units.

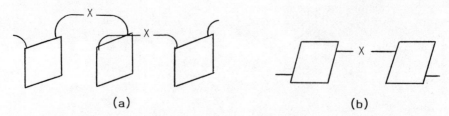

(a) (b)

Figure 4. Schematic representations of (a) an associated, and (b) a non-associated conformation of a porphyrin polymer. The porphyrin ring is represented by the square and the linking functional group by — X — . The linking groups are not necessarily adjacent or opposite to each other on the porphyrin ring system.

CONCLUSIONS

The NMR studies and molecular weight measurements combine to provide strong evidence that HPD aggregate has a polymeric structure. The same evidence also show that protoporphyrin is highly aggregated in aqueous solution but is monomeric in organic solvents. Haematoporphyrin is only slightly associated in water.

The functional group formed in the polymerisation process is either an ether or an ester with the data favouring an ether linkage. The formation of the polymer structure only in an aqueous environment is attributed to the self-association of porphyrin molecules that occurs in an aqueous solution but is not observed in organic solvents. The lability of the polymer structure means that it is likely that individual porphyrin units or small groups of porphyrins can be cleaved from the polymer chain by suitably positioned nucleophilic groups within a cell. The ability of the polymer to adopt an aggregated or open conformation depending upon the solvent or the immediate environment explains the varying fluorescence noted for HPD aggregate.

ACKNOWLEDGEMENTS

We thank the National Health and Medical Research Council of Australia and the Australian Research Grants Committee for support of this work. We also wish to thank Dr. A.A. Diamantis for the use of the vapour pressure osmometer.

REFERENCES

1. P.A. Cowled, I.J. Forbes, A.G. Swincer, V.C. Trenerry, and A.D. Ward, Separation and phototoxicity *in vitro* of some of the components of haematoporphyrin derivative, Photochem.Photobiol. 41:445 (1985).
2. R.K. Dinello and C.K. Chang, Isolation and modification of natural porphyrins, in "The Porphyrins. Vol.I. Structure and Synthesis, Part A", D.Dolphin, ed., Academic Press, New York (1978).
3. F. Ricchelli and L.I. Grossweiner, Properties of a new state of haematoporphyrin in dilute aqueous solution, Photochem.Photobiol. 40:599 (1984).
4. T.J. Dougherty, Haematoporphyrin as a photosensitizer of tumors, Photochem.Photobiol. 38:377 (1983).
5. W.I. White, Aggregation of porphyrins and metalloporphyrins, in "The Porphyrins. Vol.V. Physical Chemistry, Part C", D. Dolphin, ed., Academic Press, New York (1978).
6. H. Scheer and J.J. Katz, Nuclear magnetic resonance spectroscopy of porphyrins and metalloporphyrins, in "Porphyrins and Metallo-porphyrins", K.M. Smith, ed., Elsevier, Amsterdam (1975).
7. A.G. Swincer, A.D. Ward and G.J. Howlett, The molecular weight of haematoporphyrin derivative, its gel column fractions and some of its components in aqueous solution, Photochem.Photobiol. 41:47 (1985).
8. R. Bonnett, R.J. Ridge, P.A. Scourides and M.C. Berenbaum, On the nature of haematoporphyrin derivative, J.Chem.Soc.,Perkin Trans.I 3135 (1981).
9. T.M. Spotswood, A.G. Swincer, A.D. Ward and E.H. Williams, Solid state NMR studies on the aggregate nature of protoporphyrin in aqueous solution, Magn.Reson.Chem. in press (1985).
10. P. Daivis, I. Snook, A.G. Swincer and A.D. Ward, unpublished data.

11. T.J. Dougherty, W.R. Potter and K.R. Weishaupt, The structure of the active component of haematoporphyrin derivative, in "Progress in Clinical and Biological Research. 170. Porphyrin Localization and Treatment of Tumors", D.R. Doiron and C.J. Gomer, eds., Alan R. Liss, New York (1984).

12. T.J. Dougherty, W.R. Potter and K.R. Weishaupt, The structure of the active component of haematoporphyrin derivative, in "Porphyrins in Tumor Phototherapy", A. Andreoni and R. Cubeddu, eds., Plenum Press, New York (1984).

13. R. Bonnett and M.C. Berenbaum, HPD - a study of its components and their properties, in "Advances in Experimental Medicine and Biology. 160. Porphyrin Photosensitization", D. Kessel and T.J. Dougherty, eds., Plenum Press, New York (1983).

14. M.C. Berenbaum, R. Bonnett and P.A. Scourides, In vivo biological activity of the components of haematoporphyrin derivative, Br.J. Cancer 45:571 (1982).

15. T.J. Dougherty, D.G. Boyle, K.R. Weishaupt, B.A. Henderson, W.R. Potter, D.A. Bellnier and K.E. Wityk, Photoradiation therapy - clinical and drug advances, in "Advances in Experimental Medicine and Biology. 160. Porphyrin Photosensitization", D. Kessel and T.J. Dougherty, eds., Plenum Press, New York (1983).

16. D. Kessel, Properties of the tumor-localizing component of HPD, in "Porphyrins as Phototherapeutic Agents for Tumors and Other Diseases", Plenum Press, in press.

17. R.C. Chatelier, W.H. Sawyer, A.G. Swincer and A.D. Ward, Fluorescence studies on the interactions of porphyrins with synthetic lipid membranes, Photochem.Photobiol., in press (1985).

CHEMICAL MANIPULATION OF THE VINYL GROUPS

IN PROTOPORPHYRIN-IX

Kevin M. Smith

Department of Chemistry
University of California
Davis, California 95616

As one of only very few synthetic organic chemists at this meeting, I propose to discuss some old and some new aspects of the chemistry of "type-IX" derivatives of porphyrins. Much of our recent work has centered on total synthesis of porphyrins from acyclic precursors,[1] but today I plan to focus on various manipulations of the substituents, particularly the vinyls, on intact commercially available porphyrins since I believe that this is the type of approach which will be more useful to the present audience. Most often, the commercial porphyrin used is hemin [the iron(III) chloride complex of protoporphyrin-IX]; this is inexpensive and can be converted into protoporphyrin-IX (1) by a simple demetalation procedure.[2] Because of the constraints of the biosynthetic process leading to hemin, all porphyrins which will be discussed in this paper have the "type-IX" orientation. The type-IX orientation of substituents is that name given by Hans Fischer to the arrangement common to all natural heme and chlorophyll pigments;[3] in the case of protoporphyrin-IX (1), for example, there are four methyl groups (one on each pyrrole subunit), two vinyls (one each on of rings A and B), and two propionic acid groups (one each on rings C and D). This arrangement of substituents renders the porphyrin completely asymmetric, and this is a characteristic of all the natural plant and blood pigments. It is fitting that hematoporphyrin-IX (Hp; 2) and its "derivative" (HpD) used in phototherapy possess the natural substituent array because this may aid the general adoption of HpD for drug use, and also it makes Hp and HpD relatively readily available from blood. It must be stated, however, that Hp (2) is, in the strictest terms, not a "natural" porphyrin. It was actually the first porphyrin ever prepared (in 1867) when Thudichum[4] treated blood with concentrated sulfuric acid. This treatment caused demetalation of the heme, with concomitant hydration of the vinyl groups in the resulting protoporphyrin-IX (1), presumably by way of the sulfate esters. Though 2- and 4-(1-hydroxyethyl) groups are not common to mammalian blood pigments, a 2-(1-hydroxyethyl) group can be found in chlorophylls (e.g. 3) from green and brown sulfur bacteria.[5] Compounds of type 3, which absorb strongly above 600 nm and which can be readily obtained from culturing of bacteria, offer hope in the search for long-wavelength pigments for photoradiation therapy of tumors. They also have the 2-(1-hydroxyethyl) group required for the polymerization observed in Photofrin II.

Much of the chemistry I shall discuss in this paper was developed in order to establish synthetic methodology for introduction of deuterium and carbon-13 labels into protoporphyrin-IX derivatives. Thus, this work can simply be modified to enable preparation of samples of hematoporphyrin (and its derivatives) which bear carbon-14 or tritium labels; owing to the possibility of extraneous exchange reactions taking place, it would probably be more reliable to use carbon-14 as the radioactive tracer, and such compounds should be of considerable use for definitive assaying of porphyrin uptake in tissues.

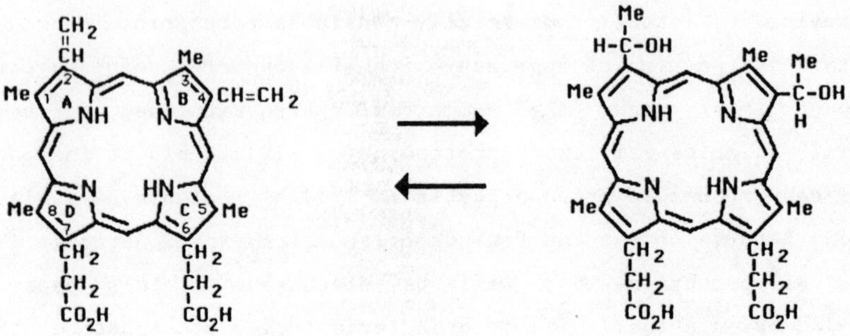

(1)   (2)

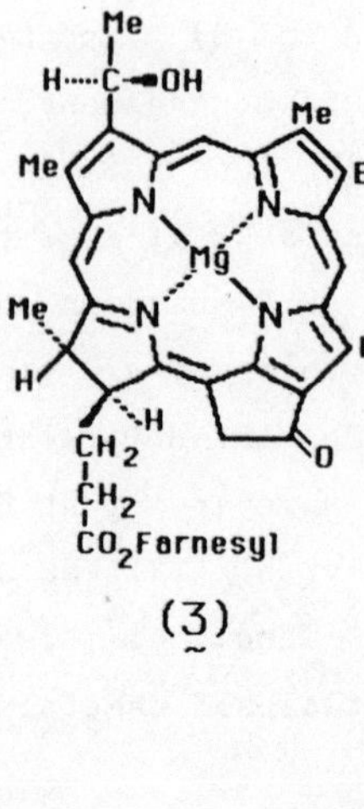

(3)

Hp (2) can be prepared readily by treatment of protoporphyrin-IX (1) with hydrogen bromide in acetic acid, followed by in situ hydrolysis of the intermediate 2,4-di-(1-bromoethyl) compound. Hp itself is not a single compound; it has two chiral centers and so exists as 2^n (n = the number of chiral centers) optical isomers. These four optical isomers are two pairs of diastereomers. On the other hand, the bacteriochlorophylls (3) described earlier, in which the 2-(1-hydroxyethyl) group is formed by enzymatic hydration of a vinyl precursor, are optically pure in many cases.[5] Hp can be simply transformed into protoporphyrin-IX (1) by heating in the presence of p-toluene sulfonic acid or by treatment with benzoyl chloride.

Both ethers and esters of Hp can be formed. Simple treatment of Hp with alcohols and acid usually gives some ether (4), though extensive dehydration to protoporphyrin-IX and other partially vinylated derivatives often occurs. Esters at the 2-(1-hydroxyethyl) can easily be prepared, as presumably occurs in the formation of HpD, by treatment with acid chlorides (or anhydrides) in base (e.g. pyridine), or with carboxylic acids in acid; in the latter case the hydroxyethyl group again undergoes extensive dehydration to vinyl.

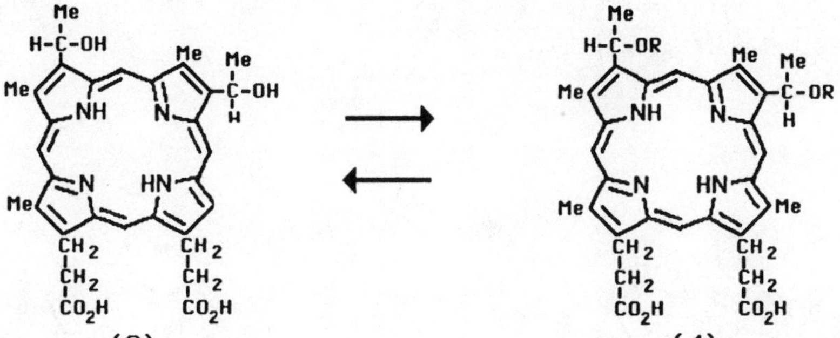

(2)

(4)

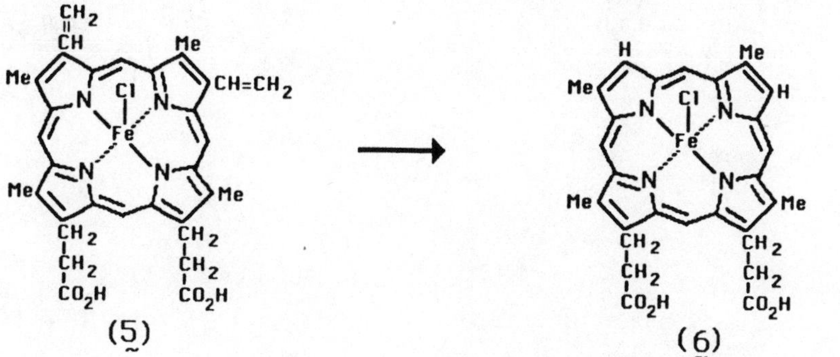

(5)

(6)

One of the key transformations in the chemistry of vinyl manipulation in porphyrins is the "Schumm" devinylation, in which the iron(III) complex of protoporphyrin-IX (i.e. hemin, 5), is heated in a melt of resorcinol. The resorcinol almost certainly plays a key role during this reaction in acceptance of the cleaved two-carbon unit.[6] However, removal of the vinyls, to give deuterohemin-IX (6), leaves a hydrogen atom at each of the 2- and 4- positions, and these sites become reactive towards other reagents, particularly electrophiles. Thus, treatment of either the iron(III) (6) or copper(II) complex of deuteroporphyrin-IX with acetic anhydride in the presence of a Lewis acid (stannic chloride) accomplishes a "Friedel-Crafts" acetylation at the 2- and 4- positions, resulting in the very efficient formation of 2,4-diacetyldeuteroporphyrin-IX (7) after demetalation. Hp can be efficiently regenerated from (7) by simple reduction with sodium borohydride, and so, if the acetyl units added in the Friedel-Crafts step were labeled with carbon-13 or carbon-14, the correspondingly labeled Hp is produced in good yield.[7] Deuterium (or tritium) can be incorporated into the (1-hydroxyethyl) groups of Hp (2) by treatment of 2,4-diacetyldeuteroporphyrin-IX with sodium borodeuteride (or sodium borotritiide) to give (8). Alternatively, acid catalyzed exchange of the acetyl methyl in the presence of a deuterium (or tritium) source gives the methyl-labeled species (9) which can be reduced with borodeuteride to give (10).

(6)   (7)

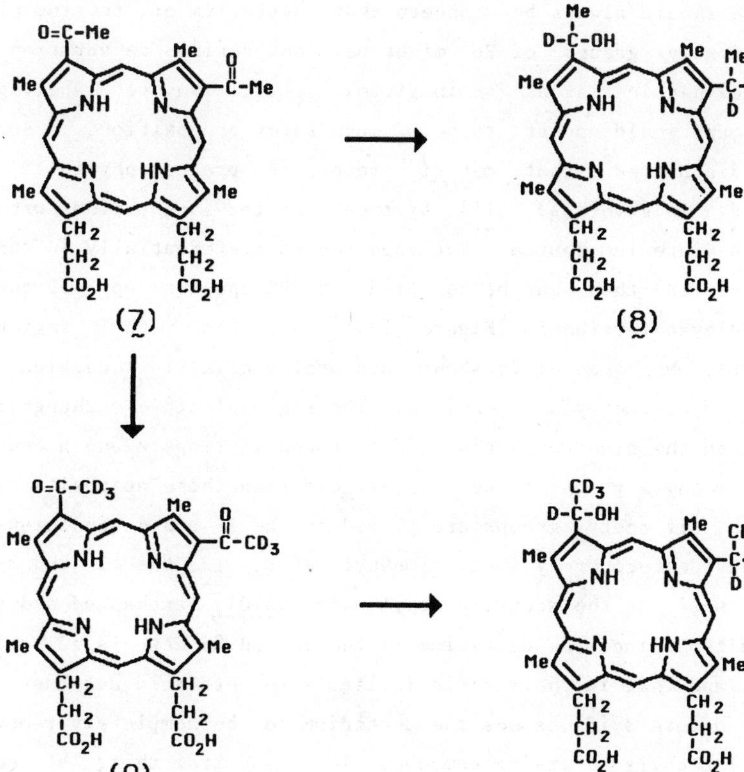
O=C-Me
Me
O
C-Me
Me
NH
N
N
HN
Me
Me
CH2
CH2
CO2H
CH2
CH2
CO2H
Me
D-C-OH
Me
Me
C-OH
D
O=C-CD3
C-CD3
CD3
D-C-OH
CD3
C-OH
D

(7) (8) (9) (10)

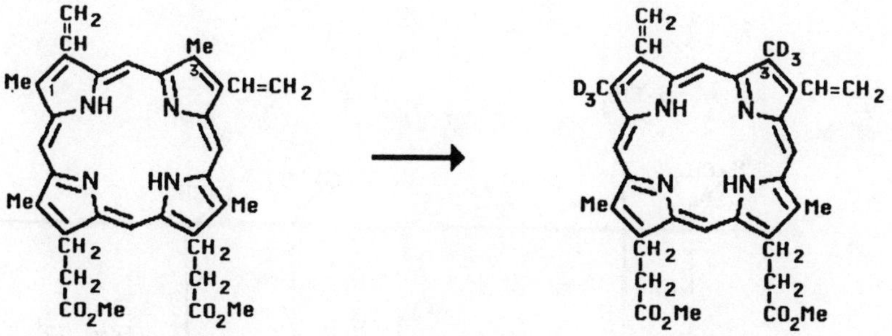
CH2
CH
Me
Me
1
3
CH=CH2
NH
N
N
HN
CO2Me
D3C
CD3

(11)

There should always be concern that deuterium or tritium placed in the hydroxyethyl groups of Hp might be lost during dehydration to give vinyls, either in tissues or in vitro. Thus, isotope labeling of the methyl groups would appear to be a much safer proposition. Some years ago we discovered[8] that methyl groups in protoporphyrin-IX can be deuterated, to give e.g. (11), by treatment for long periods with strong base and a deuterium source. Exchange occurs preferentially in the 1- and 3- methyls, and this can be verified by NMR spectroscopy of the corresponding dicyanoferrihemin (Figure 1). Only the methyl region of the paramagnetic NMR spectrum is shown, and preferential introduction of label in the 1- and 3- methyls is obvious. The regioselective exchange reaction depends upon the electronegative vinyl groups in rings A and B causing the 1- and 3- methyls groups to be more acidic than those at positions 5- and 8-. Thus, if acetyl groups are placed at the 2- and 4- positions (as in 2,4-diacetyldeuteroporphyrin-IX dimethyl ester, 12) the 1- and 3- methyl groups as well as the acetyl methyls are <u>rapidly</u> exchanged and compound (13) results. Since the deuterium in the 1- and 3- methyls is labile only to base, and that in the acetyls is labile to both acid and base, treatment with dilute acid causes the deuterium to be completely removed from the acetyl methyls, giving compound (14), and from this, by reduction, the Hp derivative (15) can be prepared. If a tritium source (e.g. T_2O) is used in place of the deuterium source, and the 2,4-diacetyldeuteroporphyrin-IX is reduced to the corresponding Hp (with $NaBH_4$) then highly tritiated material, in which the isotope is safe from extraneous exchange (except in the strongest of basic conditions) would result.

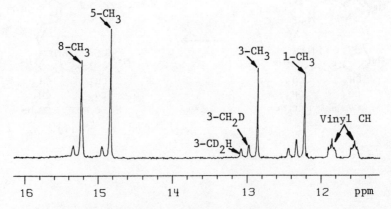

Figure 1. 360 MHz proton NMR spectrum (methyl region only) of the dicyanoferrihemin obtained by base-catalyzed exchange of protoporphyrin-IX and iron insertion. Methyl resonances are assigned; peaks to lower field of each methyl are due to CH_2D and CD_2H components.

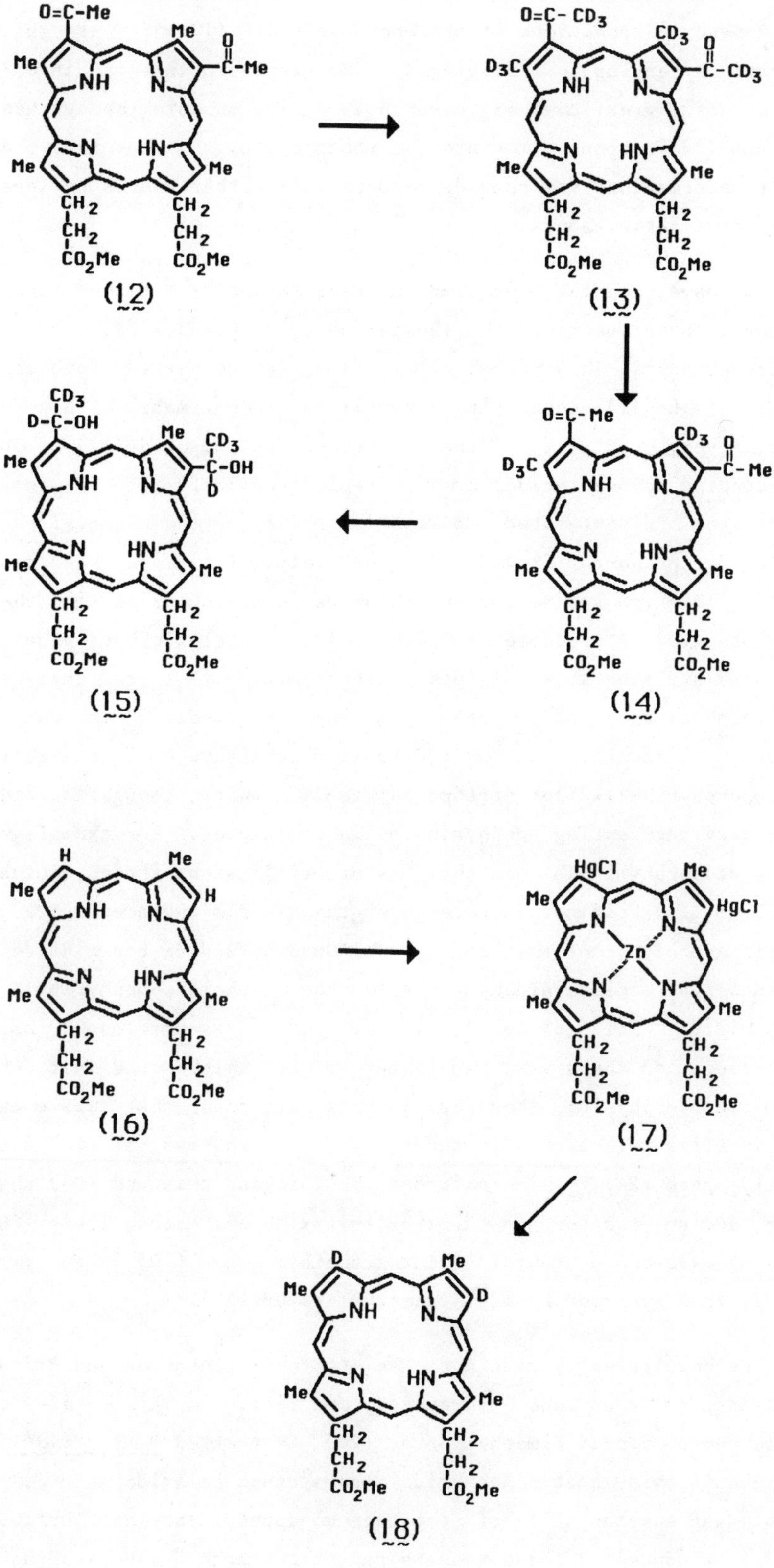
(12)
(13)
(14)
(15)
(16)
(17)
(18)

In the final part of this paper I shall deal with new chemistry at the 2- and 4- positions of protoporphyrin-IX (1) which accomplishes modifications leading to biologically important products. Since I am fully aware that a great deal of research is going on into the synthesis of Hp- and HpD-like compounds for use in phototherapy, I hope that these series of transformations may provide readers with either routes or ideas for new phototherapeutic agents.

We have recently been studying mercuration as a method for attachment of novel substituents on the the periphery of porphyrins. For example, if deuteroporphyrin-IX dimethyl ester (16) is converted into its zinc(II) complex [zinc(II) acetate in methanol] and then treated with mercuric acetate, a very high yield of the bis-mercurated compound (17) is obtained.[9] With sodium borodeuteride, or with deuterated sulfuric acid, compound (17) gives the dideuterated derivative (18), the so-called 2,4-dideuterodeuteroporphyrin-IX. Alternatively, (16) can be treated with lithium chloride and palladium chloride in acetonitrile and the 2,4-bis-acrylate (19) is produced in high yield.[9] Catalytic hydrogenation of the two acrylate substituents gives coproporphyrin-III tetramethyl ester (20), the porphyrinogen of which is a key intermediate in porphyrin and chlorophyll metabolism. We had earlier published a less direct route to coproporphyrin-III from protoporphyrin-IX[10] which, though longer, produced some very interesting porphyrins as intermediates. For example, treatment of protoporphyrin-IX dimethyl ester (21) with three equivalents of thallium(III) trifluoroacetate gave the bis-dimethylacetal (22); aqueous acidic hydrolysis of the acetal, followed by sodium borohydride reduction of the intermediate aldehyde, gave the "isohematoporphyrin-IX" (23) in high yield. This compound, for a while, assumed some importance in photoradiation therapy as an isomer of Hp which would not dehydrate to form vinylporphyrin; however, others have found that this molecule does not localize selectively in tumor tissue. Treatment of (23) with thionyl bromide gave (24), which afforded the dicyano compound (25) when treated with sodium cyanide in dimethylsulfoxide. Methanolysis (HCl/CH$_3$OH) finally gave coproporphyrin-III tetramethyl ester (20) in an overall yield of 37% from protoporphyrin-IX dimethyl ester (21).

We have recently made more use of the mercuration and Friedel-Crafts acetylation reactions discussed earlier. Thus, if copper(II) deuteroporphyrin-IX dimethyl ester (26) is treated very briefly (for ca. 20 seconds) with acetic anhydride and stannic chloride at 0°C, a mixture consisting substantially of the two monoacetyldeuteroporphyrins (27) and (28) is produced. These isomers can be separated chromatographically on

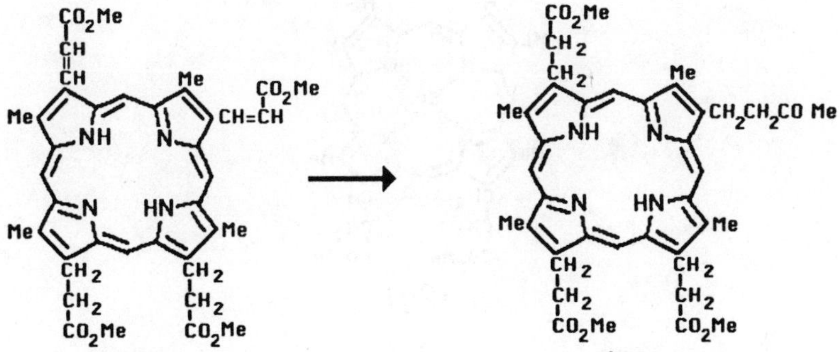

(19) (20)

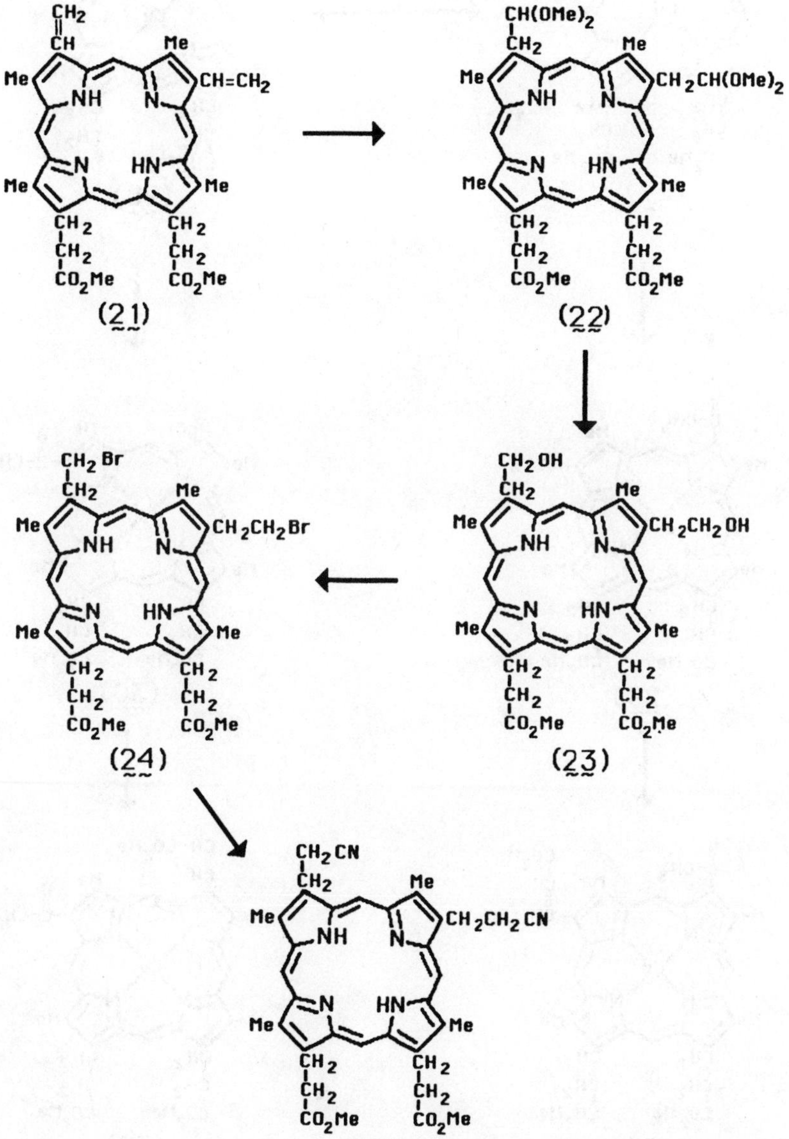

(21) (22) (23) (24) (25)

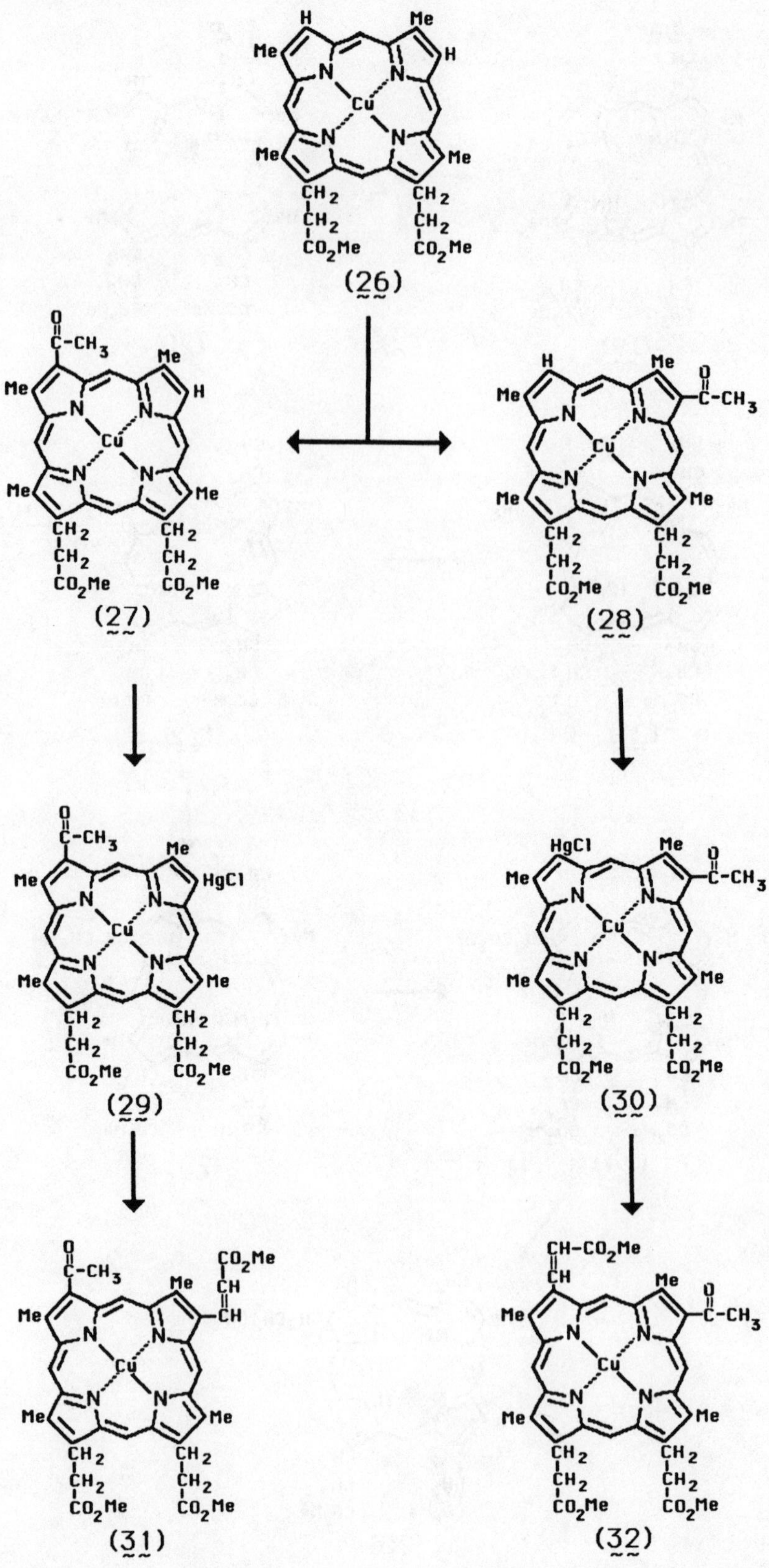
(26)
(27)
(28)
(29)
(30)
(31)
(32)

the 5 gram scale[11] and the individual isomers can be mercurated to give (29) and (30), respectively. The palladium/acrylate reaction discussed earlier leads to formation of the two isomeric monoacrylates, (31) and (32), which can be demetalated and hydrogenated (to give 33 and 34) and then reduced and dehydrated under the usual conditions to give harderoporphyrin trimethyl ester (35) and isoharderooporphyrin trimethyl ester (36). The porphyrinogens derived from the hydrolysis products of these two porphyrins were used[12] to show that in normal porphyrin metabolism between coproporphyrinogen-III and protoporphyrinogen-IX, the propionic acid group at the 2-position is modified to vinyl before the 4-propionate is touched. The hydrolyzed intermediates [(37), (38)] in the reduction of compounds (33) and (34) have 1-hydroxyethyl groups, and these compounds are a good source of porphyrins which have three carboxylic acid groups and only one hydroxyethyl, which would simplify the mixture of isomers obtained in the preparation of HpD from Hp-like compounds.

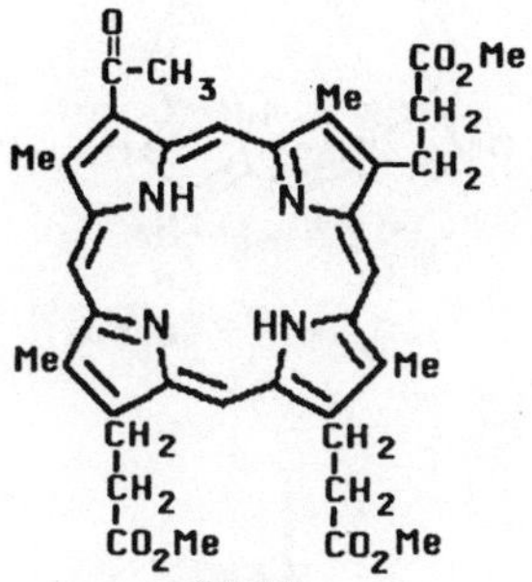

(33)

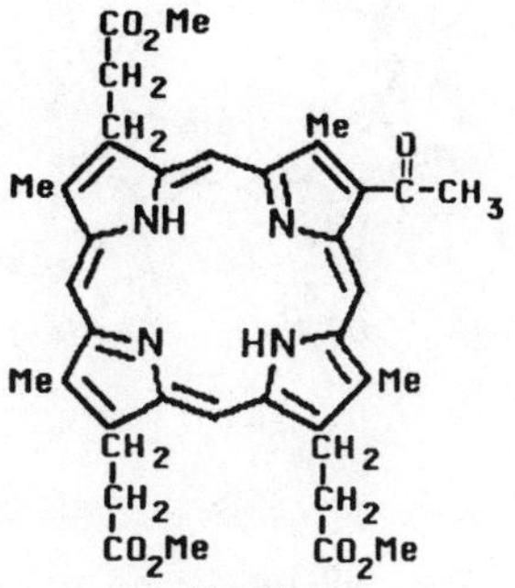

(34)

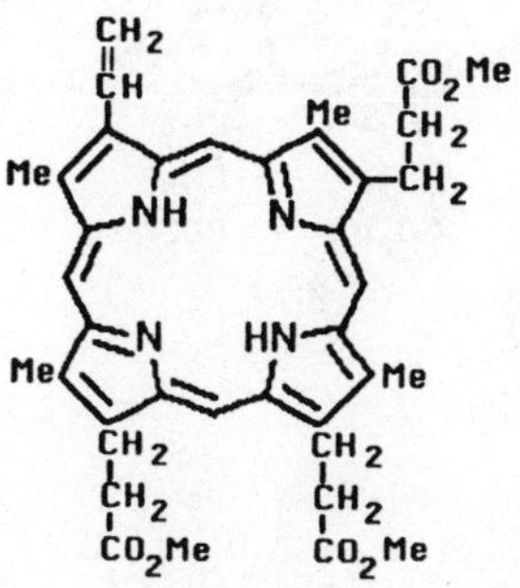

(35)

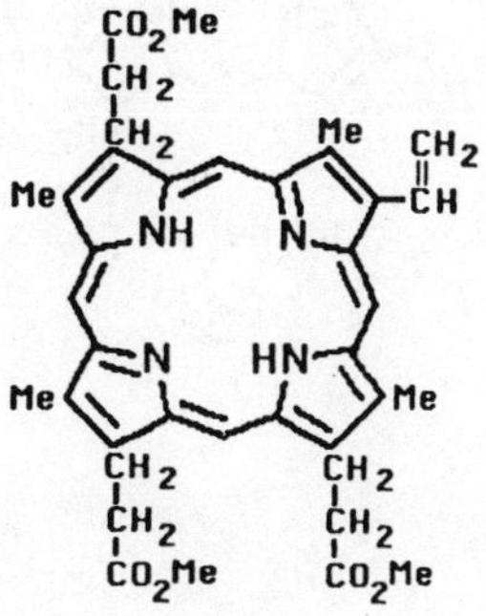

(36)

(37)

(38)

(33) → (39)

(39) → (40)

(40) → (41)

(41)

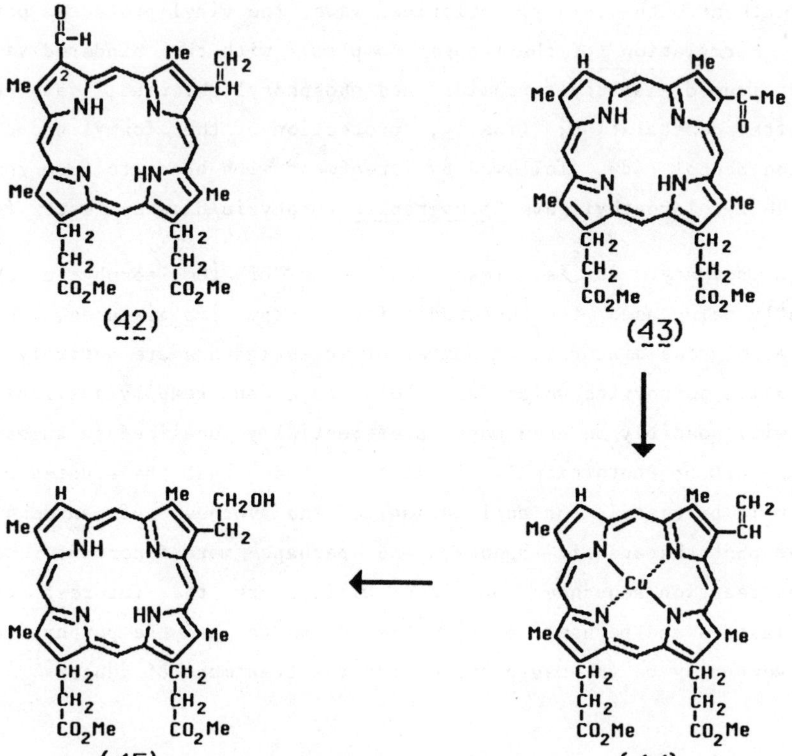

(42)   (43)

(45)   (44)

If the copper(II) complex of compound (33) is deacetylated using ethanedithiol and boron trifluoride etherate,[13] the tripropionic porphyrin (39) is obtained in excellent yield. Re-mercuration gives (40) and treatment with LiPdCl$_3$ in acetonitrile affords the tetramethyl ester (41) of S-411 porphyrin, a compound which can be isolated naturally from meconium.

The iron complex of _Spirographis_ porphyrin (42) has been isolated from the marine worm _Spirographis spallanzanii_, and the compound is interesting in that it has the vinyl group at position 2 in protoporphyrin-IX replaced with a formyl substituent; carbonyl substituents on the porphyrin periphery tend to give a long wavelength shift of all of the absorption maxima which may be useful in shifting the wavelength of phototherapy to enable deeper penetration of the radiation. We have synthesized this compound by treatment of the monoacetyldeuteroporphyrin (43) with sodium borohydride, followed by dehydration of the "mono-hemato" intermediate to give pemptoporphyrin dimethyl ester (44). Using the thallium(III) trifluoroacetate route discussed above, the vinyl in pemptoporphyrin was transformed into 2-hydroxyethyl (i.e. compound 45), and treatment with thionyl chloride gave the vinyl-protected porphyrin (46). Formylation (of the copper complex) with the hindered Vilsmeier reagent from di-isobutylformamide and phosphoryl chloride, gave compound (47) after demetalation; finally, protection of the formyl group as the ethylene acetal (48), followed by treatment with base to form the vinyl from the 2-chloroethyl gave Spirographis porphyrin dimethyl ester (49).

In summary, it is clear that most of the porphyrin compounds currently being used for photoradiation therapy are mixtures, and sometimes even gross mixtures. A number of investigators are actively seeking alternative porphyrins which are both pure and readily available, and which will possibly be even more preferentially localized in tumor tissue than are HpD or Photofrin II. It is to be hoped that the studies outlined above will be useful for application to the synthesis of radiochemically labeled phototherapeutic agents, and perhaps more important, that the various reaction sequences outlined will spark the interest of other investigators and perhaps provide the drive to prepare porphyrins which will eventually be the magic bullet for the treatment of cancers.

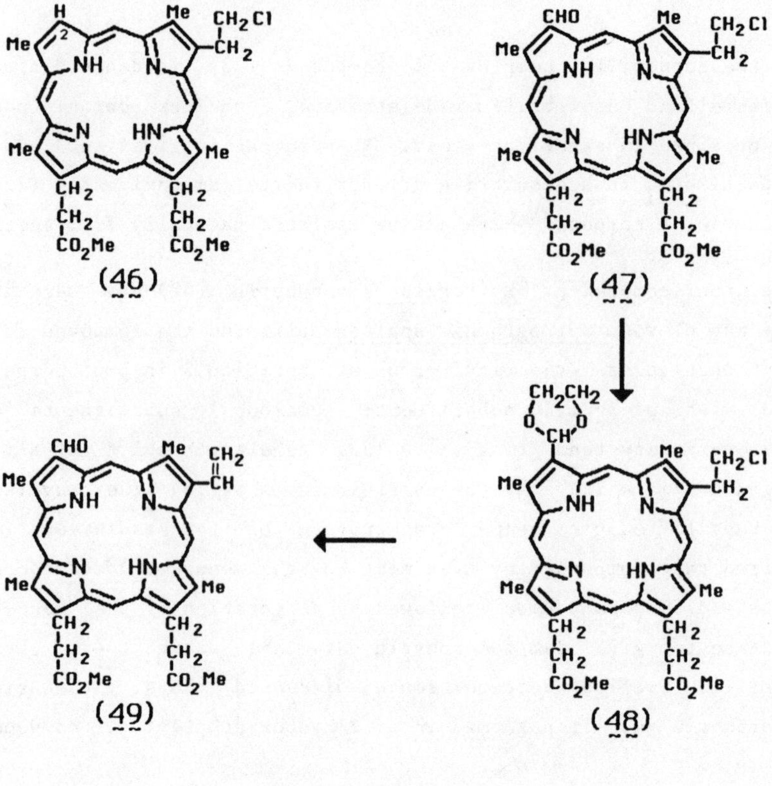

(46)

(47)

(49)

(48)

Acknowledgments. I would like to take this opportunity to thank my coworkers on the various projects described in this report for their enthusiasm, talent, and dedication; their names appear in the various literature citations. This work was supported by grants from the National Institutes of Health (HL 22252) and the National Science Foundation (CHE 81 20891).

REFERENCES

1. K. M. Smith, Protoporphyrin-IX: Some Recent Research, _Accounts Chem. Res._, 12: 374 (1979). K. M. Smith, Synthetic and Spectroscopic Studies of Hemes and Heme Proteins, _Rev. Port. Quim._, 25: 138 (1983).

2. J. -H. Fuhrhop and K. M. Smith, Laboratory Methods, _in_ "Porphyrins and Metalloporphyrins", K. M. Smith, ed., Elsevier, Amsterdam, (1975), p. 800.

3. H. Fischer and H. Orth, "Die Chemie des Pyrrols", Akademische Verlag., Leipzig, Vol. II part 1, (1937), p. 390.

4. J. L. W. Thudichum, _Report Med. Off. Privy Council,_ 10: 152 (1867).

5. K. M. Smith, L. A. Kehres and H. D. Tabba, Structures of the Bacteriochlorophyll-c Homologues: Solution to a Longstanding Problem, _J. Am. Chem. Soc.,_ 102: 714 (1980). K. M. Smith, D. A. Goff, J. Fajer and K. M. Barkigia, Chirality and Structures of the Bacteriochlorophylls d. _J. Am. Chem. Soc.,_ 104: 4337 (1982).

6. E.g. G. W. Kenner, J. M. E. Quirke and K. M. Smith, Transformations of Protoporphyrin-IX into Harderoporphyrin, Pemptoporphyrin, Chlorocruoroporphyrin and Their Isomers, _Tetrahedron,_ 32: 2753 (1976).

7. K. M. Smith, E. M. Fujinari, K. C. Langry, D. W. Parish and H. D. Tabba, Manipulation of Vinyl Groups in Protoporphyrin-IX: Introduction of Deuterium and Carbon-13 Labels for Spectroscopic Studies, _J. Am. Chem. Soc.,_ 105: 6638 (1983).

8. B. Evans, K. M. Smith, G. N. La Mar and D. B. Viscio, Regioselective Base-Catalyzed Exchange of Ring Methyl Protons in Protoporphyrin-IX: A New Facet of Porphyrin Chemistry, _J. Am. Chem. Soc.,_ 99: 7070 (1977).

9. K. M. Smith and K. C. Langry, Electrophilic Mercuration Reactions of Derivatives of Deuteroporphyrin-IX: New Syntheses of Coproporphyrin-III, Harderoporphyrin, Isoharderoporphyrin and S-411 Porphyrin, J. Org. Chem., 48: 500 (1983).

10. G. W. Kenner, S. W. McCombie and K. M. Smith, Protection of Porphyrin Vinyl Groups. A Synthesis of Coproporphyrin-III from Protoporphyrin-IX, Justus Liebigs Ann. Chem., 1329 (1973).

11. D. W. Parish, Ph.D. Dissertation, University of California, Davis, 1985.

12. J. A. S. Cavaleiro, G. W. Kenner and K. M. Smith, Biosynthesis of Protoporphyrin-IX from Coproporphyrinogen-III, J. Chem. Soc., Perkin Trans. I, 1188 (1974).

13. K. M. Smith and K. C. Langry, Protiodeacetylation of Porphyrins and Pyrroles: A New Partial Synthesis of Dehydrocoproporphyrin (S-411 Porphyrin), J. Chem. Soc., Chem. Commun., 283 (1981).

AGGREGATION EFFECTS ON THE PHOTOPHYSICAL PROPERTIES OF PORPHYRINS IN RELATION TO MECHANISMS INVOLVED IN PHOTODYNAMIC THERAPY

R.W. Redmond[1], E.J. Land[2] and T.G. Truscott[1]

[1] Department of Chemistry, Paisley College of Technology
High Street, Paisley, PA1 2BE, Scotland
[2] Paterson Laboratories, Christie Hospital and Holt Radium
Institute, Wilmslow Road, Manchester, M20 9BX, England

INTRODUCTION

The use of porphyrin preparations in conjunction with visible light in Photodynamic Therapy (PDT) is dependent on two important properties exhibited by these compounds, i.e. their preferential uptake and localisation in malignant tissue and the fact that such porphyrins are very efficient photosensitisers. By direct absorption of light they are electronically excited to higher energy levels from which they can, by energy transfer processes, donate this energy to acceptor molecules giving porphyrin deactivation and elevation of acceptor to excited states possibly not attainable by direct absorption under these conditions.

In addition porphyrins exhibit a tendency towards aggregation in aqueous systems (Brown et al., 1976; Andreoni et al., 1982), the extent of aggregation being largely dependent on the polarity of the tetrapyrrole ring substituents. This paper is concerned with the effects of porphyrin aggregation state on the photophysical properties of certain porphyrin species involved in PDT. Haematoporphyrin itself (HP), haematoporphyrin derivative (HPD), hydroxyethyl vinyl deuteroporphyrin (HVD) and the recently introduced di-haematoporphyrin ether (DHE) were all studied for reasons given in the text.

Many reports have been published proposing the involvement of activated singlet oxygen $^1O_2$, in the inactivation of cellular systems (see review by Kessel 1984). The $^1O_2$ is formed via energy transfer from the lowest excited triplet state of the porphyrin to the triplet ground state of molecular oxygen ($^3\Sigma_g^-$). $^1O_2$ has been indirectly implicated by the use of chemical quenchers such as diazo bicyclo octane (DABCO) (e.g. Torinuki et al., 1980) and dimethylisobenzofuran (e.g. Reddi et al., 1983), although such quenchers are not thought to be 100% specific for $^1O_2$. In this paper we report $^1O_2$ yields from the various porphyrin photosensitisers measured by direct detection of $^1O_2$ luminescence emission at 1270 nm thus giving unambiguous results. The effect of porphyrin aggregation state on $^1O_2$ yields is shown and in combination with photophysical data gives an insight into the mechanism of photodynamic inactivation of cellular systems.

MATERIALS AND METHODS

Deuterium oxide (D_2O), HP free base and Human Serum Albumin (HSA) (fraction V) were obtained from Sigma. Photofrin I (PFI) (a commercial preparation of HPD) and Photofrin II (PFII) (a non-polar fraction separated from PFI) were supplied by Oncology Research and Development. HVD was obtained from Porphyrin Products. A purified sample of DHE was kindly supplied by Dr. David Kessel. Cetyl trimethyl ammonium bromide (CTAB) and sodium dodecyl sulphate (SDS) were obtained from British Drug Houses and Triton X100 from Merck. All solvents used were of Analar spectroscopic grade.

The main technique used in this study was laser flash photolysis. Porphyrin samples were excited by the frequency doubled 347 nm emission from a J.K. Lasers System 2000 solid state ruby laser. A 250 watt pulsed Xenon lamp was used as the source of monitoring light through a 1 cm path length flow through sample cell at right angles to the direction of the laser beam. The transmitted light was wavelength selected by use of a monochromator and finally detected by a photomultiplier tube with the current passed through a 50 Ω load resistor and resultant voltage displayed on a storage oscilloscope. Changes in transmission of light through the sample after laser irradiation are thus detected and recorded as voltage against time.

Difference extinction co-efficients, $\Delta\varepsilon_T$, were obtained by the complete conversion method which consists of measuring the change in optical density (ΔOD) with increase in laser intensity until no further increase in ΔOD is observed. At this point all molecules are assumed to be converted into transient species and $\Delta\varepsilon_T$ is simply given by $\Delta OD^{max} = \Delta\varepsilon_T \times c \times \ell$ where ℓ = path length and c is the concentration of transients produced which, of course, is equal to the ground state concentration under these conditions.

Quantum yields of triplet formation, Φ_T, were measured using the comparative technique (Amand and Bensasson, 1975). This involves comparison of the concentration of triplet species produced by a compound with unknown Φ_T to that formed by a standard molecule with accurately known Φ_T and $\Delta\varepsilon_T$

values, under conditions where the number of photons incident on the solutions are identical. Then Φ_T is given by

$$\Phi_T^U = \Phi_T^S \; \frac{\Delta OD^U}{\Delta OD^S} \times \frac{\Delta\varepsilon_T^S}{\Delta\varepsilon_T^U} \times \frac{OD_{G.S.}^S \; (347 \text{ nm})}{OD_{G.S.}^U \; (347 \text{ nm})} \qquad\qquad \begin{array}{l} S = \text{standard} \\[4pt] U = \text{unknown} \end{array}$$

$OD_{G.S.}$ (347 nm) is measured for both standard and unknown and included to correct for unequal photon absorption by both solutions at the laser wavelength.

Relative yields of $^1O_2$ were measured by direct observation of the phosphorescence emission (1270 nm) from $^1O_2$ following irradiation of oxygen saturated D_2O solutions of various porphyrins by the frequency tripled 355 nm output from a J.K. Lasers System 2000 Nd/YAG laser with Nd/glass amplifier. Samples were contained in a $1 \times 1 \times 4$ cm quartz cell in an aluminium holder. $^1O_2$ luminescence was detected at 90° to the laser beam with an I.R. Germanium diode close to the sample cell. The diode current was passed through a variable load resistor and the resultant voltage applied to a Tektronix 7A22 differential amplifier in a Tektronix 7912D digitiser connected to a Commodore Pet 2001 computer system with signal averaging capability. Initial $^1O_2$ emission intensities were taken as a measure of $^1O_2$ yield. Ground state OD values of 1.0 at 355 nm were used for all samples to be irradiated.

RESULTS AND DISCUSSION

The porphyrin samples chosen for this study were considered relevant to possible mechanisms involved in PDT. HPD in the commercial form PFI was studied. As this is a mixture of various porphyrins it was decided to study HP itself and also HVD which are known to be major components present in HPD (Kessel and Chou, 1983; Bonnett et al., 1981). Finally DHE in the form of PFII, and also as isolated by Dr. Kessel, thought to be the active component of HPD in PDT (Dougherty et al., 1984) was also included in this work.

Aggregation is evident for all these samples in aqueous solution as shown by the nature of the absorption spectrum in each case. For HP, HVD and to a lesser extent PFI and PFII addition of detergent such as CTAB or using organic solvent such as methanol, produces a bathochromic shift of λ_{max} value of the Soret peak from 370-380 nm to 400 nm with concomitant sharpening and intensification of the band. This behaviour is well known but it is also particularly interesting to study aggregation effects on the photophysical properties of the species involved.

Detergent micelles and organic solvents are known to produce complete monomerisation of HP. It is instructive to look at the relatively simple situations of HP and HVD in a variety of environments where the extent of aggregation will differ. Table 1 gives Φ_T and $\Delta\varepsilon_T$ values obtained for both HP and HVD in various environments.

For both HP and HVD the effect of a disaggregating environment is that of increasing $\Delta\varepsilon_T$ values, i.e. 17250 $M^{-1}cm^{-1}$ for HP in acetone compared to 8300 $M^{-1}cm^{-1}$ in aqueous P.B.S. Micellar systems are produced when detergent is present above its critical micelle concentration (C.M.C.) and disaggregate porphyrins by solubilisation of porphyrin molecules within the micelle structure as the hydrocarbon chains provide a hydrophobic environment for the non-polar regions of porphyrins. HVD itself lies structurally between HP and the very hydrophobic protoporphyrin (PP) and will be intermediate in polarity and be significantly more aggregated in aqueous systems than HP. This is shown by $\Delta\varepsilon_T$ values of 3300 $M^{-1}cm^{-1}$ for HVD in P.B.S. compared to 8300 $M^{-1}cm^{-1}$ for HP in the same environment. This variation of $\Delta\varepsilon_T$ with environment may serve as a guide to the extent of aggregation of porphyrins in solution (Craw et al., 1984) and suggests that only disaggregated species are capable of producing triplet states, an observation which has to be taken into consideration when postulating mechanisms of photodynamic damage.

The most notable difference between HP and HVD lies in the significantly lower Φ_T values obtained for HVD than HP in all environments studied. On photophysical grounds alone HP would then be expected to have a more efficient role at lower cellular concentrations than that for the same effect produced by HVD.

Figure 1 shows, for comparison, the triplet difference spectra for HP

Table 1. Photophysical Properties of HP and HVD in various Environments

Environment	HP		HVD	
	$\Delta\varepsilon_T^{max}$(nm)/$M^{-1}cm^{-1}$	Φ_T	$\Delta\varepsilon_T^{max}$(nm)/$M^{-1}cm^{-1}$	Φ_T
PBS	8,300 (420)	0.94	3,300 (420)	0.34
7×10^{-2}M SDS	10,200 (440)	0.74	10,500 (440)	0.44
1.5×10^{-2}M CTAB	14,100 (440)	0.90	13,600 (440)	0.63
3.3×10^{-2}M TX100	14,600 (440)	0.90	13,250 (440)	0.70
9:1 MeOH H_2O	15,100 (440)	0.80	13,200 (440)	0.69
9:1 Acetone H_2O	17,250 (440)	1.00	13,200 (440)	0.53

Concentration of HP and HVD in each case was either 8 or 10 μm.

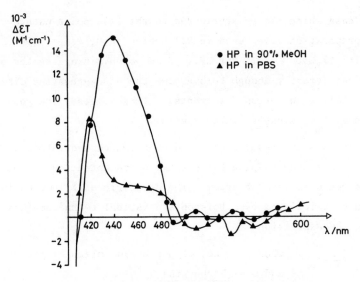

Fig. 1. Triplet Difference Spectra for
HP in PBS and 9:1 Methanol:H_2O

in P.B.S. and in 9:1 Methanol:H_2O. HVD shows essentially similar spectra
to HP in all systems. The λ_{max} value shifts from 420 nm to 440 nm on going
from aggregated to monomeric form with a sharpening and intensification of
the band; a behaviour which reflects changes in the ground state absorption
spectra.

PFI (HPD) is known to contain a larger proportion of highly aggregated
porphyrin species than HP which has been proposed to account for its
superior tumour localisation efficiency (Moan et al., 1982). Again
absorption spectra show shifts produced by disaggregating systems as
described except that a significant shoulder remains on the Soret peak
around 370 nm indicating that not all porphyrin is in monomeric form.

Table 2 summarises the laser flash photolysis data for PFI in the same
systems as HP and HVD. $\Delta\varepsilon_T$ values are much lower than those exhibited by

Table 2. Photophysical Properties of Photofrin I in various Environments

Environment	$\Delta\varepsilon_T^{max}$ (nm)/$M^{-1}cm^{-1}$	Φ_T
PBS	2300	0.74
$7\times10^{-2}M$ SDS	4430	0.79
$1.5\times10^{-2}M$ CTAB	5400	0.84
$3.3\times10^{-2}M$ TX100	5900	0.80
9:1 MeOH:H_2O	6000	0.72
9:1 Acetone:H_2O	7000	0.61

HP in each case which can be attributed to the less polar nature of the porphyrin preparation, i.e. even in 9:1 acetone:H_2O, $\Delta\varepsilon_T$ = 7,000 $M^{-1}cm^{-1}$ compared with 17,250 $M^{-1}cm^{-1}$ for HP. Some porphyrin aggregates must be bound together strongly enough to overcome the disaggregating effect of detergent micelles and organic solvents. This suggests the possibility of the existence of covalently linked species.

Table 3 shows the results obtained when the concentration of TX100 micelles was increased keeping the porphyrin concentration constant. $\Delta\varepsilon_T$ values were measured in each case. The porphyrin distribution in such systems can be calculated by a Poisson statistical technique (Chauvet et al., 1981) using the expression

$$P^k = \frac{Q^k e^{-Q}}{k!}$$ where k = no. of porphyrin molecules per micelle

and $Q = \dfrac{[Porphyrin]}{[Micelles]}$

Thus the percentage of micelles which are vacant, singly, doubly or multiply occupied may be calculated as shown in Table 3.

The calculations predict an increase in monomeric porphyrin as the concentration of TX100 is increased. The trend of increase in $\Delta\varepsilon_T$ up to a concentration of $1.65\times10^{-2}M$ TX100 follows predictions. However further increase in TX100 concentration has no effect on $\Delta\varepsilon_T$ values suggesting that the extent of porphyrin aggregation is unchanged, i.e. some strongly-bound aggregated component remains and is not monomerised by the solubilisation effect of the micelles.

Laser flash photolysis studies show negligible triplet formation by both DHE and PFII in aqueous solution. Introduction of detergent gives rise to a production of triplet species but $\Delta\varepsilon_T$ values as obtained by complete conversion are low, e.g. 1600 $M^{-1}cm^{-1}$ and 2100 $M^{-1}cm^{-1}$ for DHE in 2% Triton X100 and $10^{-2}M$ CTAB solutions respectively. This confirms the

Table 3. Effect of [TX100] on Porphyrin Distribution and $\Delta\varepsilon_T$

Environment	P^0	P^1	P^2	$P^{3-\infty}$	$\Delta\varepsilon_T^{max}(M^{-1}cm^{-1})$
PBS	–	–	–		2300
$0.33\times10^{-2}M$ TX100	44.81	35.97	14.44	4.78	4300
$0.82\times10^{-2}M$ TX100	74.01	22.28	3.35	0.36	5400
$1.65\times10^{-2}M$ TX100	86.29	12.72	0.94	0.05	5900
$3.3\times10^{-2}M$ TX100	92.96	6.78	0.25	0.01	5900
$8.25\times10^{-2}M$ TX100	97.14	2.82	0.04	–	5900

existence of a large proportion of highly aggregated species in these
preparations.

One further experiment confirms the differences in aggregation and
strength of aggregate binding between HP, HPD and DHE.   When a porphyrin
preparation is injected into the patient it is thought to be carried through
the bloodstream to the tumour site bound to plasma serum proteins (Jori et
al., 1984).   It is important not only to study the clinical preparation
itself but also any changes produced by serum binding.   By keeping the
porphyrin concentration fixed and increasing the concentration of the serum
protein HSA we can study aggregation by observing the changes in initial
triplet optical density at the monomer maximum of 440 nm (as $\Delta OD \propto \Delta\varepsilon_T$)
with variation in [HSA]/[Porphyrin] ratio.   Figure 2 shows the corresponding
plots for HP, PFI and PFII in PBS as the porphyrin preparation.

ΔOD increases dramatically for HP with increase in [HSA]/[HP] ratio to
reach a maximum at a 1:1 ratio, above which there is no further increase in
ΔOD.   All HP molecules were then in monomeric form with HSA exhibiting one
site for porphyrin binding as has been determined elsewhere by other methods
(Reddi et al., 1981).   PFI shows limited disaggregation again reaching a
maximum at a 1:1 ratio but ΔOD^{max} is far lower than that achieved by HP.
PFII however shows virtually no triplet formation either in the presence or
absence of HSA.   'In vivo' we would then expect HP molecules to be in

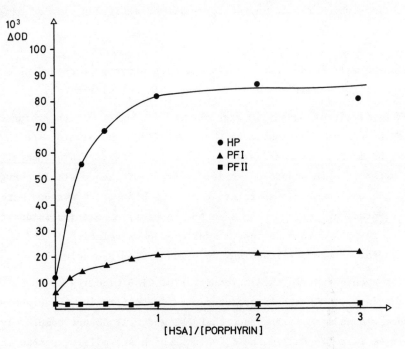

Fig. 2. Effect of [HSA] on Porphyrin Aggregation

Table 4. Singlet Oxygen Data for various Porphyrins

Porphyrin	$[CTAB]/mol \ \ell^{-1}$	Φ_T	Relative $\Phi\Delta$	Relative SΔ
HP	-	0.94	1.00	1.00
HP	0.01	0.90	2.27	2.43
HVD	-	0.34	0.43	1.19
HVD	0.01	0.63	1.51	2.25
PFI	-	0.74	0.36	0.46
PFI	0.01	0.84	1.63	1.87
DHE	-	-	0.11	-
DHE	0.01	-	1.51	-

monomeric form on reaching the tumour site whereas PFI may have some monomer but significant amounts of bound aggregated species and in the case of PFII all will be bound in aggregated form.

Knowing the state of the porphyrin preparation on entering the tumour cells we can go on to look at mechanisms of photodamage. Singlet oxygen is produced by energy transfer from the porphyrin triplet state. The actual yield of singlet oxygen, Φ_Δ, is dependent both on the quantum yield of sensitiser triplet formation, Φ_T, and on the fraction of oxygen quenching interactions which result in the formation of activated singlet oxygen, S_Δ

$$\Phi_\Delta = S_\Delta \cdot \Phi_T$$

Photodamage will be due then to sensitiser photoproperties and also to the location of the sensitiser near to vulnerable cellular sites.

The effects of porphyrin aggregation on singlet oxygen production were studied by the same rationale as for laser flash photolysis with the porphyrins in systems where aggregation will differ. Unfortunately, there is no water-soluble standard sensitiser available hence singlet oxygen yields can only be taken as relative to each other. Φ_Δ values were obtained from the gradients of the slopes on the respective emission intensity vs. laser energy plots in each case. Table 4 shows relative Φ_Δ and S_Δ values obtained for HP, HVD, PFI and DHE in D_2O and CTAB/D_2O solutions.

For each porphyrin the introduction of CTAB micelles produces an increase in Φ_Δ. HP itself produces most $^1O_2$ reflecting its higher amount of monomeric species. The fact that PFI and PFII do not completely monomerise is reflected in the Φ_Δ values which are all lower than that for

HP. DHE and PFI in aqueous systems show little $^1O_2$ production indicating only a very small proportion of species capable of forming triplet excited states. These results must be considered in view of possible mechanisms of photodamage.

PFI and PFII are used in cancer phototherapy rather than HP due to their increased tumour localising ability. However it is shown here that in pre-injection form these preparations would be inefficient for a type II mechanism involving $^1O_2$. HP and HVD are taken up by tumour cells in addition to the non-polar, high molecular weight, porphyrin species. 'In vitro' studies (Moan and Sommer, 1983) have shown that the less hydrophobic monomer species begin to leak out from the cell relatively quickly eventually leaving only the non-polar hydrophobic species within the cell. Since clinical practice is to have a period of up to five days between drug injection and tumour irradiation (Dahlman et al., 1983) it would be expected that all monomeric species originally taken up by tumour cells would subsequently have leaked out. Aggregated species are shown to be ineffective in $^1O_2$ production in the form studied, thus some change in their nature must occur to render them photoactive 'in vivo'. DHE is proposed to have a unit structure of two molecules of Hp joined by an ether bridge through the hydroxyethyl substituent groups although many such links may be formed resulting in a covalently bound polymeric species. This change may be a slow disaggregation and/or polymer ether bond cleavage to give monomers of HP, HVD or DHE itself. Modification of the aggregated species on localisation is supported by the fact that tumours will exhibit a strong red fluorescence which persists long after injection when the initial monomer would be absent from the cell. Aggregated PFII has a fluorescence yield considerably smaller than HP itself, the small fluorescence probably being due to the small amount of monomeric species in the sample. Thus some monomeric species may gradually be produced from the localised aggregate.

We conclude that only monomeric species are likely to be photoactive, i.e. produce $^1O_2$ and evidence suggests that some modification of the localised non-polar aggregated species must occur to facilitate the photodynamic effect of porphyrin preparations in conjunction with visible light to give the desired tumour destruction.

ACKNOWLEDGEMENTS

We thank Dr. D. Kessel for the sample of DHE. Also we thank SERC, CRC and MRC for financial support and RWR acknowledges an SERC CASE studentship.

REFERENCES

Amand, R. and Bensasson, R.V., 1975, Determination of Triplet Quantum Yields by Laser Flash Absorption Spectroscopy, Chem. Phys. Lett., 34:44.

Andreoni, A., Cubeddu, R., De Silvestri, S., Laporta, P., Jori, G. and Reddi, E., 1982, Hematoporphyrin Derivative: Experimental Evidence for Aggregated Species, Chem. Phys. Lett., 88:33.

Bonnett, R., Ridge, R.J., Scourides, P.A. and Berenbaum, M.C., 1981, On the Nature of Haematoporphyrin Derivative, J. Chem. Soc., Perkin I, 3135.

Brown, S.B., Shillcock, M. and Jones, P., 1976, Equilibrium and Kinetic Studies of the Aggregation of Porphyrins in Aqueous Solution, Biochem J., 153:279.

Chauvet, J.P., Viovy, R., Santus, R. and Land, E.J., 1981, One-Electron Oxidation of Photosynthetic Pigments in Micelles. Bacteriochlorophyll a, Chlorophyll a, Chlorophyll b and Pheophytin a, J. Phys. Chem., 85:3449.

Craw, M., Redmond, R. and Truscott, T.G., 1984, Laser Flash Photolysis of Haematoporphyrins in some Homogeneous and Heterogeneous Environments, J. Chem. Soc., Faraday Trans. I, 80:2293.

Dahlman, A., Wile, A.G. Burns, R.G., Mason, G.R. Johnson, F.M. and Berns, M.W., 1983, Laser Photoradiation Therapy of Cancer, Cancer Res., 43:430.

Dougherty, T.J., Potter, W.R. and Weishaupt, K.R., 1983, The Structure of the Active Component of Hematoporphyrin Derivative, in: "Porphyrins in Tumour Phototherapy", A. Andreoni and R. Cubeddu, eds., Plenum Press, 23.

Jori, G., Beltramini, M., Reddi, E., Salvato, B., Pagnan, A., Ziron, L., Tomio, L. and Tsanov, T., 1984, Evidence for a Major Role of Plasma Lipoproteins as Hematoporphyrin Carriers in Vivo, Cancer Letters, 24:291.

Kessel, D. and Chou, T.-H., 1983, Tumor-Localizing Components of the Porphyrin Preparation Hematoporphyrin Derivative, Cancer Res., 43:1994.

Kessel, D., 1984, Hematoporphyrin and HPD: Photophysics, Photochemistry and Phototherapy, Photochem. Photobiol., 39:851.

Moan, J., Christensen, T. and Sommer, S., 1982, The Main Photosensitizing Components of Hematoporphyrin Derivative, Cancer Letters, 15:161.

Moan, J. and Sommer, S., 1983, Uptake of the Components of Hematoporphyrin Derivative by Cells and Tumours, Cancer Letters, 21:167.

Reddi, E., Ricchelli, F. and Jori, G., 1981, Interaction of Human Serum Albumin with Hematoporphyrin and its Zn^{2+} and Fe^{3+} Derivatives, Int. J. Peptide Protein Res., 18:402.

Reddi, E., Jori, G., Rodgers, M.A.J. and Spikes, J., 1983, Flash Photolysis Studies of Hemato- and Copro-Porphyrins in Homogeneous and Microheterogeneous Aqueous Dispersions, Photochem. Photobiol., 38:639.

Torinuki, W., Miura, T. and Seiji, M., 1980, Lysosome Destruction and Lipoperoxide Formation Due to Active Oxygen Generated from Hematoporphyrin and U.V. Irradiation. Br. J. Dermatol., 102:17.

STUDIES ON THE SOLVOPHOBIC NATURE OF PORPHYRIN SELF AGGREGATION VS. THE

LIPOPHYLIC NATURE OF PORPHYRIN-MEMBRANE BINDING

Rimona Margalit, Smadar Cohen and Michal Rotenberg

Department of Biochemistry, The George S. Wise Faculty of
Life Sciences, Tel-Aviv University, Tel-Aviv 69978, Israel

INTRODUCTION

Fractionation of hematoporphyrin derivative by chromatography has been
performed by several methods such as TLC, HPLC and gel exclusion (Dougherty,
1983; Dougherty et al., 1983; Grossweiner et al., 1983; Kessel, 1983; Evensen et al., 1984). The pattern, number and nature of the fractions depend
obviously on the method of separation. Yet, common to all methods, not all
fractions retain the ability of the drug to act as a photodynamic agent for
tumor treatment and there are also indications that the active and nonactive
fractions differ in terms of aggregation (Dougherty et al., 1983; Kessel,
1984; Evensen et al., 1984).

To gain more understanding on the linkage between the types of hematoporphyrin derivative fractions and their aggregation properties, we have
probed the forces driving porphyrins to aggregate in aqueous phases. Considering the molecular factors which might be involved, favorable contribution
to self-aggregation could result from $\pi - \pi$ interactions between the aromatic moieties of the porphyrins. However, such interactions could not
account alone for the highly favorable aqueous aggregation and could occur
independent of the solvent system (Abraham et al., 1966; Abraham et al.,
1976; Janson and Katz, 1972). Thus, in assessing the forces driving porphyrin aggregation in aqueous phases, the solvent involvement was taken into
consideration. To that end we have studied porphyrins at three states -
monomers, dimers and higher-order aggregates, focusing on fractionation of
fresh vs. aged hematoporphyrin derivative and DP systems by gel-exclusion
chromatography, dimerization equilibrium and its temperature dependence
and on the binding of porphyrin monomers to membranes. The hematoporphyrin
derivative we investigated was from two sources. To differentiate between
the drug in general and the specific products, we will refer to the former
by the full name and to the latter by the abbreviations, HPD for one product, Pf for the other.

EXPERIMENTAL

Materials

DP, MP and HPD were purchased from Porphyrin Products, Logan (Utah,
U.S.A.). HP and PP were purchased from Sigma Chemical Co. Pf (in saline

solution) was purchased from Photofrin Med. Inc. Fluorescence spectra were
recorded on a Perkin-Elmer model MPF-44B fluorescence spectrophotometer.

Methods

1. Solution preparation: Porphyrin stocks of mM range were prepared
by dissolving porphyrin in 0.1 N NaOH bringing the solution to the final
volume and pH (7.2) by phosphate-buffered saline typical ratios being
porphyrin:base:PBS 1:125:875 (mg,ul,ul). Phosphate-buffered saline contained 3.3mM KH_2PO_4, 3.3mM Na_2HPO_4, 143mM NaCl and 4.7mM KCl, adjusted to
pH 7.2.

II. Gel-exclusion chromatography of fresh and aged porphyrins: Column
chromatography was performed essentially according to Dougherty (1983), on
BioGel P-10 (25x1 cm). 0.1 ml samples of concentrated porphyrin solution in
PBS (5mg/ml) were loaded on the column and eluted with water at pH = 8.
Chromatography of fresh porphyrin systems was immediately after solution
preparation. Chromatography of aged porphyrin systems was performed on
samples from the same initial stocks as the fresh ones, which have been
kept in the dark at 25°C for 24 hours. The absorbance of the eluante was
read at 370 nm and at 390 nm.

III. Determination of porphyrin dimerization constant (equilibrium);

IIIa. Experimental design: The increase of fluorescence intensity with the
increase in total porphyrin concentration, in an aqueous medium (PBS) was
followed, over the concentration ranges of .01 - .20 uM for MP and PP, .05-
1.5 uM for Dp, HP and HPD.

IIIb. Data processing: For the concentration region in which the fraction
of aggregates of higher order than dimers is negligible, the following relationship should hold (See Margalit et al., 1983 for further mathematical
details):

$$[T] /F = 2Kdk^2F + k \qquad (2)$$

Where [T] is the total porphyrin concentration in the system (expression
per mol monomer), F is the monomer fluorescence at a given wavelength
(usually peak emission was taken), k is the coefficient relating to fluorescence to monomer molar concentration (which can vary from one experiment
to the other) and Kd the dimerization equilibrium constant. Plotting [T]/F
vs. F should give a linear relationship and the magnitude of Kd can be obtained from the slope and intercept.

IIIc. Thermodynamic parameters of dimerization: The dimerization equilibrium constants were determined for a given porphyrin over a range of temperatures, the linear relationship of the logarithm of the dimerization constant (ln Kd) to the reciprocal of temperature was verified and the enthalpy
change obtained from the slope.

IV. Binding of porphyrins to liposomes:

IVa. Liposomes: Large unilamellar liposomes (LUV) were prepared by the
method of reverse phase evaporation (Szoka and Papahadjopoulos, 1980). The
liposome preparation was filtered through a polycarbonate filter with a
pore size of 0.6 um and only the filtrate was collected and used for the
binding experiments. Fiant unilamellar liposomes (GUV) were prepared according to the dialysis method of Oku et al. (1982). Quantitative determination
of the lipids was according to the methods of Yoshida et al. (1979) for PC
and Zlatkis and Zak (1968) for cholesterol.

IVb. Binding experiments: The experimental design was to have a constant porphyrin concentration (in the 10nM – 1uM range) in a series of reaction mixtures and vary the liposome concentration over the 0.01–2 mg/ml lipid. The reaction mixtures were incubated for short (2 hours) or long (24 hours) periods, at 37°C, in the dark. Separation was by ultracentrifugation for 1 hour at 225000g and 25°C, the pellet suspended in PBS and aliquots were taken into Triton X-100, to a final concentration of 0.6%. Quantitative determination of the liposome-bound' porphyrin was by fluorimetry of the samples in Triton X-100, using suitable calibration curves.

IVc. Data processing: The data were processed according to a model assuming that only porphyrin monomers bind to the membrane, the dimers participating indirectly, through dissociation into monomers, each porphyrin monomer binding no more than one liposome. The equilibrium binding constant, for this model, is given by (see Margalit and Cohen, 1983 for further mathematical details):

$$K_{BM} = \left([T_b]/[L] \right) \left(\frac{-1+\sqrt{1+8\ K_D([T]-[T_b])}}{4K_D} \right)^{-1} \qquad (2)$$

Where K_{BM} is the monomer-liposome equilibrium binding constant, $[L]$ is the lipid concentration, $[T]$ and $[Tb]$ the total porphyrin in the reaction mixture and the total liposome-bound porphyrin, respectively, expressed per mol monomer.

RESULTS

1. <u>Gel-exclusion chromatography of fresh and aged aggregated porphyrin systems</u>

Chromatographing a concentrated, freshly-prepared, solution of DP on a BioGel P-10 column which separates according to size, eluting at low ionic strength, gives two major fractions. A typical elution profile is illustrated in figure 1, the fractions denoted A' and B'. Dissociation (into monomers) of samples from fractions A' & B' was attempted by dilution into buffer or by the addition of detergent. Such dissociation can be monitored by fluorescence (increase) or absorbance (a red shift in the soret soret peak) spectra. As shown by the data listed in Table 1, there is an increase in the red-shift with the increase in dilution into buffer, for samples from both fractions of DP. Furthermore, the red-shift increase is similar, close to 40% for a sixteen-fold increase in the factor of dilution. (Obviously, for such concentrated solutions this dilution is not sufficient to drive to complete monomerization). Subjecting a similar, but aged, DP solution to chromatography under the same conditions did not show any significant difference from the fresh system, neither in elution pattern nor in the effects of the type listed in Table 1. Thus, the fractionation of DP seems to give two dominant size-populations of the same type of aggregate, to which we will assign the working term type-II. We propose this type includes the "micelle-like" polymers and dimers previously reported for nonchromatographed solutions of DP (Brown et al., (1976; 1980).

Going through the same chromatographic procedure for freshly prepared solutions of HPD also results in two major fractions, denoted A and B in figure 1 (but notice the differences in the elution volumes compared to the corresponding DP fractions). Samples from the B fraction were found to be similar to both fractions of DP, as exemplified in Table 1, indicating this

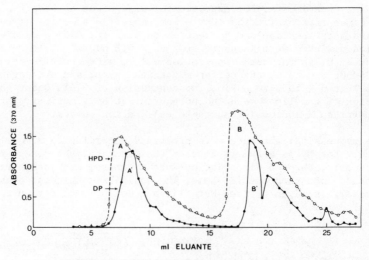

Figure 1: Elution profiles of DP and HPD on BioGel P-10 columns

fraction has a high share of type-II aggregates. Samples from the A fraction differed: Dissociation attempts by dilution or by detergent treatment were unsuccessful. As exemplified by the data in Table 1, there was no significant change in the red-shift with dilution. I. Another demonstration of the differences between the two HPD fractions can be seen in figure 2 - the extent of aggregation as measured by the ratio of absorbances at 370 to 390 (i.e. the reciprocal of the red-shift) is higher for the A fraction. A third difference between the HPD fractions is in the effect of time: We found the weight of fraction A to increase with time, in a slow process on the order of days, at the expense of fraction B. This was accompanied by a further increase in the extent of fraction A aggregation, as illustrated by the data in figure 2 (compare the fresh vs. aged HPD systems). Thus, the first fraction of HPD differs enough from the second one and from both DP fractions, to merit another assignment in terms of types of aggregates. We denoted the aggregates populating the first fraction of HPD "type-I". Similar chromatographies of Pf showed it to contain a type-I rich fraction A and a type-II rich fraction B also (see for example Table 1) and to undergo time-effects similar to those of HPD.

The time effects indicate a slow aggregation process is taking place, which we propose is associated with type-I aggregates rather than type-II. This proposition is supported by our finding that DP, which has only the latter type was not affected by time, and on the fast aggregation kinetics reported by Brown et al. (1976; 1980).

II. Temperature-dependence of porphyrin dimers

To obtain direct evidence for the forces driving porphyrin aggregation we determined the temperature dependence of the dimerization equilibrium for a series of porphyrin, the dimers being the smallest and the only well-defined porphyrin aggregates. As shown by the results listed in Table 2, for all porphyrin species studied the dimerization decreases with increasing temperature, indicating the involvement of a negative enthalpy change. A detailed temperature study conducted for two porphyrin species, from which the magnitudes of the thermodynamic parameters could be determined, lends further support: The data listed in Table 2 clearly show the dimerization to be driven by a highly favorable enthalpy change with an unfavorable but negligible contribution of the entropy change.

306

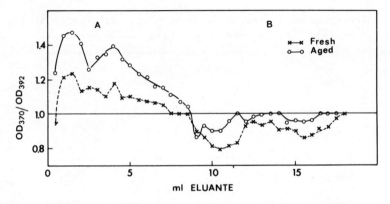

Figure 2: Elution profiles of fresh and aged HPD on BioGel P-10

III. The lipophylic nature of porphyrin monomers.

The ability of porphyrins to be extracted from aqueous into organic solutions undergoing monomerization in the course of this process (Kessel, 1977; Kessel and Rossi, 1982) is accepted as experimental evidence for the lipophylic nature of porphyrin monomers. The porphyrin solvation and mono-merization in detergent-containing solutions is another similar case. As we will show later in this report this monomerization can be interpreted differently, yet the lipophylic nature of porphyrin monomers can be clearly supported by other experimental data.

Direct support comes from studies of porphyrin binding to membranes. For example from our studies on the binding of porphyrins to unilamellar liposomes, for porphyrin ranges of concentration in which the fraction of aggregates larger than the dimer is negligible. We have previously ascert-tained that the dimers do not participate directly in the binding to the liposomes, only indirectly through their dissociation into the monomers (Margalit and Cohen, 1983; see also the "methods" section here). Thus, the binding constants obtained, listed in Table 3, are those of the respective monomer units. The data clearly show that the porphyrin monomers have affinity to the lipid, some of considerable magnitude. Both the type of liposome and the lipid composition seem to affect the magnitudes of the binding constants. However, these effects are minor and do not mask the overall sequence of affinity, DP > Pf ⪈ HPD > HP, which correlates with the order of increasing hydrophylic nature of the variant side-chain substit-uents (which are the only structural differences between these species): Hydroxy residues in HP; hydroxy, vinyl-hydroxy and vinyl-vinyl residues in HPD and Pf (which are reported to contain HP, HVD, DHE and PP); hydrogens only in DP.

DISCUSSION

The theory of the hydrophobic effect (Kauzmann 1959; Tanford 1973; Nemethy and Scheraga, 1961; Paterson et al., 1981) and that of the solvo-phobic force (Sinanoglu and Abdulnur, 1965; Sinanoglu, 1968, 1981) are two approaches (not mutually exclusive) accounting for the role of the solvent in driving self-aggregation of solute molecules.

Two experimental observations of our studies show the lipophylic

307

Table 1: The increase in the ratio of porphyrin absorbance at 390 nm to that at 370 nm with the increase in porphyrin dilution.

| Dilution ratio | OD 390 / OD 370 | | | | | |
| | DP | | HPD | | Pf | |
	A'	B'	A	B	A	B
1:2	.924	.951	.808	1.033	.989	1.052
1:4	1.008	.975	.770	.994	.807	1.132
1:8	1.085	1.006	.803	1.042	.791	1.260
1:16	1.185	1.113	.850	1.152	.806	1.386
1:32	1.257	1.355	.880	1.284	.753	1.536

(a) Cohen and Margalit, in preparation

nature of porphyrin monomers: The affinity of the monomers to the liposomal membranes which is high enough to compete with the self-aggregation. The ranking of the affinity among porphyrins. These observations indicate that the hydrophobic effect could be the dominant factor in aggregation. For such systems, the aggregation process is driven by a favorable entropy change with little or no role for the enthalpy change, aggregation increasing with increasing temperatures. Yet, at least for type-II aggregates, this is not the case. Brown et al. (1976; 1980) have shown the aggregation of the micelle-like polymers (in DP) to decrease with increasing temperatures, similar to our observations for the dimers (which are the smallest type-II aggregates) indicating the involvement of a negative enthalpy change. Moreover, the thermodynamic parameters we have determined for dimerization clearly show the process to be enthalpy driven. The magnitudes determined for the enthalpy change are large enough to account for the favorable free energy change, the entropy change having a small (but negligible) unfavorable contribution. These data fit well with the expectations of the solvophobic force theory (Sinanoglu, 1968, 1981), according to which aggregation should decrease with increasing temperatures.

For systems in which the solvophobic force is dominant in driving aggregation, the surface tension of the solvent is the major factor involved. Work against the surface tension, an input of enthalpy mainly, is done in increasing the surface area of the solvent in the course of dissolving solute molecules, forming a cavity for each molecule. Some of this investment in enthalpy is regained when molecules aggregate in the solvent, and the aggregation is accompanied by a decrease in the cavity size (aggregate vs. individual molecules), which is a decrease in the solvent surface area. Thus, the higher the surface tension of the solvent, the higher the potential gain in enthalpy, making the aggregation more favorable. Lowering the surface tension should, accordingly, reduce the energy gain in aggregation making it less favorable and at the limit - unfavorable (Sinanoglu, 1968). This expectation fits with the experimental observations: Dimerization is sufficiently favorable in aqueous phases, water having an exceptionally high surface tension, at extremely low concentrations (nM for PP to uM for HP). Yet, lowering the surface

Table 2: Dimerization equilibrium constants and thermodynamic parameters at neutral pH

Porphyrin	K_D (M^{-1}x10^{-5})		ΔG^o (Kc/mol)	ΔH^o (Kc/mol)	ΔS^o (e.u.)
	25°C	37°C			
HP[b]	4.0(±.4)	2.8(±.4)			
DP[1,b]	23(±8)	11(±4.5)	−8.7	−11	−7.7
HPD[b]	23(±7)	19(±9)			
MP[b]	54(±8)	17(±5)			
PP[a]	190(±40)				
ZnHP[2,c]	2.7(±.5)		−7.4	−9.3	−6.4

(1) Temperature-independence of enthalpy change verified for the range of 22°-37°C. Free-energy and entropy changes are given for 37°C.

(2) Temperature-independence of enthalpy change verified for the range of 19°-31°C. Free-energy and entropy changes are for 25°C.

(a) Margalit and Cohen, 1983. (b) Margalit and Rotenberg, 1984. (c) Cohen and Margalit, in preparation.

tension of pure water by the addition of detergents causes monomerization (Margalit and Cohen, 1983; Margalit and Rotenberg, 1984). Another demonstration of the surface tension effect is the observation that at these same concentrations there is no aggregation in organic solvents which have surface tension much lower than water (Brown et al., 1976; 1980; Margalit et al., 1983). Similar effects, showing the aggregation to become less favorable with the decrease in solvent surface tension (compared to water) were reported for aggregates of higher order than the dimers also (Brown et al., 1976; 1980; Kessel and Rossi, 1982). Thus, the stability of type-II aggregates in aqueous solutions, marked by a favorable enthalpy change, is due to the high surface tension of water as well as to structural features of the aggregate formed (i.e. the cavity size occupied by the aggregate vs. those of the monomeric units) rather than to the lipophylic nature of the monomers.

As to type-I aggregates, those are not defined enough, presently, to allow a thermodynamic study as has been done for the dimers of type-II. We and others have found type-I species (or HPD fractions rich in type-I) to be more stable than type-II. This is demonstrated by the observation that lowering the concentration by simple dilution is not sufficient to affect even partial dissociation (recall Table 1). Another is the finding that with time the weight of type-I aggregates and the extent of aggregation in the type-I rich HPD and PF fractions, increase at the expense of type-II. Thus, even though the growth of type-I species is seen to be slow, the system moves with time towards dominance of type-I (recall figure 2).

Table 3: Binding of porphyrins to liposomes - equilibrium constants, at neutral pH and 37°C.

Porphyrin	Liposome type	lipid	K_{BM} $(M^{-1}\text{x}10^{-3})$	Reference
HP	LUV	PC	1.8($\pm$.3)	(a)
HPD	LUV	PC	4.2($\pm$.8)	(a)
HPD [1]	LUV	PC		
f1			9	(b)
f2			1.5	(")
HPD [1]	LUV	PC/CH (1:1)		
f1			5	(b)
f2			.3	
Pf	LUV	PC	6.1($\pm$.9)	(c)
DP	LUV	PC/CH (3:2)	49($\pm$15)	(d)
DP	GUV	PC/CH (3:2)	24($\pm$6)	(d)
DP	GUV	PC	37($\pm$18)	(d)

(1) Incubation of reaction mixtures was for 16 hours. All other data are for incubation of 2 hours.

(a) Margalit and Cohen, 1983. (b) Cohen and Margalit, 1985. (c) Cohen and Margalit, in preparation. (d) Rotenberg and Margalit, in preparation.

Were hydrophobic effects the dominant factor in type-I aggregation, then its high stability would call for it to contain a substantial share of the more-hydrophobic porphyrins. The experimental data showing type-I rich HPD fractions to contain a dominant share of the more-hydrophylic porphyrins (Dougherty et al. 1983) is contrary to this expectation, leading us to propose that if the solvent has a dominant role in driving type-I aggregation, it would be in the direction of the solvophobic effect. Then, on the macroscopic level, if solvent is involved in both type-I and type-II aggregations, there would be no qualitative differences between these species, only quantitative. This would also lead to a conclusion on the microscopic level, predicting that the more-stable aggregate should have the more-compact structure affording a higher gain in enthalpy when the aggregate is formed from the monomeric units. If, on the other hand, the solvent is not dominant in driving type-I aggregation this in itself would be a qualitative difference between the two species, but would not give insight into the nature of the differences. To understand the latter we suggest the effort should be directed towards the microscopic level. This would include not only structural features but also the kinetics of type-I association and dissociation, which we are currently investigating.

We found type-I aggregates in hematoporphyrin derivative preparations from two sources, HPD and Pf, but not in a regular porphyrin species such as DP. This affirms the conclusion that type-I species are an integral component of the drug "hematoporphyrin derivative", independent of the source of preparation. Furthermore, it allows extending data and understanding on self-aggregation and membrane-interaction obtained from studies on HPD to Pf - the product for clinical use and investigation.

REFERENCES

Abraham, R.J., P.A. Burbridge, A.H. Jackson and D.B. Macdonald (1966) J. Chem. Soc. B, 620-626.

Abraham, R.J., F. Eivazi, H. Pearson and K.G. Smith (1976) J. Chem. Soc. Commun., 698-701.

Brown, S.B., H. Hatzikonstantinou and D.G. Herries (1980) Int. J. Biochem. 12, 701-707.

Brown, S.B., M. Schillcock and P. Jones (1976) Biochem. J. 153, 279-285.

Cohen, S. and R. Margalit (1985) Biochim. Biophys. Acta 813, 307-312.

Dougherty, T.J. (1983) Photochem. Photobiol. 38, 377-379.

Dougherty, T.J., D.G. Boyle, K.R. Weishaupt, B.A. Henderson, W.R. Potter, D.A. Bellnier and K.E. Wityk (1983) Porphyrin Photosensitization (Kessel, D. and T.J. Dougherty, eds.), pp. 3-13, Plenum Press, New York.

Evensen, F.J., S. Sommer, J. Moan and T. Christensen (1984) Cancer Research 44, 482-486.

Grossweiner, L.I. and G.C. Goyal (1983) Photochem. Photobiol. 37, 529-532.

Janson, T.R. and J.J. Katz (1972) J. Magn. Reson. 6, 209-220.

Kauzmann, W. (1959) Adv. Protein Chem. 14, 1-63.

Kessel, D. (1977) Biochemistry 16, 3443-3449.

Kessel, D. (1984) Photochem. Photobiol. 40, 851-859.

Kessel, D. and T.-H Chou (1983) Cancer Research 43, 1994-1999.

Kessel, D. and E. Rossi (1982) Photochem. Photobiol. 35, 37-41.

Margalit, R. and S. Cohen (1983) Biochem. Biophys. Acta 736, 163-170.

Margalit, R. and Rotenberg, M. (1984) Biochem. J. 219, 445-450.

Margalit, R., N. Shaklai and C. Cohen (1983) Biochem. J. 209, 547-552.

Nemethy, G. and H.A. Scheraga (1962) J. Chem. Phys. 36, 3401-3417.

Oku, N., D.A. Kendall and R.C. MacDonald (1982) Biochim. Biophys. Acta 692, 384-388.

Paterson, Y., G. Nemethy and H.A. Scheraga (1981) Ann. N.Y. Acad. Sci. 367, 132-150.

Sinanoglu, O. (1968) Molecular Associations in Biology (Pullman, B., ed.), pp. 427-445, Academic Press, New York.

Sinanoglu, O. (1981). Int. J. Quant. Chem. 18, 381-392.

Sinanoglu, O. and S. Abdulnur (1965) Fed. Proc. Fed. Am. Soc. Expl. Biol. 14, Suppl. III, 512-523.

Szoka, F.C. and D. Papahadjoupoulos (1980) Ann. Rev. Bioenerg. 9, 467-508.

Tanford, C. (1973) The Hydrophobic Effect, John Wiley and Sons, New York.

Zlatkis, A. and B. Zak (1968) Anal. Biochem. 29, 143.

PHOTODYNAMIC THERAPY

Thomas J. Dougherty

Head, Division of Radiation Biology, Department of
Radiation Medicine
Roswell Park Memorial Institute, 666 Elm Street
Buffalo, NY 14263

INTRODUCTION

Photodynamic therapy (PDT) involves the action of light on a
photosensitizer retained in malignant or other diseased tissues.
Application to treatment of cancer in man depends upon relative selective
retention in the tumor, low systemic toxicity and the ability of the
activating light to reach the diseased site.

The sensitizers in clinical use include hematoporphyrin derivative
(Hpd) and a purified form tentatively identified as dihematoporphyrin
ether (DHE). These porphyrins are of low systemic toxicity and are
retained in most malignant tissue at levels sufficient to elicit a
localized photosensitized reaction when activated by light near 630 nm.
In many cases the tissues overlying or surrounding the tumors have lower
levels of drug allowing for a relatively selective destruction of the
malignant tissue. The main disadvantage of these materials is their long
retention in skin, which can lead to severe phototoxic reaction if
patients are exposed to bright light.

Mechanism of Action

Questions regarding Hpd and DHE uptake and distribution in animal
tissue, cellular target sites, tissue target sites, and mechanism of
tissue distribution following PDT have been addressed for several years by
several groups. However, until recently little definitive data were
available.

In a study by Bugelski, et al.[1] the distribution of ^{3}H-Hpd matched
the distribution of labeled albumin. At longer periods (over 6 hr) the
porphyrin was cleared from most normal tissue with certain exceptions. No
change was seen in the tumor level from 3 hr to over a several day period
although dilution occurred presumably due to cell proliferation. Within
the tumor the porphyrin was found to be primarily associated with vascular
stroma (macrophages, mast cells, possibly fibroblasts). The relative
density of porphyrin in stroma to that in tumor cells was approximately
5:1. In the liver, hepatocytes appeared to retain little if any
porphyrin, but Kupfer cells retained high levels. The porphyrin appears

to be associated with reticuloendothelial cells in general in all tissues. Fluorescence microscopy of experimental mouse tumors as well as biopsy samples from patients indicates that fluorescence occurs intracellularly in cells associated with the vascular stroma as well as in the tumor cells themselves.

As indicated above, the porphyrin in the experimental mouse tumors appears to be associated primarily with the vascular stroma. In fact, the earliest histological effects of Hpd + light occur in this area. Bugelski, et al. examined tumors by electron microscopy at various times following treatment.[1] The earliest change which could be observed was blebbing of tumor cells adjacent to the microvasculature. The endothelium appeared to be intact at this point. At longer periods tumor cells more remote from the vessels demonstrated blebbing and endothelial cells were seen to be retracted and damaged and red cells had extravasated. Finally, massive extravasation and tumor cell destruction was observed.

Such data raised the question of the role of direct tumor cell destruction, such as that studied in vitro (see above), and destruction of the vasculature in the overall process of tissue destruction following PDT.

Henderson, et al. have studied this question by examining tumor cell survival following PDT in vivo, using the EMT-6 tumor system, an in vivo to in vitro colony formation survival assay.[2] Tumors were treated using DHE and 630 nm light with doses which produced destruction of tumor bulk in all animals and 52% long-term cures. No significant tumor cell inactivation was found following PDT in vivo when tumors were explanted immediately after completion of treatment. However, when tumors were allowed to remain in situ for varying lengths of time (1-10 hr) following treatment, tumor cell death was found to occur rapidly and progressively. This delayed cell death was similar to that caused by shut down of the tumor blood circulation by simply killing the mouse. Thus, at least in this system, death of tumor cells seemed to be the result of vascular damage by PDT rather than of direct photodynamic damage. It is likely that sublethal damage to tumor cells combined with vascular destruction is necessary for cure of the tumor.[3]

Singlet oxygen has been shown to be likely involved in in vitro destruction of cells containing porphyrins[4] (see above). While no comparable experiments have been carried out in vivo, it is likely that singlet oxygen may be involved at least as an initially formed cytotoxic species. It has been demonstrated that if tumors in the legs of mice are shut off from oxygen by leg-clamping that the usual destructive effect of Hpd + light is prevented.[5] Recently, in our laboratory, Miller has demonstrated the involvement of oxygen radicals in tumor cell death by PDT (unpublished results).

Hematoporphyrin Derivative (Hpd) and Its Active Component (DHE)

Numerous porphyrins have been tested as in vivo photosensitizers of experimental SMT-F animal tumors, e.g. hematoporphyrin, protoporphyrin, uroporphyrin, etc.[6] Among these, only tetraphenylporphine sulfonate (TPPS), hematoporphyrin derivative (Hpd), and its active component DHE have been found to be active in this system (this is not necessarily true if activity is measured in vitro; see below). Hematoporphyrin itself has been reported to be active in vivo by some investigators[7] and inactive by others.[8] It is now known that various preparations of hematoporphyrin may contain material similar to the active material in Hpd and when this is removed, in vivo activity disappears.[8]

314

While TPPS was found to be at least as active as Hpd, its low rate of serum clearance[9] and reported neurotoxocity[10] precluded it from being tested clinically.

Hpd, most extensively used to date in clinical trials, was first prepared by Schwartz.[11] When Hp is acetylated, in the first step to preparing Hpd, it yields monoacetate (20-30%), di-acetate (50-60%), and unchanged Hp (5-20%). In addition the mono- and di-dehydration products of Hp are formed, hydroxyethylvinyldeuteroporphyrin (HVD) and protoporphyrin (proto) respectively. None of the components of this acetate mixture are active in photosensitizing tumors.[12] However, when dissolved for injection in dilute sodium hydroxide solution or slightly acidic saline, the acetate groups are rapidly hydrolyzed producing a mixture of porphyrins very active as a tumor localizer and photosensitizer and termed here hematoporphyrin derivative (Hpd), or by some authors, alkali-treated Hpd. Since none of the known components of Hpd, e.g. Hp, HVD and proto, were active tumor photosensitizers, it was clear that the activity was a result of a component not identified. In 1981 Dougherty separated Hpd by gel exclusion chromatography and identified a new fraction representing approximately 45% of the mixture. This material was found to be responsible for the tumor photosensitizing ability of Hpd.[13] In addition, it appeared (in animal testing) to provide a higher therapeutic ratio (tumor vs skin) than the Hpd mixture. Using a combination of NMR spectroscopy, fast-atom bombardment mass spectroscopy, and various chemical reactions, this material has been tentatively identified as bis-1-[3(1-hydroxyethyl)deutero-porphyrin-8-yl] ethyl ether (I) and/or other isomers of this structure. This ether apparently is formed during hydrolysis of the acetate(s) provided the hydrolysis is carried out under conditions where self-aggregation of the acetate occurs. Under dilute conditions where self-aggregation does not occur, hydrolysis produces none of this ether, which conceptually can be considered as the ether derived from hematoporphyrin by elimination of water from hydroxyethyl groups on two separate Hp molecules. A common name suggested for this ether is dihematoporphyrin ether (DHE). Recently it has been suggested that DHE may be esters rather than ethers.[14]

DHE has a strong tendency to self-associate in aqueous solutions even in the presence of albumin at a concentration similar to that found in serum (~30 mg/ml). Grossweiner has estimated that 15-20% of this material remains unbound to serum protein in patients.[15] In fact, spectroscopy of sera taken from patients 0.5 to 3 hr after receiving DHE indicated considerable self-association of the porphyrin whereas at longer times it appeared to be mainly dissociated (probably bound to serum protein). It is this strong tendency for self-association that may be responsible for long periods of retention of Hpd and DHE in tumors. Phagocytosis of DHE by macrophages, Kupfer cells, mast cells, and cells of the reticuloendothelial system has been demonstrated.[1] Swincer, et al.[16] has shown that the nominal molecular weight of DHE in aqueous solution is approximately 20,000. However, in organic solvents containing tetrahydrofuran, only low molecular weight fractions can be found (Kessel, D., private communication).

DHE in aqueous solutions demonstrates no fluorescence. However, in tetrahydrofuran, a solvent that causes disaggregation, the typical red porphyrin fluorescence was seen. Further, bright fluorescence was seen in tumors in animals or humans, similar to that found for Hpd. The actual fluorescent yield of DHE has not yet been determined; for Hpd it is 2-3%.[17]

Also, when aggregated, DHE does not generate singlet oxygen, being less than one-quarter as efficient as Hp at similar concentrations in water.[13] Therefore, it is reasonable to assume that DHE aggregates, while they may be important in tumor uptake and retention, are not responsible as such for the fluorescence or photosensitizing effect of DHE. Such aggregates apparently must first disaggregate (and/or metabolize) within the tissue to be effective. The uptake of DHE and Hpd in the DBA/2 Ha mouse tissue indicates that compared to Hpd, only half the injected amount of DHE is necessary to achieve similar tissue levels, e.g. 3-5 μg/g tumor (SMT-F) 3 to 24 hr following 10 mg/kg Hpd or 5.0 mg/kg DHE intraperitoneally.[18]

The distribution of Hpd in various organs and tissues of the mouse delineated by visual fluorescence is not consistent with the distribution determined using radioactivity tagged porphyrins. For example, while little fluorescence is apparent in liver, kidney, and spleen after systemic Hpd injection, these organs take up more ^{3}H- or ^{14}C-Hpd than any other tissues.[19] While in principle it could be argued that porphyrin metabolites are responsible for this discrepancy, animals receiving Hpd and illumination of the peritoneal cavity (red light) die of liver damage (see below). It is more likely that fluorescence in these organs cannot be observed due to the high absorptivity by the blood for both the activating light (usually blue) and the fluorescent light (630, 690 nm). Thus, when tumors are treated in areas near or involving these organs, great care should be exercised to prevent their exposure to the therapeutic light.

In Vivo Action Spectrum

Since tunable dye lasers are most commonly used for PDT, it is important to determine the wavelength most effective in producing tumoricidal effects with Hpd or DHE. The reason for using red light for activation rather than shorter wavelengths which actually are absorbed more strongly by these porphyrins is that the red is considerably more penetrating through tissue (see below). In solution the maximum wavelength for absorption of Hpd is near 620 nm whether or not serum albumin is added. However, when absorbed onto or into cells, the maximum red absorption is near 630 nm. The action spectrum for control of the transplanted SMT-F tumor in DBA/2 mice was determined by Dougherty and colleagues.[20] Using standard conditions (75 mW/cm^2, 30 min, 1 day post 7.5 mg/kg Hpd or 5.0 mg/kg DHE) and varying only the wavelength, it was determined that maximum response occurred at 630 nm. In addition, in an earlier study this group had determined that maximum normal skin response of Hpd in albino mice occurred at the same wavelength. Recently Gomer has demonstrated the same action spectrum for cultured cells.[21]

Shortly after PDT (immediate to one day, depending on conditions) tumors in experimental animals became hemorrhagic (e.g. subcutaneous SMT-F tumor in the mouse) or in some cases totally blanched (e.g. Green melanoma in rabbit eye). When effectively treated, the tumor mass becomes non-palpable or non-observable usually within a matter of a few days. Complete eradication is possible depending on treatment conditions. It has been demonstrated that oxygen levels in experimental tumors drop rapidly during PDT treatment.[22,23] When examined histologically, the earliest changes occur in and around the tumor vasculature with red cell congestion and some enucleation of cells adjacent to the vessels. At later times obvious red cell extravasation is seen along with apparent total destruction to tumor cells.[1]

It has been shown autoradiographically that tritium-labeled Hpd with the SMT-F tumor or the Sarcoma 180 mouse tumor is located in the vascular stroma at a level five times that in or around the tumor cells.[1] A similar pattern can be observed by fluorescence microscopy of frozen sections of these tumors. However, a common characteristic of these experimental animal tumors is their poor vascularity. It is unclear if similar drug distribution and vascular damage occur in tumors in humans.

In Vitro Studies

Several groups have studied the uptake of Hpd in normal cells in culture for comparison with cells of malignant tissue origin. Moan, et al.[24] examined four types of mouse embryo fibroblast cells of different oncogenic potential for their ability to take up Hpd measured cytofluorometrically. It was found that there was no increased Hpd uptake or photosensitivity with increasing tendency for these cells to produce tumors in syngeneic immunosuppressed mice. Also, Chang and Dougherty[25] demonstrated no differences in uptake or clearance of Hpd or TPPS between normal cells (CHO, L929) and cells of malignant character (HeLa, PC-1). Similar results were found for a variety of cultured cells by Henderson, et al.[26] On the other hand, Christensen, et al.[27] have demonstrated 25-50% more Hpd uptake in transformed cells than in the normal counterparts (10T1/2). Also, Andreoni, et al.[28] have demonstrated enhanced destruction of a transformed epithelial rat thyroid cell compared to the normal cell following exposure to Hpd and light.

Berns, et al. have examined a kangaroo epithelial cell line (PTK_2) following exposure to Hpd and have shown a rapid uptake for the first 2 hr followed by slower accumulation over the following two days.[29] They used a fluorescence detection system for individual cells. Excretion was measured after Hpd removal from the medium which demonstrated a residual amount of Hpd which could not be removed, the residual increasing with time of initial exposure. Also in this study, the fluorescence in myocardial cells was seen to be primarily associated with mitochondria located near the nuclear membrane. This observation correlates with early damage to mitochondria observed by Coppola, et al. in lymphoma cells treated in vitro with hematoporphyrin and light.[30]

Much of the data indicate the central role of cellular membrane as sites of damage resulting from porphyrins and light. This damage may be a result of cross-linking of membrane protein reported by Dubbelman, et al.[31] and Girotti, et al.[32] However, Hilf, et al. have demonstrated inactivation of mitochondrial membrane enzymes as an early event in cells exposed to light in vitro following Hpd uptake in vivo.[33]

Recently several investigators have demonstrated a "tightly bound" porphyrin fraction in cells exposed to Hpd or DHE. Its fraction of total porphyrin increases with time of incubation to a maximum of 50-60% at 24 hr. Henderson, et al.[26] and Bellnier and Lin[34] have demonstrated increased photosensitivity of cells containing "tightly bound" porphyrin compared to those with "loosely bound" (i.e. removed by washing in media) porphyrin. It is expected that these in vitro conditions may be better correlated with the in vivo situation. Kessel, et al.[35] and Christensen, et al.[27] have demonstrated enhanced uptake in vitro of DHE compared to the other components of Hpd. The DHE appears to correspond to the "tightly bound" fraction of Hpd since it is the only Hpd component not removed from cells by washing following even long-term exposure.[27]

While very little of the Hpd in cells is taken up by the nucleus, effects of photoradiated porphyrins on DNA can be measured. Single strand breaks have been demonstrated by Moan and Boye in NHIK 3025 cells exposed to hematoporphyrin and light, but the frequency is low compared to ionizing radiation and does not correlate with cell survival.[36] Also, sister chromatid exchange has been demonstrated by Moan, et al.[37] and by Gomer, et al.[38] On the other hand, Hpd + light has been found not to produce mutants in CHO cells using light doses sufficient to produce survival similar to that produced by X-irradiation which produced a high degree of mutation in these cells.[38] It has also been reported that Hpd + light is negative in the Ames test for mutagenicity in bacteria.[10]

Dosimetry of Light in Tissue

An important parameter determining the extent of tumor necrosis is the penetrability of the necessary visible light through tissue. Svaasand, et al.[39,40] have considered this question theoretically and made measurements of penetration of various wavelengths in various tissues in experimental animals and tissues. For a diffusion dominant, one dimensional case (surface illumination)

$$(1) \quad \Phi = \Phi_0 e^{-\alpha X}$$ where Φ = space irradiance
Φ_0 = irradiance at surface
α = total attenuation coefficient $(\beta/\xi)^{1/2}$
where β = absorption coefficient and
ξ = diffusion coefficient

X = distance

Measured values for δ (mm) = I/α, where δ is the attenuation distance or the depth at which the incident intensity falls off to I/ϵ or approximately 37% range from 1 to 4 mm. When fibers are inserted into the tumor for interstitial PDT, the equation for space irradiance in the diffusion dominant case is

$$(2) \quad \phi = \phi_0 \frac{(a)}{(r)} e^{-\alpha(r-a)}$$

where r = radial distance in tissue
a = radius of fiber

The pathology of tumors removed after treatment by PDT indicates necrosis of 5 to 10 mm generally. Coagulation necrosis of tumor obstructing bronchi following PDT has been demonstrated by inserting special fibers which allow for lateral light distribution (see below). Such fibers have been demonstrated to cause necrosis within a radius of 9-10 mm from the point of insertion.[41]

While many of the early clinical studies were carried out with a filtered xenon arc lamp with a fairly broad spectral output of 600-700 nm, such systems are rarely used today, although they may have some application for superficial skin lesions. Most clinical studies use lasers in order to take advantage of the high coupling efficiency to optical fibers which allow light to be delivered conveniently to many areas.

A variety of delivery fibers are available: those with an optically flat end with lenses to produce a spot of homogeneous intensity generally used to treat cutaneous or subcutaneous tumors; fibers with various lenses on the ends to further expand the spot, and those with diffusers of various lengths which produce a cylindrical pattern along the diffuser length and are generally used within a lumen or for interstitial PDT. Fibers without lenses or diffusers should not be used since they produce inhomogeneous spots or hot spots when used interstitially.

While the PDT treatment is a photochemical process rather than a thermal ablation or coagulative process as with the CO_2 or Nd-YAG laser, considerable heat can be generated at the tips or broken surfaces of the fibers, particularly if they come in contact with absorbing tissue or blood.

There has been one reported case of a fire during an intratracheal PDT treatment under 100% oxygen when a hot spot on the fiber came in contact with tissue and/or blood. No incidences have been reported for treatments carried out in air, although it must be borne in mind that while a fire is not likely in air, the fiber can nonetheless be destroyed if high power is applied when in contact with blood or other strongly absorbing tissue.

PDT in Clinical Trials

Since 1976 it is estimated that approximately 3,000 patients have been treated by PDT worldwide. In most cases these patients had a long history of cancer and had exhausted many other treatments prior to PDT. In spite of this, all investigators have reported tumor response rates of 50-90%. This should not be taken to indicate patient benefit of this order. Because of the advanced state of most of the patients, only a small portion of the patient's cancer was treated by PDT. The current trend is to use PDT at much earlier stages of disease where patient benefit may be expected to be higher.

The table summarizes the results of published clinical data on PDT up to 1984. In addition to these data are numerous as yet unpublished reports and data on other sites. The Japanese experience in addition to that included in the table has been summarized elsewhere.[42]

In treating tumors of the skin by PDT it is important to carefully select a drug-light dose combination which causes selective tumor necrosis and little normal skin damage. With 3.0 mg/kg Hpd or 2.0 mg/kg DHE, 48-72 hr prior to light treatment the upper limit of the light dose is 25-35 Joules/cm^2 for Hpd and 35-50 Joules/cm^2 for DHE. However, various situations may require downward adjustment of the light dose, e.g. previous adriamycin or highly traumatized skin or ulcers. Isolated, large tumors (>1 cm thick) which can be treated with little or no normal skin in the treatment field may receive many times higher light doses (e.g. 100-200 Joules/cm^2) in order to obtain sufficient depth of treatment. However, in general these tumors are best treated by interstitial methods perhaps in combination with superficial treatment. In such cases, the fibers should have a diffuser of 1-3 cm length. In general a radius of necrosis of 5-10 mm can be expected. Therefore tumors larger than this must receive multiple treatments, preferably simultaneously. The depth of necrosis obtained for superficial lesions receiving 25-50 J/cm^2 ranges from 3 to 5 mm. Measured values of are generally 2 to 4 mm (Potter and Dougherty, unpublished results).

Recently we have treated a few patients with a dose of 1.0 mg/kg DHE and initiated treatment from 3 to 72 hr post injection at various light doses. The incentive for this is to reduce skin photosensitivity and compensate for the lower tumor drug level with increasing amounts of light. While the data are preliminary, there are several observations:

TABLE 1

SUMMARY OF RESPONSE OF VARIOUS TUMORS TO PDT (TO 1984)

Tumor Type	No. Patients or (Sites)	Response CR	Response PR	Response NR	Longest Followup to Date
Skin (met. breast, basal cell, squamous, melanoma)	(219)	(147)	(29)	(43)	4 years
Endobronchial - late stage	262	106*	136	20	1.5 years
Endobronchial - early stage	32	17	10	-	4.5 years
Bladder - superficial	(70)	(38)	(16)	(13)	2 years
Head/neck - recurrent	49	9	26	14	2 years
Gyn - recurrent	11	4	5	2	2 years
Esophagus - advanced	17	0	11**	6	1 year

*Complete opening to wall of lumen following physical removal after PDT
**Partial relief of obstruction

(1) the skin sensitivity to sunlight appears to be considerably reduced judging from patient information (uncontrolled). The light dose at 630 nm necessary to cause skin damage is greater than 144 Joules/cm^2. Thus, at this drug level there is clearly a non-reciprocal relationship between drug and light dose; (2) treatment at 3 hr is sub-optimal compared to treatment at 24 hr or longer; (3) tumor necrosis to >5 mm has been measured following a light dose of 108 Joules/cm^2. We are continuing these studies for further definition of the variables.

Finally, it should be noted that while the data are sparse, basal cell tumors appear to be especially responsive to PDT even when previously treated by other modalities. In addition, the cosmetic effect following healing appears to be excellent. If a topical drug preparation is developed, these patients may be good candidates for PDT.

Endobronchial tumors are highly responsive to PDT and in early stages may be a curative treatment.[43,44] In addition, bulky obstructive tumors can be removed to the wall following PDT. In a series of over 100 patients Balchum[45] has used the following protocol: the patient receives 3.0 mg/kg Hpd or 2.0 mg/kg DHE. At 72 hr post injection light at 630 nm was delivered endoscopically by inserting the fibers with 0.5 to 1.5 cm diffusers directly into the lesion when possible. The light dose was 400 mW per cm of diffuser to deliver 200 Joules/cm diffuser, requiring 8.3 min. Three days later the necrotic tumor and debris were removed endoscopically. While this procedure does not affect survival, palliation may be achieved in many cases. While the Nd-YAG laser is capable of similar tumor ablation, PDT can be also used in the peripheral airways reachable by the bronchoscope. These two modalities may well be complimentary or used effectively in an adjuvant setting (i.e. rapid removal of the bulk of the tumor using the Nd-YAG laser followed by PDT to complete tumor destruction to the wall). As pointed out in an earlier study by Vincent, et al.[46] removal of the necrotic tumor and debris is important in order to avoid potentially serious complications.

Superficial bladder cancer, especially CIS, may provide an ideal application of PDT. Because of the multi-focal nature of this disease, it is desirable to treat the entire bladder, which can now be accomplished with special light diffusers. In a determination of the ablative capabilities of PDT, Hisazumi, et al.[47] determined that complete response is likely only in tumors 2 cm or smaller and requires light doses of at least 100 Joules/cm^2 given 48-72 hr post Hpd (2.0 to 3.2 mg/kg). Benson[48] has demonstrated selective uptake and retention of Hpd in CIS or dysplasia of the bladder in a series of patients scheduled for cystectomy. Early results indicate that CIS lesions are especially responsive to PDT although bladder shrinkage has been seen in a few patients who had PDT following other therapies, especially radiation therapy. It is now thought that overdistention of the bladder contributed to this complication. A distended volume of not over 200 cc during treatment is recommended. Controlled, randomized clinical trials are underway currently to compare PDT to thiotepa for ablative and prophylactic application in superficial bladder tumors. Whole bladder light doses will be 25-35 Joules/cm^2, 48-72 hr post 2.0 mg/kg DHE.

Application of PDT to Cancers of the Head and Neck. Wile and co-workers studied PDT for treatment of various head and neck tumors.[49] Nineteen sites in 16 patients were treated for recurrent squamous cell carcinoma. Five of these demonstrated complete response to treatment (up to one year currently) and 10 partial responses. In some cases an endoscope was used to visualize the lesion which was then treated through the scope. In most cases patients were given 3.0 mg/kg Hpd, 72 hr prior to treatment. Light doses were in the range of 25-90 Joules/cm^2. Among the patients benefitting from treatment was a patient with a tongue cancer which persisted following radiation therapy but responded completely to PDT with improvement in eating and speaking. A second patient with a recurrent tumor of the vocal cord which recurred after radiation therapy also responded completely to PDT. Wile is currently carrying out a more extensive study of these types of tumors. Also, Taketa and Imakiire[50] have reported six cases of squamous cell carcinoma including laryngeal carcinoma, cancer of the nasopharynx, tongue and vocal cords. In most cases a reduction in tumor mass was found but generally the biopsies remained positive in the deeper portion of the tumors involving areas beneath the mucosa.

Cancer of the Esophagus. The rationale for use of PDT in esophageal cancer is similar to that for treatment of lung cancer, i.e., early stage disease can be treated locally with little risk and advanced cases may receive palliation. It should be emphasized that PDT provides the clinician with another modality which should be appropriately worked into the overall management of the patient along with other modalities. The guiding principle is that PDT is more localized and selective than some of the other treatments but cannot be expected to entirely eliminate large, bulky tumor, especially outside the lumen or in lymph nodes.

To date approximately 30 patients with esophageal cancer have received PDT. Since in most instances, like lung cancer, esophageal cancer is not often detected in an early stage (except in China), most patients have had advanced disease and previously had received therapy (surgery, radiation therapy). A case described by Aida indicates the benefit, however, even for some advanced stage patients.[51] A patient had undergone total gastrectomy, total resection of the thoracic esophagus and right hemicolectomy. Two early stage lesions were found in the esophagus as well as the residual cervical esophagus for which he received two courses of radiation therapy (>15,000 rads). Further surgery had been ruled out. However, within a few months of radiation the biopsy remained positive. He received 2.5 mg/kg Hpd and two days later endoscopic PDT was performed followed by a second treatment the next day. Two years later he remains free of disease.

Tian of the Henan Tumor Institute[52] has treated 13 cases of early and Stage I esophageal cancer detected in a mass screening program. This group used a procedure similar to that described by Hayata for treatment of early bronchial tumors by PDT; i.e., 630 nm light delivered endoscopically to the lesion 72 hr post 2.5 mg/kg Hpd. The dose was 100-200 Joules/cm^2 delivered from a fiber without any type of lens or diffuser. In four cases (Stage I) the patients went on to operation. In three of these, residual tumor nests were found in the submucosal layers. This may have been a result of inadequate light delivery, a common problem with these types of fibers since they produce an inhomogeneous spot. Fibers with diffusing ends are recommended. The remaining cases have shown no recurrences up to one year post treatment. McCaughan[53] has reported on seven cases with near or total esophageal obstructions. PDT was used in order to relieve symptoms of dysphagia and to restore the ability to take solid foods. In these cases the fiber was inserted into the tumor or remaining lumen. Light delivery was 200-300 Joules per cm of diffuser. All patients responded to treatment with massive sloughing of tumor within a few days of treatment. In all cases dysphagia was relieved and normal diet resumed. McCaughan points out that while residual tumor may remain after treatment, palliation was achieved up to nine months so far in one patient. Similar results have been obtained by Aida[51] and Nava,[54] although these authors noted that patients complained of severe mediastinal pain for several days following treatment. Also, one patient developed an esophageal-tracheal fistula presumably due to slough of invading tumor. One patient in this group of five remained free of symptoms more than one year after treatment.

PDT is currently being investigated for treatment of a variety of other types of tumors, but results are still very preliminary. For example, Laws has reported a feasibility study for PDT treatment of brain tumors,[55] and McCulloch has applied PDT in combination with surgery and ionizing radiation to treatment of at least 17 glioblastoma patients.[56]

One of the most promising new areas for application of PDT is in treatment of intraocular tumors such as malignant melanoma of the choroid and retinoblastoma. Early results of Murphree, et al. have shown that malignant melanoma of the choroid responds to PDT (3.0 mg/kg Hpd) when the light is applied both through the cornea and through the sclera.[57] Similar results have been obtained by Bruce, et al..[58] Long-term followup in these patients is not yet available although there is one case of recurrence at approximately one year after PDT of a choroidal melanoma treated by L'Esperance and Dougherty.[59] The proper light dose and means of delivery are still under investigation. In general patients treated by Bruce, et al. have been treated by high power densities (up to 600 mW/cm^2) whereas Murphree, et al. have limited the dose rate to 200 mW/cm^2 where they have demonstrated that little damage occurs to ocular structures in rabbits.[60] The high degree of pigmentation of some of the melanoma lesions raises the possibility of a significant contribution by thermal effects to the overall results.

The main toxicity of the drug remains the skin photosensitivity. While it appears that this can be minimized by reducing the drug dose, the major means of avoidance remains patent compliance with instructions to remain out of sunlight for 30 or more days. Treatment toxicity includes pain, especially when skin necrosis occurs in treating large areas of the chest wall, irritated symptoms in the bladder following PDT lasting a few weeks, and possible increased incidence of hemoptysis in certain advanced cancer patients.

CONCLUSION

In order for PDT to be made available widely and accepted as a useful cancer modality, it is necessary to carry out randomized comparative trials. Such trials are currently underway for treatment of superficial bladder cancer (versus thiotepa) and in relief of airway obstruction (versus radiation therapy). Future research should focus on new photosensitizers which do not induce skin photosensitivity and which absorb at more tissue-penetrating wavelengths in the far red and near infrared, since limited depth of necrosis remains a limitation of the therapy. It will also be helpful to devise more reliable light delivery fibers and to improve upon current methods of measuring dosimetry of light delivery to various tumor sites and surrounding tissues.

REFERENCES

1. P. J. Bugelski, C. W. Porter, and T. J. Dougherty, Autoradiographic distribution of hematoporphyrin derivative in normal and tumor tissue of the mouse. Cancer Res. 41:4606 (1981).

2. B. W. Henderson, T. J. Dougherty, and P. B. Malone, Studies on the mechanism of tumor destruction by photoradiation therapy, in: "Porphyrin Localization and Treatment of Tumors," D. R. Doiron and C. J. Gomer, eds., Alan R. Liss, Inc., New York (1984).

3. B. W. Henderson, S. M. Waldow, T. S. Mang, and T. J. Dougherty, Photodynamic therapy and hyperthermia, in: "Proc. of Porphyrins as Phototherapeutic Agents for Tumors and Other Diseases," G. Jori, ed., (1985), in press.

4. K. R. Weishaupt, C. J. Gomer, and T. J. Dougherty, Identification of singlet oxygen as the cytotoxic agent in photodynamic inactivation of a murine tumor. Cancer Res. 36:2326 (1976).

5. K. R. Weishaupt and T. J. Dougherty, Unpublished results (1983).

6. T. J. Dougherty, Photodynamic therapy (PDT) of malignant tumors. CRC Critical Reviews in Oncology/Hematology 2(2):83 (1984).

7. L. Tomio, P. L. Zorat, L. Corti, F. Calzavara, E. Reddi, and G. Jori, Effect of hematoporphyrin and red light on AH-130 solid tumors in rats. Acta Radiolog. 12:1 (1982).

8. T. J. Dougherty, Hematoporphyrin as a photosensitizer of tumors. Photochem. Photobiol. 38:377 (1983).

9. D. A. Bellnier, In vitro photoradiation - Hematoporphyrin derivative accumulation and interaction with ionizing radiation. Ph.D. Thesis, State University of New York at Buffalo (1982).

10. J. Kennedy, Unpublished results (1983).

11. R. L. Lipson, The photodynamic and fluorescent properties of a particular hematoporphyrin derivative and its use in tumor detection. Master's Thesis, University of Minnesota (1960).

12. M. C. Berenbaum, R. Bonnett, and P. A. Scourides, In vivo biological activity of components of hematoporphyrin derivative. Br. J. Cancer 45:571 (1982).

13. T. J. Dougherty, D. G. Boyle, K. R. Weishaupt, B. A. Henderson, W. R. Potter, D. A. Bellnier, and K. E. Wityk, Photoradiation therapy - Clinical and drug advances, in: "Porphyrin Photosensitization," D. Kessel and T. J. Dougherty, eds., Plenum Press, New York/London (1983).

14. D. Kessel, Properties of the tumor-localizing component of Hpd, in: "Proc. of Porphyrins as Phototherapeutic Agents for Tumors and Other Diseases," G. Jori, ed., (1985), in press.

15. L. Grossweiner, Personal communication (1984).

16. A. G. Swincer, A. D. Ward and G. J. Hawlett, The molecular weight of haematoporphyrin derivative, its gel column fractions and some of its components in aqueous solution. Photochem. Photobiol. 41:47-50 (1985).

17. D. R. Doiron, E. Profio, R. G. Vincent, and T. J. Dougherty, Fluorescence bronchoscopy for detection of lung cancer. Chest 76:27 (1979).

18. T. J. Dougherty, W. R. Potter and K. R. Weishaupt, The structure of the active component of hematoporphyrin derivative, in: "Porphyrin Localization and Treatment of Tumors," D. R. Doiron and C. J. Gomer, eds., Alan R. Liss, Inc., New York (1984).

19. C. J. Gomer and T. J. Dougherty, Determination of ^{3}H- and ^{14}C-hematoporphyrin derivative distribution in malignant and normal tissue. Cancer Res. 39:146 (1979).

20. T. J. Dougherty, K. E. Wityk and P. B. Malone, Unpublished results (1983).

21. C. J. Gomer, D. R. Doiron, N. Rucker, N. J. Razum, and S. W. Fountain, Action spectrum (620 nm-640 nm) for hematoporphyrin derivative induced cell killing. Photochem. Photobiol. 39:365 (1984).

22. H. I. Bicher, F. W. Hetzel, P. Vaupel, and T. S. Sandhu, Microcirculation modifications by localized microwave hyperthermia and hematoporphyrin phototherapy. Biblthca. Anat. 20:628 (1981).

23. F. W. Hetzel and H. Farmer, Dose effects relationships in a mouse mammary tumor, in: "Porphyrin Localizaion and Treatment of Tumors," D. R. Doiron and C. J. Gomer, eds., Alan R. Liss, Inc., New York (1984).

24. J. Moan, H. B. Steen, K. Feren, and T. Christensen, Uptake of hematoporphyrin derivative and sensitized photoinactivation of C_3H cells with different oncogenic potential. Cancer Lett. 14:291 (1981).

25. C. Chang and T. J. Dougherty, Photoradiation therapy: Kinetics and thermodynamics of porphyrin uptake and loss in normal and malignant cells in culture. Radiat. Res. 74:498 (1978).

26. B. W. Henderson, D. A. Bellnier, B. Ziring, and T. J. Dougherty, Aspects of the cellular uptake and retention of hematoporphyrin derivative and their correlation with the biological response to PRT in vitro, in: "Porphyrin Photosensitization," D. Kessel and T. J. Dougherty, eds., Plenum Press, New York (1983).

27. T. Christensen, T. Sandquist, K. Feren, H. Waksvik, and J. Moan, Retention and photodynamic effects of hematoporphyrin derivative in cells after prolonged cultivation in the presence of porphyrin. Br. J. Cancer 48:35 (1983).

28. A. Andreoni, R. Cubeddu, S. DeSilvestri, P. Laporta, F. S. Ambesi-Impiombato, M. Esposito, M. Mastrocinque, and D. Tramontano, Effects of laser irradiation on hematoporphyrin-treated normal and transformed thyroid cells in culture. Cancer Res. 43:2076 (1983).

29. M. W. Berns, A. Dahlman, F. M. Johnson, R. Burns, D. Sperling, M. Guiltinan, A. Siemens, R. Walter, W. Wright, M. Hammer-Wilson, and A. Wile, In vitro cellular effects of hematoporphyrin derivative. Cancer Res. 42:2325 (1982).

30. A. Coppola, E. Viggiani, L. Salzarulo, and G. Rasile, Ultrastructural changes in lymphoma cells treated with hematoporphyrin and light. Amer. J. Path. 99:175 (1980).

31. T. M. A. R. Dubbelman, A. F. P. M. DeGoeij, and J. VanSteveninck, Protoporphyrin-induced photodynamic effects on transport processes

across the membrane of human erythrocytes. Biochem. Biophys. Acta 595:33 (1980).

32. A. W. Girotti, Photodynamic action of protoporphyrin. IX. On human erythrocytes: Cross-linking of membrane proteins. Biochem. Biophys. Res. Comm. 72:1367 (1976).

33. R. Hilf, D. B. Smail, R. S. Murant, P. B. Leakey, and S. L. Gibson, Hematoporphyrin derivative-induced photosensitivity of mitochondrial succinate dehydrogenase and selected cytosolic enzymes of R3230AC mammary adenocarcinomas of rats. Cancer Res. 44:1483 (1984).

34. D. Bellnier and C. Lin, Photodynamic destruction of cultured human bladder tumor cells by hematoporphyrin derivative: Effects of porphyrin molecular aggregation. Photobiochem. Photobiophys. 6:357 (1983).

35. D. Kessel and T. Chou, Tumor-localizing components of porphyrin preparation hematoporphyrin derivative. Cancer Res. 43:1994 (1983).

36. J. Moan and E. Boye, Photodynamic effect on DNA and cell survival of human cells sensitized by hematoporphyrin. Photobiochem. Photobiophys. 2:301 (1981).

37. J. Moan, H. Waksvik, and T. Christensen, DNA single-strand breaks and sister chromatid exchanges induced by treatment with hematoporphyrin and light or by x-rays in human NHIK 3025 cells. Cancer Res. 40:2915 (1980).

38. C. J. Gomer, N. Rucker, A. Banerjee, and W. F. Benedict, Comparison of mutagenicity and induction of sister chromatid exchange in Chinese hamster cells exposed to hematoporphyrin derivative photoradiation, ionizing radiation or ultraviolet radiation. Cancer Res. 43:2622 (1983).

39. L. O. Svaasand, D. R. Doiron, and T. J. Dougherty, Temperature rise during photoradiation therapy of malignant tumors. Med. Phys. 10:10 (1983).

40. L. O. Svaasand and R. Ellingsen, Calibration of applications of non-ionizing electromagnetic radiation. Part 2. Experimental results. Report #RB/PE029, Division of Physical Electronics, University of Trondheim, Norway, Norwegian Institute of Technology (1983).

41. T. J. Dougherty, Unpublished results (1983).

42. Y. Hayata and T. J. Dougherty, eds., "Lasers and Hematoporphyrin Derivative in Cancer," Igaku-Shoin, Tokyo/New York (1983).

43. Y. Hayata, H. Kato, R. Amemiya, and J. Ono, Indications of photoradiation therapy in early stage lung cancer on the basis of post-PRT histological findings, in: "Porphyrin Localization and Treatment of Tumors," D. R. Doiron and C. J. Gomer, eds., Alan R. Liss, Inc. New York (1984).

44. D. A. Cortese and J. H. Kinsey, Hematoporphyrin derivative phototherapy in the treatment of bronchogenic carcinoma. Chest 86:8 (1984).

45. O. Balchum, D. R. Doiron, and G. Huth, Photoradiation therapy of endobronchial lung cancer using the photodynamic action of hematoporphyrin derivative. Lasers Surg. Med. 4:13 (1984).

46. R. G. Vincent, T. J. Dougherty, U. Rao, D. G. Boyle, and W. R. Potter, Photoradiation therapy in advanced carcinoma of the trachea and bronchus. Chest 85(1):29 (1984).

47. H. Hisazumi, T. Misaki, and N. Miyoshi, Photodynamic therapy of bladder tumors, in: "Lasers and Hematoporphyrin Derivative in Cancer," Y. Hayata and T. J. Dougherty, eds., Igaku-Shoin, Tokyo/New York (1983).

48. R. C. Benson, The use of hematoporphyrin derivative (HpD) in the localization and treatment of transitional cell carcinoma (TCC) of the bladder, in: "Porphyrin Localization and Treatment of Tumors," D. R. Doiron and C. J. Gomer, eds., Alan R. Liss, Inc., New York (1984).

49. A. G. Wile, J. Movotny, G. R. Mason, V. Passy, and M. W. Berns, Photoradiation therapy of head and neck cancer, in: "Porphyrin Localization and Treatment of Tumors," D. R. Doiron and C. J. Gomer, eds., Alan R. Liss, Inc., New York (1984).

50. C. Taketa and M. Imakiire, Cancer of the ear, nose and throat, in: "Lasers and Hematoporphyrin Derivative in Cancer," Y. Hayata and T. J. Dougherty, eds., Igaku-Shoin, Tokyo/New York (1983).

51. M. Aida and T. Hirashima, Cancer of the esophagus, in: "Lasers and Hematoporphyrin Derivative in Cancer," Y. Hayata and T. J. Dougherty, eds., Igaku-Shoin, Tokyo/New York (1983).

52. M. Tian, Unpublished results (1983).

53. J. S. McCaughan, W. Hicks, L. Laufman, E. May, and R. Roach, Palliation of esophageal malignancy with photoradiation therapy. Cancer 54:2905 (1984).

54. H. Nava, Unpublished results (1983).

55. E. R. Laws, D. A. Cortese, J. H. Kinsey, R. T. Eagan, and R. F. Anderson, Photoradiation therapy in the treatment of malignant brain tumors: A phase I (feasibility) study. Neurosurg. 9:672, (1981).

56. G. A. J. McCulloch, I. J. Forbes, K. L. See, P. A. Cowled, F. J. Jacka, and A. D. Ward, Phototherapy in malignant brain tumors, in: "Porphyrin Localization and Treatment of Tumors," D. R. Doiron and C. J. Gomer, eds., Alan R. Liss, Inc., New York (1984).

57. A. L. Murphree, D. R. Doiron, C. J. Gomer, B. Szirth, and S. Fountain, Hematoporphyrin derivative photoradiation treatment of ophthalmic tumors. Presented at the Clayton Foundation Symposium on Porphyrin Localization and Treatment of of Tumors, Santa Barbara, CA, April 24-28, 1983 (abstract).

58. R. A. Bruce, An approach to the treatment of ocular malignant melanomas, in: "Porphyrin Localization and Treatment of Tumors," D. R. Doiron and C. J. Gomer, eds., Alan R. Liss, Inc., New York (1984).

59. F. L'Esperance and T. J. Dougherty, Unpublished results (1983).

60. C. J. Gomer, D. R. Doiron, J. V. Jester, B. C. Szirth, and A. L. Murphree, Hematoporphyrin derivative photoradiation therapy for the treatment of intraocular tumors: Examination of acute normal ocular tissue toxicity. <u>Cancer Res.</u> 43:721 (1983).

BIBLIOGRAPHY ON PORPHYRIN PHOTOSENSITIZATION AND RELATED TOPICS

David Kessel

Wayne State University School of Medicine
Harper-Grace Hospitals
Detroit MI 48201

Abhold, R.H., Leid, R.W., Magnuson, J.A., and Hegreberg, G.A. (1986) Inhibition of rat peritoneal mast cell histamine secretion by protopor-
phyrin and incandescent light. Int. Arch. Allergy Appl. Immunol.
76, 126-132.
Aida, M., and Kawaguchi, M. (1983) In Lasers and Hematoporphyrin Deriva-
tive in Cancer Therapy. Y Hayata and TJ Dougherty, eds, Igako-Shoin,
Tokyo, New York. pp 65-69.
Amagasa, J. (1981) Dye binding and photodynamic action. Photochem.
Photobiol. 33, 947-955.
Anderson, S.M. and Krinsky, N.I. (1973) Protective action of carotenoid
pigments against photodynamic damage to liposomes. Photochem.
Photobiol. 18, 403-408.
Andreoni, A., R. Cubeddu, S. De Silvestri, G. Jori, P. Laporta and
E. Reddi (1983) Time-resolved fluorescence studies of hematoporphyrin
in different solvents. Zeit. Natur. 38C, 83-89.
Andreoni, A., Cubeddu, R., De Silvestri, S., and Laporta, P. (1982) Hema-
toporphyrin derivative: photophysical properties and laser activa-
tion. Med. Biol. Environ. 10, 261-268.
Andreoni, A., Cubeddu, R., De Silvestri, S., Laporta, P., Ambesi-
Impiombato, S., Esposito, M., Mastrocinque, M., and Tramontano,
D. (1983) Effects of laser irradiation on hematoporphyrin-treated
normal and transformed thyroid cells in culture. Cancer Res. 43,
2076-2080.
Andreoni, A., Cubeddu, R., De Silvestri, S., Laporta, P. and Svelto,
O. (1982) Two-step laser activation of hematoporphyrin deriva-
tive. Chem. Phys. Lett. 88, 37-39.
Andreoni, A., De Silvestri, S., Laporta, P., Jori G. and Reddi E. (1982)
Hematoporphyrin derivative: experimental evidence for aggregated
species. Chem. Phys. Lett. 88, 33-36.
Ankerst, J., Montan, S., Svanberg, K., and Svanberg, S. (1984) Laser-
induced fluorescence studies of hematoporphyrin derivative (HPD) in
normal and tumor tisseue of rat. Applied Spectroscopy 38, 890-896.
Baghdassarian, R., Wright, M.W., Vaughan, S.A., Berns, M.W., Martin,
D.C., and Wile, A.G. (1985) The use of lipid emulsion as an in-
travesical medium to disperse light in the potential treatment of
bladder tumors. J. Urol. 133, 126-130.
Balchum, O.J., Doiron, D.R., and Huth, G.C. (1983) HPD photodynamic
therapy for obstructing lung cancer. In Porphyrin Localization and
Treatment of Tumors. (Edited by D.R. Doiron and C.J. Gomer). Alan
R. Liss, New York, pp. 727-746.

Barker, D.S., Henderson, R.W., and Storey, E. (1970) The in vivo localization of porphyrins. Br. J. Exp. Path. 51, 628-638

Bellnier, D.A. and T.J. Dougherty (1982) Membrane lysis in chinese
hamster ovary cells treated with hematoporphyrin derivative plus
light. Photochem. Photobiol. 36, 43-47.

Bellnier, D.A., and Lin, C-W. (1983) Photodynamic destruction of cultured
human bladder tumor cells by hematoporphyrin derivative: effects of
porphyrin molecular aggregation. Photobiochem. Photobiophys. 6, 357-
366.

Bellnier, D.A., and Lin, C-W. (1984) Giant cell formation in bladder
tumor cells following hematoporphyrin derivative-sensitized photoirradiation. Photochem. Photobiol. 39, 425-428.

Bellnier, D.A., and Lin, C-W. (1985) Photosensitization and split-dose
recovery in cultured human urinary bladder carcinoma cells containing
nonexchangeable hematoporphyrin derivative. Cancer Res., 45, 2507-
2511.

Bellnier, D.A., Prout Jr., G.R., and Lin, C-W. (1985) Effect of 514.5-
nm argon ion laser radiation on hematoporphyrin derivative-treated
bladder tumor cells in vitro and in vivo. J. Natl. Cancer Inst., 74,
617-626.

Ben-Hur, E. and Rosenthal, I. (1985) The phthalocyanines: a new class of
mammalian cell photosensitizers with a potential for cancer phototherapy. Int. J. Radiobiol. Relat. Stud. Phys. Chem. Med. 47:145-
147.

Benson, R., Farrow, G., Kinsey, J., Cortese, D., Zinke, D. and Utz D.C.
(1982) Detection and localization of in situ carcinoma of the bladder
with hematoporphyrin derivative. Mayo Clin. Proc. 57, 548-555.

Benson, R., Kinsey, J., Cortese, D., Farrow, G., and Utz, D. (1983)
Treatment of transitional cell carcinoma of the bladder with hematoporphyrin derivative phototherapy. J. Urol. 130, 1090-1095.

Berenbaum, M., Bonnett R. and Scourides P. (1982) In vivo biological activity of the components of haematoporphyrin derivative. Br. J.
Cancer 45, 571-581.

Berns, M., Dahlman, A., Johnson, F., Burns, R., Sperling, D., Guiltinan,
M., Siemans, A., Walter, R., Wright, W., Hammer-Wilson M. and Wile
A. (1982) In vitro cellular effects of hematoporphyrin derivative.
Cancer Res. 42, 2325-2329.

Berns, M., Hammer-Wilson, M., Walter, R., Wright, Chow, M., Nahabedian M.
and Wile A. (1984) Uptake and localization of HPD and 'active fractions' in tissue culture and in serially-biopsied human tumors. In
Porphyrin Localization and Treatment of Tumors. (Edited by
D.R. Doiron and C.J. Gomer). Alan R. Liss, New York, pp. 501-520.

Berns, M., Wilson, M., Rentzepis, P., Burns R. and Wile A. (1983) Cell
biology of hematoporphyrin derivative (HPD). Lasers in Med. Surg. 2,
261-266.

Bertoloni, G., Dall'Aqua, M., Vazzoler, M., Salvato B. and Jori G. (1982)
Photosensitizing action of hematoporphyrin on some bacterial strains.
Med. Biol. Environ. 10, 239-242.

Bertoloni, G., Noventa, B., Viel, A., Dall'Aqua, M., and Vazzoler,
M. (1983) Photosensitization of mollicutes by hematoporphyrin. Med.
Biol. Environ. 11, 369-376.

Bickers, D.R., Dixit, R. and Mukhtar, H. (1982) Hematoporphyrin photosensitization of epidermal microsomes results in destruction of
cytochrome P-450 and decreased monooxygenase activities and heme content. Biochem. Biophys. Res. Commun. 108, 1032-1039.

Blazek, E.R., and Hariharan, P.V. (1984) Alkaline elution studies of
hematoporphyrin-derivative photosensitized DNA damage and repair in
Chinese hamster ovary cells. Photochem. Photobiol. 40, 5-13.

Blum, A., and Grossweiner, L.I. (1985) Singlet oxygen generation by
hematopoprhyrin IX, uroporphyrin I and hematoporphyrin derivative at

546 nm in phosphate buffer and in the presence of egg phosphatidylcholine liposomes. Photochem. Photobiol. 41, 27-32.

Bodaness, R.S., and Chan, P.C. (1977) Singlet oxygen as a mediator in the
hematoporphyrin-catalyzed photooxidation of NADPH to NADP⁺ in
deuterium oxide. J. Biol. Chem. 252, 8554-8560.

Boggan, J.E., Walter, R., Edwards, M., Borcich, J.K., Davis, R.L.,
Koonce, M., and Berns, M.W. (1984) Distribution of hematoporphyrin
derivative in the rat 9L gliosarcoma brain tumor analyzed by digital
video fluorescence microscopy. J. Neurosurg. 61, 1113-

Bonnett, R., and Berenbaum, M.C. (1983) HPD- a study of its components
and their properties. photosensitization. In Porphyrin Photosensitization (Edited by D. Kessel and T.J. Dougherty). Plenum Press, New
York, pp 241-250.

Bonnett, R., C.R. Lambert, E.J. Land, P.A. Scourides, R.S. Sinclair and
T.G. Truscott, (1983) The triplet and radical species of haematoporphyrin and some of its derivatives. Photochem. Photobiol. 38, 1-8.

Bonnett, R., Ridge, R.J., and Scourides, P.A. (1980) Haematoporphyrin
derivative. J.C.S. Chem Comm. 24, 1198-1199.

Bonnett, R., R.J. Ridge and P.A.Scourides (1981) On the nature of
haematoporphyrin derivative. J.C.S Perkin I, 3135-3140.

Bottiroli, G., Freitas, I., Docchio, F., Ramponi, R., and Sacchi,
C.A. (1984) The time-dependent behavior of hematoporphyrin-derivative
in saline: a study of spectral modifications. Chem. Biol. Interact.
49, 1-11.

Bottiroli, G., Freitas, I., Docchio, F., Ramponi, R., and Sacchi,
C.A. (1984) Spectroscopic studies of hematoporphyrin-derivative in
culture medium. Chem. Biol. Interact. 50, 153-157.

Boye, E., and Moan, J. (1980) The photodynamic effect of hematoporphyrin
on DNA. Photochem. Photobiol. 31, 223-228.

Breitbart, H., and Malik, Z. (1982) The effects of photoactivated protoporphyrin on reticulocyte membranes, intracellular activities and
hemoglobin precipitation. Photochem. Photobiol. 35, 365-369.

Broderson, R. (1979) Bilirubin: solubility and interaction with albumin
and phospholipid. J. Biol. Chem. 254, 2364-2369.

Brookfield, R., Craw, M., Lambert, Land, E., Redmond, R., Sinclair R. and
Truscott T.G. (1984) Excited state properties of haematoporphyrin.
In Porphyrins in Tumor Therapy. (Edited by A. Andreoni and R. Cubed-
du). Plenum Press, New York, pp. 3-10.

Brown, S.B., Hazikonstantinou H. and Herries D. (1980) The structure of
porphyrins and haems in aqueous solution. Int. J. Biochem. 12, 701-
707.

Brown, S.B. and Shillcock M. (1976) Equilibrium and kinetic studies of
the aggregation of porphyrins in aqueous solution. Biochem. J. 153,
279-285.

Brun, A., Høvding G. and Romslo I. (1981) Protoporphyrin-induced
photohemolysis: differences related to the subcellular distribution
of protoporphyrin in erythropoietic protoporphyria and when added to
normal red cells. Int. J. Biochem. 13, 225-228.

Buettner, G.R. (1984) Thiyl free radical production with hematoporphyrin
derivative, cysteine and light: a spin-trapping study. FEBS Lett.,
177, 295-299.

Buettner, G.R. and Need, M.J. (1985) Hydrogen peroxide and hydroxyl free
radical production by hematoporphyrin derivative, ascorbate and
light. Cancer Lett. 25, 297-304.

Buettner, G.R., and Oberly, L.W. (1980) The apparent production of superoxide and hydroxyl radicals by hematoporphyrin and light as seen by
spin-trapping. FEBS Lett. 121, 161-164.

Bugelski, P., Porter C. and Dougherty T. (1981) Autoradiographic distribution of hematoporphyrin derivative in normal and tumor tissue of
the mouse. Cancer Res. 41, 4060-4612.

Cannistraro, S., Jori, G. and Van de Vorst, A. (1982) Quantum yield of electron transfer and of singlet oxygen production by porphyrins: an ESR study. Photobiochem. Photobiophys. 3, 353-363.

Cannistraro, S., and Van de Vorst, A. (1977) Photosensitization by hematoporphyrin: ESR evidence for free radical induction in unsaturated fatty acids and for singlet oxygen production. Biochem. Biophys. Res. Commun. 74, 1177-1185.

Cannistraro, S., Van de Vorst, A., and Jori, G. (1978) EPR studies on singlet oxygen production by porphyrins. Photochem. Photobiol. 28, 257-259.

Canti, G., Franco, P., Marelli, O., Ricci, L., and Nicolin, A. (1984) Hematoporphyrin derivative rescue from toxicity caused by chemotherapy or radiation in a murine leukemia model (L1210). Cancer Res. 44, 1551-1556.

Carpenter, B.J., Ryan, R.J., Neel III, H.B., and Sanderson, D.R. (1977) Tumor fluorescence with hematoporphyrin derivative. Ann. Otol. 86, 661-666.

Carruth, J.A.S., and McKenzie, A.L. (1985) Preliminary report of a pilot study of photoradiation therapy for the treatment of superficial malignancies of the skin, head and neck. Eur. J. Surg. Oncol. In Press.

Cauzzo, G., Gennari, D., Jori, G., and Spikes, J.D. (1977) The effect of chemical structure on the photosensitizing efficiencies of porphyrins. Photochem. Photobiol. 25, 389-395.

Chang, C. and Dougherty, T. (1978) Photoradiation therapy: kinetics and thermodynamics of porphyrin uptake and loss in normal and malignant cells in culture. Rad. Res. 74, 498-506.

Cheli, R., Addis, F., Mortellaro, C.M., Fonda, D., Andreoni, A., and Cubeddu, R. (1984) Hematoporphyrin derivative photochemotherapy of spontaneous animal tumors: clinical results with optimized drug dose. Cancer Lett. 23, 61-66.

Cheng, M-K., McKean, J., Boisvert, D., Tulip, J., and Mielke, B.W. (1984) Effects of photoradiation therapy on normal rat brain. Neurosurg. 15, 804-810.

Christensen, T. (1981) Multiplication of human NHIK 3025 cells exposed to porphyrins in combination with light. Br. J. Cancer 44, 433-439.

Christensen, T., Feren, K., Moan, J., and Pettersen, E. (1981) Photodynamic effects of haematoporphyrin derivative on synchronized and asynchronous cells of different origin. Br. J. Cancer 44, 717-724.

Christensen, T., and Moan, J. (1979) Photodynamic inactivation of synchronized human cells in vitro in the presence of hematoporphyrin. Cancer Res. 39, 3735-3737.

Christensen, T., Moan, J. McGhie, J.B., Waksvik, H. and Stigum H. (1983) Studies of HPD: chemical composition and in vitro photosensitization. In Porphyrin Photosensitization (Edited by D. Kessel and T.J. Dougherty). Plenum Press, New York, pp 63-76.

Christensen, T., Moan, J., Sandquist T. and Smedshammer L. (1984) Multicellular spheroids as an in vitro model system for photoradiation therapy in the presence of HPD. In Porphyrin Localization and Treatment of Tumors. (Edited by D.R. Doiron and C.J. Gomer). Alan R. Liss, New York, pp. 381-390.

Christensen, T., Moan, J., Wibe, E., and Ofteboro, R. (1979) Photodynamic effect of haematoporphyrin through the cell cycle of the human cell line NHIK 3025 cultivated in vitro. Br. J. Cancer 39, 64-68.

Christensen, T., Sandquist, T., Feren, K., Waksvik H. and Moan, J. (1983) Retention and photodynamic effects of haematoporphyrin derivative in cells after prolonged cultivation in the presence of porphyrin. Br. J. Cancer 48, 35-43.

Christensen, T., Volden, G., Moan, J., and Sandquisyt, T. (1982) Release of lysosomal enzymes and lactate dehydrogenase due to hematoporphyrin derivative and light irradiation of NHIK 3025 cells in vitro. Ann. Clin. Res. 14, 46-52.

Christensen, T., Wahl, A., and Smedshammer, L. (1984) Effects of haematoporphyrin derivative and light in combination with hyperthermia on cells in culture. Br. J. Cancer 50, 85-89.

Clezy, P., Hai, T., Henderson R. and van Thuc L. (1980). The chemistry of pyrrolic compounds. XLV Haematoporphyrin derivative: Haematoporphyrin diacetate as the main product of the reaction of haematoporphyrin with a mixture of acetic and sulfuric acids. Aust. J. Chem. 33, 585-597.

Codd, J.E. (1983) A histological study of the uptake of porphyrins by normal and malignant tissue. Med Lab. Sci. 40, 401.

Cohen, S, and Margalit, R. (1985) Binding of hematoporphyrin derivative to membranes. Expression of porphyrin heterogeneity and effects of cholesterol studied in large unilamellar liposomes. Biochim. Biophys. Acta 813, 307-312.

Coppola, A. Viggiana, E., Salzarulo, L., and Rasile, G. (1980) Ultrastructural changes in lymphoma cells treated with hematoporphyrin and light. Am. J. Path., 99, 175-192.

Cortese, D. and Kinsey J. (1982) Endoscopic management of lung cancer with hematoporphyrin derivative phototherapy. Mayo Clin. Proc. 57, 543-547.

Cortese, D., and Kinsey, J. (1982) Hematoporphyrin derivative phototherapy for local treatment of cancer in the tracheobronchial tree. Ann. Otol. Rhinol. Laryngol. 91, 652-655.

Cowled, P., Grace J. and Forbes I. (1984) Comparison of the efficacy of pulsed and continuous-wave red laser light in induction of phototoxicity by haematoporphyrin derivative. Photochem. Photobiol. 39, 115-117.

Cowled, P., Forbes, I., Swincer, A., Trenerry, V. and Ward, A.D. (1985) Separation and phototoxicity in vitro of some of the components of haematoporphyrin derivative. Photochem. Photobiol 41, 445-451.

Cox, G., Krieg, M., and Whitten, D. (1982) Self-sensitized photooxidation of protoporphyrin IX derivatives in aqueous surfactant solutions: product and mechanistic studies. J. Amer. Chem. Soc. 104, 6930-6937.

Cox, G. and Whitten, D. (1983) Excited state interactions of protoporphyrin IX and related porphyrins with molecular oxygen in solutions and in organized assemblies. In Porphyrin Photosensitization (Edited by D. Kessel and T.J. Dougherty. Plenum Press, New York, pp. 279-292.

Cozzani, I., Jori, G., Bertoloni, G., Milanesi, C., Carlini, P., Siouoro, T and Ruschi, A. (1985) Efficient photosensitization of malignant human cells in vitro by liposome-bound porphyrins. Chem-Biol. Interact. 53, 131-144.

Cozzani, I., Jori, G., Reddi, E., Fortunato, A., Bruno, G., Felice, M., Tomio, L., and Zorat, P. (1981) Distribution of endogenous and injected porphyrins at the subcellular level in rat hepatocytes and in ascites hepatoma. Chem.-Biol. Interact. 37, 67-75.

Cozzani, I. and Spikes, J. (1982) Photodamage and photokilling of malignant human cells in vitro by hematoporphyrin-visible light: molecular basis and possible mechanisms of phototoxicity. Med. Biol. Enviorn. 10, 271-276.

Creekmore, S. and Zaharko, D. (1983) Modification of chemotherapeutic effects on L1210 cells using hematoporphyrin and light. Cancer Res. 43, 5252-5257.

Dahlman, A., Wile, A., Burns, R., Mason, G., Johnson F. and Berns, M. (1983) Laser photoradiation therapy of cancer. Cancer Res., 43, 430-434.

Das, M., Dixit, R., Mukhtar, H., and Bickers, D.R. (1985) Role of active oxygen species in the photodestruction of microsomal cytochrome P-450 and associated monooxygenases by hematoporphyrin derivative in rats. Cancer Res. 45, 608-615.

De Goeij, A., Van Straalen, R. and Van Steveninck, J. (1976) Photodynamic modification of proteins in human red blood cell membranes, induced by protoporphyrin. Clin. Chim. Acta 71, 485-494.

De Goeij, A., Vevergaert, P. and Van Steveninck, J. (1975) Photodynamic effects of protoporphyrin on the architecture of erythrocyte membranes in protoporphyria and in normal red cells. Clin. Chem. Acta, 62, 287-292.

De Paolis, A., Chandra, S., Charalambides, A., Bonnett, R. and Magnus, I.A. (1985) The effect on photohemolysis of variation in the structure of the porphyrin photosensitizer. Biochem. J. 226, 757-766.

Deziel, M. and Girotti, A. (1982) Lysis of resealed erythrocyte ghosts by photoactivated tetrapyrroles: estimation of photolesion dimensions. Int. J. Biochem., 14, 263-266.

Diamond, I., Granelli, S. and McDonagh, A. (1977) Photochemotherapy and photodynamic toxicity: simple methods for identifying potentially active agents. Biochem. Med. 17, 121-127.

Diamond, I., Granelli, S., McDonagh, A., Nielsen, S., Wilson C. and Jaenicke R. (1973) Photodynamic therapy of malignant tumors. Lancet 2, 1175-1177.

Diezel, W., Sonnichsen, N., and Meffert, H. (1983) Treatment of psoriasis with hematoporphyrin derivative and long-wave ultraviolet light. Stud. Biophys. 94, 45-46.

Docchio, F., Ramponi, R., Sacchi, S., Bottiroli G. and Freitas, I. (1982) Fluorescence studies of biological molecules by laser irradiation. Lasers in Surg. and Med. 2, 21-28.

Docchio, F., Ramponi, R., Sacchi, C., Bottiroli, G. and Freitas, I. (1982). Time-resolved fluorescence microscopy of hematoporphyrin-derivative in cells. Lasers in Med. Biol. 2, 21-28.

Doiron, D., Svaasand, L. and Profio, A. (1983) Light dosimetry in tissue: application to photoradiation therapy. In Porphyrin Photosensitization (Edited by D. Kessel and T.J. Dougherty). Plenum Press, New York, pp. 63-76.

Dougherty, T.J. (1974) Activated dyes as antitumor agents. J. Natl. Cancer Inst. 52, 1333-1336.

Dougherty, T. (1981) Photoradiation therapy for cutaneous and subcutaneous malignancies. J. Invest. Dermatol. 77, 122-124.

Dougherty, T.J. (1982) Photoradiation therapy for bronchogenic cancer. Chest, 81, 265-266.

Dougherty, T., Weishaupt, K. and Boyle, D. (1982) Photoradiation therapy of malignant tumors. in Principals and Practice of Oncology ed. V. De Vita, S. Hellman and S. Rosenberg. J.P. Lippincott, Philadelphia. pp 1836-1844.

Dougherty, T.J. (1983) Hematoporphyrin as a photosensitizer of tumors. Photochem. Photobiol. 38, 377-379.

Dougherty, T.J. (1984) The structure of the active component of hematoporphyrin derivative. In Porphyrin Localization and Treatment of Tumors. (Edited by D.R. Doiron and C.J. Gomer). Alan R. Liss, New York, pp. 301-314.

Dougherty, T.J. (1984) Photoradiation therapy (PRT) of malignant tumors. CRC Critical Reviews in Oncology/Hematology, CRC Publ. Co., New York.

Dougherty, T.J. (1984) An overview of the status of photoirradiation therapy. In Porphyrin Localization and Treatment of Tumors. (Edited by D.R. Doiron and C.J. Gomer). Alan R. Liss, New York, pp. 75-87.

Dougherty, T., Boyle, D., Weishaupt, K., Henderson, B., Potter, W., Bellnier, D. and Wityk, K. (1983) Photoradiation therapy - clinical and drug advances. In Porphyrin Photosensitization (Edited by D. Kessel and T.J. Dougherty) Plenum Press, New York. pp. 3-13.

Dougherty, T., Boyle, D., Weishaupt, K., Potter, W. and Thoma, R. (1983) Photoradiation therapy of malignant tissues. in New Frontiers in Laser Medicine and Surgery, ed. K. Atsumi, Excerpta Medica (Amsterdam-Oxford-Princeton) pp. 161-165.

Dougherty, T., Gomer, C. and Weishaupt, K. (1976) Energetics and efficiency of photoinactivation of murine tumor cells containing hematoporphyrin. Cancer Res. 36, 2330-2333.

Dougherty, T., Grindey, G., Weishaupt, K. and Boyle, D. (1975)
Photoradiation therapy II. J. Natl. Cancer Inst. 55, 115-120.
Dougherty, T., Kaufman, J., Goldfarb, A., Weishaupt, K., Boyle, D. and
Mittleman, A. (1978) Photoradiation therapy for the treatment of
malignant tumors. Cancer Res. 38, 2628-2635.
Dougherty, T., Lawrence, G., Kaufman, J., Boyle, D., Weishaupt, K., and
Goldfarb, A. (1979) Photoradiation in the treatment of recurrent
breast carcinoma. J. Natl. Cancer Inst. 62, 231-237.
Dougherty, T., Thoma, R., Boyle, D. and Weishaupt, K. (1981) Interstitial
photoradiation therapy for primary solid tumors in pet cats and dogs.
Cancer Res. 41, 401-404.
Dougherty, T., Weishaupt, K. and Boyle, D.G (1982) Photosensitizers. In
Cancer: Principals and Practice of Oncology (Edited by V. DeVita and
S.A. Rosenberg) J.P. Lippincott, Philadelphia, pp. 1836-1844.
Dubbelman, T., De Bruijne, A. and Van Steveninck, J. (1977) Photodynamic
effects of protoporphyrin on red blood cell deformability. Biochem.
Biophys. Res. Commun., 77, 811-817.
Dubbelman, T., De Goeij, A., Christiansen, K. and Van Steveninck, J.
(1981) Protoporphyrin-induced photodynamic effects on band 3 protein
of human erythrocyte membranes. Biochim. Biophys. Acta 649, 310-316.
Dubbelman, T., De Goeij, A. and Van Steveninck, J. (1978) Photodynamic
effects of protoporphyrin on human erythrocytes. Nature of the cross-
linking of membrane proteins. Biochim. Biophys. Acta 511, 141-151.
Dubbelman, T., De Goeij, A. and Van Steveninck, J. (1978) Protoporphyrin-
sensitized photodynamic modification of proteins in isolated human
red blood cell membranes. Photochem. Photobiol. 28, 197-204.
Dubbelman, T., De Goeij, A. and Van Steveninck, J. (1978) Photodynamic
effects of protoporphyrin on human erythrocytes. Nature of the
cross-linking of membrane proteins. Biochim. Biophys. Acta 511, 141-
151.
Dubbelman, T., De Goeij, A. and Van Steveninck, J. (1980) Protoporphyrin-
induced photodynamic effects on transport processes across the membrane of human erythrocytes. Biochim. Biophys. Acta 595, 133-139.
Dubbelman, T., Haasnoot, C. and Van Steveninck, J. (1980) Temperature dependence of photodynamic red cell membrane damage. Biochim. Biophys.
Acta 601, 220-227.
Dubbelman, , Leenhouts J. and Van Steveninck, J. (1984) Photodynamic in-
activation of L929 cells after treatment with hematoporphyrin deriva-
tive. In Porphyrins in Tumor Therapy. (Edited by A. Andreoni and
R. Cubeddu). Plenum Press, New York, pp. 167-176.
Dubbelman, T., Van Steveninck, A. and Van Steveninck, J. (1982)
Hematoporphyrin-induced photo-oxidation and photodynamic cross-
linking of nucleic acids and their constituents. Biochim. Biophys.
Acta 719, 47-52.
Dubbelman, T., Van Steveninck A. and Van Steveninck, J. (1982)
Hematoporphyrin-induced photo-oxidation and photodynamic cross-
linking of nucleic acids and their constituents. Biochim. Biophys.
Acta. 719, 47-52.
Dubbelman, T. and Van Steveninck, J. (1984) Photodynamic effects of
hematoporphyrin-derivative on transmembrane transport systems of
murine L929 fibroblasts. Biochim. Biophys. Acta 771, 201-207.
Ebert, P., Hess, R. and Tschudy, D. (1985) Augmentation of hematoporphy-
rin uptake and in-vitro growth inhibition of L1210 leukemia cells by
succinylacetone. J. Natl. Cancer Inst., 74, 603-608.
Ebert, P., Smith, P., Bonner, R., Hess, R., Costa, J. and Tschudy,
D. (1983) Effect of defined wavelength and succinylacetone on the
photoinactivation of leukemia L1210 cells in vitro by hematoporphyrin. Photobiochem. Photobiophys. 6, 165-175.
Ehrenberg, B., Malilk, Z. and Nitzan, Y. (1985) Fluorescence spectral
changes of hematoporphyrin derivative upon binding to lipid vesicles,

Staphylococcus aureus and Escherichia coli cells. Photochem.
Photobiol. 41, 429-435.

El-Far, M. and Pimstone, N. (1983) Tumour localization of uroporphyrin
isomers I and III and their correlation to albumin and serum protein
binding. Cell Biochem. Funct. 1, 156-160.

El-Far, M. and Pimstone, N. (1983) Uroporphyrin I is a better tumor
localizer than hematoporphyrin derivative in mouse mammary carcinoma.
Clin. Res. 31, 2.

Emiliani, C. and Delmelle, M. (1983) The lipid solubility of porphyrins
modulates their phototoxicity in membrane models. Photochem.
Photobiol. 37, 487-490.

Emiliani, C., Delmelle, M., Cannistraro, S. and Van de Vorst, A. (1983)
Solubility of hematoporphyrin and photodynamic damages in liposomal
systems: optical and electron spin resonance studies. Photobiochem.
Photobiophys. 5, 119-128.

Evensen, J. and J. Moan (1982) Photodynamic action and chromosomal damage
of haematoporphyrin derivative (HpD) and light with X-irradiation.
Br. J. Cancer 45, 456-465.

Evensen, J., Moan, J.,Hindar A. and Sommer, S. (1984) Tissue distribution
of ^{3}H-hematoporphyrin derivative and its main components, ^{67}Ga and
^{131}I-albumin in mice bearing Lewis lung carcinoma. In Porphyrin
Localization and Treatment of Tumors. (Edited by D.R. Doiron and
C.J. Gomer). Alan R. Liss, New York, pp. 541-562.

Evensen, J., Somer, S., Moan, J. and Christensen, T. (1984) Tumor-
localizing properties of the main components of hematoporphyrin
derivative. Cancer Res. 44, 482-486.

Facchini, V., Cela, M., Gadducci, A., Tau, A. and Cozzani, I. (1984)
Phototherapy of tumors sensitized by liposome-bound porphyrins locally infiltered in neoplastic areas. Med. Biol. Environ. 12, 469-474.

Felix, C., Reszka, K. and Sealy, R. (1983) Free radicals from photoreduction of hematoporphyrin in aqueous solution. Photochem. Photobiol.
37, 141-147.

Fiel, R., Datta-Gupta, N., Mark E. and Howard, J. (1981) Induction of DNA
damage by porphyrin photosensitizers. Cancer Res. 41, 3543-3545.

Figge, F., Wieland, G. and Mangiello, L. (1948) Cancer detection and
therapy. Affinity of neoplastic, embryonic and traumatized tissue
for porphyrins and metalloporphyrins. Proc. Soc. Exptl. Biol. Med.
68, 640-641.

Forbes, I., Cowled, P., Leong, A., Ward, A.D., Black, R., Blake, A. and
Jacka, F. (1980) Phototherapy of human tumours using haematoporphyrin
derivative. Med. J. Aust. 2, 489-493.

Forbes, I., Ward, A.D., Jacka, F., Blake, A., Swincer, A.G., Wilksch, P.,
Cowled, P. and See, K. (1984) Multidisciplinary approach to photo-
therapy of human cancers. In Porphyrin Localization and Treatment of
Tumors. (Edited by D.R. Doiron and C.J. Gomer). Alan R. Liss, New
York, pp. 693-708.

Franco, P., Mirelli, S., Sarra, F. and Nicolin, A. (1981) Inhibition of
cellular DNA synthesis and lack of antileukemic activity by non-
photoactivated hematoporphyrin derivative. Tumori 67, 183-189.

Ganguly, T. and Bhattacharjee, S. (1983) Photodynamic inactivation of
Chinese hamster cells. Photochem. Photobiol. 38, 65-69.

Gibson, S., Cohen, H. and Hilf, R. (1984) Evidence against the production of superoxide by photoradiation of hematoporphyrin derivative.
Photochem. Photobiol. 40, 441-448.

Gibson, S. and Hilf, R. (1983) Photosensitization of mitochondrial
cytochrome C oxidase by hematoporphyrin derivative and related porphyrins in vitro and in vivo. Cancer Res. 43, 4191-4197.

Giloh, H. and Sedat, J. (1982) Fluorescence microscopy: reduced
photobleaching of rhodamine and fluorescein protein conjugates by n-
propyl gallate. Science 217, 1252-1254.

Girotti, A. (1976) Photodynamic action of protoporphyrin IX on human erythrocytes: cross-linking of membrane protein. Biochem. Biophys. Res. Commun. 72, 1367-1374.

Girotti, A. (1979) Protoporphyrin-sensitized photodamage in isolated membranes of human erythrocytes. Biochem. 18, 4403-4411.

Girotti, A. (1980) Photosensitized cross-linking of erythrocyte membrane proteins. Evidence against participation of amino groups in the reaction. Biochim. Biophys. Acta 602, 45-56.

Girotti, A. and Deziel, M. (1983) Photodynamic action of protoporphyrin on resealed erythrocyte membranes: mechanism of release of trapped markers. In Porphyrin Photosensitization (Edited by D. Kessel and T.J. Dougherty). Plenum Press, New York, pp. 213-225.

Goldstein, B. and Harber, L. (1972) Erythropoietic protoporphyria: lipid peroxidation and red cell membrane damage associated with photohemolysis. J. Clin. Invest. 51, 892-902.

Gomer, C.J. (1980) DNA damage and repair in CHO cells following hemato-porphyrin photoradiation. Cancer Lett. 11, 161-167.

Gomer, C., Doiron, D., Jester, J., Szirth B. and Murphree, A. (1983) Hematoporphyrin derivative photoradiation therapy for the treatment of intraocular tumors: examination of acute normal ocular tissue toxicity. Cancer Res. 43, 721-727.

Gomer, C., Doiron, D., Rucker, N., Razum, N. and Fountain, S. (1984) Action spectrum (620-640 nm) for hematoporphyrin derivative induced cell killing. Photochem. Photobiol. 39, 365-368.

Gomer, C., Doiron, D., While, L., Jester, J., Dunn, S., Szirth, B., Razum, N. and Murphree, A. (1984) Hematoporphyrin derivative photoradiation induced damage to normal and tumor tissue of pigmented rabbit eye. Current Eye Res. 3, 229-237.

Gomer, C. and T. Dougherty (1979) Determination of [^{3}H]-and [^{14}C]hematoporphyrin derivative distribution in malignant and normal tissue. Cancer Res. 39, 146-151.

Gomer, C. and Razum, N. (1984) Acute skin response in albino mice following porphyrin photosensitization under oxic and anoxic conditions. Photochem. Photobiol. 40,435-439.

Gomer, C., Rucker, N., Banerjee A. and Benedict, W. (1983) Comparison of mutagenicity and induction of sister chromatid exchange in Chinese hamster cells exposed to hematoporphyrin derivative photoradiation, ionizing radiation or ultraviolet radiation. Cancer Res. 43, 2622-2627.

Gomer, C., Rucker, N., Razum, N and Murphree, A.L. (1985) In vitro and in vivo light dose rate effects related to hematoporphyrin derivative photodynamic therapy. Cancer Res. 45, 1973-1977.

Gomer, C. and Smith, D. (1980) Photoinactivation of Chinese hamster cells by hematoporphyrin derivative and red light. Photochem. Photobiol. 32, 341-348.

Gouterman, M. (1959) Study of the effects of substitution on the absorption spectra of porphin. J. Chem. Phys. 30, 1139-1160.

Goyal, G., Blum, A. and Grossweiner, L. (1983) Photosensitization of liposomal membranes by hematoporphyrin derivative. Cancer Res. 43, 5826-5830.

Granelli, S., Diamond, I., McDonagh, A., Wilson, C. and Nielsen, S. (1975) Photochemotherapy of glial cells by visible light and hemato-porphyrin. Cancer Res. 35, 2567-2570.

Gregorie, H., Horder, E., Ward, J., Green, J., Richards, T., Robertson, H.C., and Stevenson, T.B. (1968) Hematoporphyr-derivative fluores-cence in malignant neoplasms. Ann. Surg. 167, 820-828.

Grey, M., Lipson, M., Mack, J., Parker, L. and Rombyn, D. (1967) Use of hematoporphyrin derivative in detection and management of cervical cancer. Am. J. Obstet. Gynecol. 99, 766-770.

Grossweiner, L.I. (1984) Membrane photosensitization by hematoporphyrin and hematoporphyrin derivative. In Porphyrin Localization and Treat-ment of Tumors. (Edited by D.R. Doiron and C.J. Gomer). Alan R. Liss, New York, pp. 391-404.

Grossweiner, L. and Goyal, G. (1983) Photosensitized lysis of liposomes by hematoporphyrin derivative. Photochem. Photobiol. 37, 529-532.

Grossweiner, L. and Goyal, G. (1984) Photosensitization of liposomes by porphyrins. J. Photochem. 2, 253-265.

Grossweiner, L. and Goyal, G. (1984) Binding of hematoporphyrin derivative to human serum albumin. Photochem. Photobiol. 40, 1-4.

Grossweiner, L. and Grossweiner, J. (1982) Hydrodynamic effects in the photosensitized lysis of liposomes. Photochem. Photobiol. 35, 583-586.

Grossweiner, L., Patel, A. and Grossweiner, J. (1982) Type I and type II mechanisms in the photosensitized lysis of phosphatidylcholine liposomes by hematoporphyrin. Photochem. Photobiol. 36, 159-167.

Gruner, S., Meffert, H., Vold, H., Grunow, R. and Jahn, S (1985) The influence of hematoporphyrin derivative and visible light on murine skin graft survival, epidermal Langerhans cells, and stimulation of the allogenic mixed leukocyte reaction. Scand. J. Immnol. 21, 267-274.

Gutter, B., Speck, W. and Rosenkranz, H. (1978) The photodynamic modification of DNA by hematoporphyrin. Biochim. Biophys. Acta, 475, 307-314.

Hariharan, P., Courtney, J. and Eleczko, S. (1980) Production of hydroxyl radicals in cell systems exposed to hematoporphyrin and red light. Int. J. Rad. Biol. 37, 691-694.

Hayata, H. (1983) Laser equipment and technique. In Lasers and Hematoporphyrin Derivative in Cancer Therapy. Y Hayata and TJ Dougherty, eds, Igako-Shoin, Tokyo, New York. pp 11-20.

Hayata, Y. and Kato, H. (1983) Applications of laser phototherapy in the diagnosis and treatment of lung cancer. Jap. Ann. Thoracic Surg. 3, 203-210.

Hayata, Y., Kato, H., Konaka, C., Amemiya, R., Ono, J., Ogawa, I., Kinoshita, K., Sakai, H. and Takahashi, H. (1984) Photoradiation therapy with hematoporphyrin derivative in early and stage 1 lung cancer. Chest 86, 169-177.

Hayata, Y., Kato, H., Konaka, C., Hayashi, N., Tahara, M., Saito T. and Ono, J. (1983) Fiberoptic bronchoscopic photoradiation in experimentally induced canine lung cancer. Cancer 51, 50-56.

Hayata, Y., Kato, H., Konaka, C., Ono, J., Matsushima, Y., Yoneyama, K. and Nishimiya, K. (1982) Fiberoptic bronchoscopy laser photoradiation for tumor localization in lung cancer. Chest 82, 10-14.

Hayata, Y., Kato, H., Konaka, C., Ono J. and Takizawa, N. (1982) Hematoporphyrin derivative and laser photoradiation in the treatment of lung cancer. Chest 81, 269-277.

Hayata, H., Kato, H., Konaka, C., Ono, J. and Takizawa, N (1982) Hematoporphyrin derivative and laser photoradiation therapy in the treatment of lung cancer. Chest 81, 269-277.

Hayata, Y., Kato, H., Ono, J., Matsushima, Y., Hayashi, N., Saito T. and Kawate, N. (1982) Fluorescence fiberoptic bronchoscopy in the diagnosis of early stage lung cancer. Recent Results Cancer Res. 82, 121-130.

Henderson, B., Bellnier, D., Zirling, B. and Dougherty, T. (1983) Aspects of the cellular uptake and retention of hematoporphyrin derivative and their correlation with the biologic response to PRT in vitro. In Porphyrin Photosensitization (Edited by D. Kessel and T.J. Dougherty). Plenum Press, New York, pp. 279-292.

Henderson, R., Christie, G., Clezy, P. and Lineham, J. (1980) Haematoporphyrin diacetate: a probe to distinguish malignant from normal tissue by selective fluorescence. Br. J. Exp. Pathol. 61, 345-350.

Henderson, B. and Dougherty, T. (1984) Studies on the mechanism of tumor destruction by photoradiation therapy. In Porphyrin Localization and Treatment of Tumors. (Edited by D.R. Doiron and C.J. Gomer). Alan R. Liss, New York, pp. 601-612.

Henderson, B., Waldow, S., Mang, T., Potter, W., Malone, P. and Dougherty, T.J. (1985) Tumor destruction and kinetics of cell death in two experimental mouse tumors following photodynamic therapy. Cancer Res. 45, 572-576.

Hilf, R., Leakey, P., Sollott, S. and Gibson, S. (1983) Photodynamic inactivation of R3230AC mammary carcinoma in vitro with hematoporphyrin derivative: effects of dose, time and serum on uptake and phototoxicity. Photochem. Photobiol. 37, 633-642.

Hilf, R., Smail, D., Murant, R., Leakey, P. and Gibson, S. (1984) Hematoporphyrin derivative-induced photosensitivity of mitochondrial succinate dehydrogenase and selected cytosolic enzymes of R3230C mammary adenocarcinoma of rats. Cancer Res. 44, 1483-1488.

Hilf, R., Warne, N., Smail, D. and Gibson, S. (1984) Photodynamic inactivation of selected intracellular enzymes by hematoporphyrin derivative and their relationship to tumor cell viability in vitro. Cancer Lett. 24, 165-172.

Hisazumi, H. (1984) A trial manufacture of a motor-driven laser light scattering optic for whole bladder wall irradiation. In Porphyrin Localization and Treatment of Tumors. (Edited by D.R. Doiron and C.J. Gomer). Alan R. Liss, New York, pp. 239-248.

Hisazumi, H., Mikasi, T. and Miyoshi, N. (1983) Photoradiation therapy of bladder tumours. J. Urol. 130, 685-687.

Hisazumi, H., Miyoshi, N., Ueki, O., Nishino, A. and Nakajima, K. (1984) Cellular uptake of hematoporphyrin derivative in KK-47 bladder cancer cells. Urol. Res. 12, 143-146.

Holt, S., Tulip, J., Hamilton, D., Cummins, J., Fields, A. and Dick, C. (1985) Experimental laser phototherapy of the Morris 7777 hepatoma in the rat. Hepatology (Baltimore) 5:175-180.

Ito, T. (1981) Photodynamic action of hematoporphyrin on yeast cells - a kinetic approach. Photochem. Photobiol. 34, 521-524.

Ito, T. (1983) Photodynamic agents as tools for cell biology. In Photochemical and Photobiological Reviews, Vol. 7. (Edited by K.C. Smith). Plenum Publ. Co., New York, pp. 141-186.

Ito, T. (1983) Mode of photodynamic action with special reference to the sensitizer-cell interaction: porphyrins and membrane damage. Studia Biophys. 94, 1-6.

Jacka, F. and Blake, A. (1983) A lamp for cancer phototherapy. Aust. J. Phys. 36, 221-226.

Jones, G., Kinsey, J., Neel, H. and Cortese, D. (1984) The effect of cooling in the photodynamic action of hematoporphyrin derivative during interstitial phototherapy of solid tumors. Otolaryngol. Head Neck Surg. 92, 532-536.

Jori, G., Beltramini, M., Reddi, E., Salvato, B., Pagnan, A., Ziron, L., Tomio, L. and Tsanov, T. (1984) Evidence for a major role of plasma lipoproteins as hematoporphyrin carriers in vivo. Cancer Lett. 24, 291-297.

Jori, G., Pizzi, G., Reddi, E., Tomio, L., Salvato, B., Zorat, P. and Calzavara, F. (1979) Time dependence of hematoporphyrin distribution in selected tissues of normal rats and in ascites hepatoma. Tumori 65, 425-434.

Jori, G., Reddi, E., Rossi, E., Cozzani, I., Tomio, L., Zorat, P.L., Pizzi, G.B. and Calzavara, F. (1980) Porphyrin-sensitized photoreactions and their use in cancer phototherapy. Med. Biol. Environ. 8, 141-154.

Jori, G., Reddi, E., Tomio, L. and Calzavara, F. (1983) Factors governing the mechanism and efficiency of porphyrin-sensitized photooxidations in homogeneous and organized media. In Porphyrin Photosensitization (Edited by D. Kessel and T.J. Dougherty). Plenum Press, New York, pp. 193-212.

Jori, G. and Spikes, J. (1984) Photobiochemistry of Porphyrins. In

Topics in Photomedicine, Ed K.C. Smith, Plenum Press, New York,
pp. 183-318.

Jori, G., Tomio, L., Reddi, E., Rossi, E., Corti, L., Zorat P. and Calzavara, F. (1983) Preferential delivery of liposome-incorporated
porphyrins to neoplastic cells in tumour-bearing rats. Br. J. Cancer
48, 307-309.

Karns, G., Gallagher, W. and Elliott, W. (1979) Dimerization constants of
water-soluble porphyrins in aqueous alkali. Bioorg. Chem. 8,69-81.

Kelly, J. and Snell, M. (1976) Hematoporphyrin derivative: a possible aid
in the diagnosis and therapy of carcinoma of the bladder. J. Urol.
115, 150-151.

Kelly, J., Snell, M. and Berenbaum, M. (1975) Photodynamic destruction of
human blader carcinoma. Br. J. Cancer 31, 237-244.

Kennedy, J. (1983) HPD photoradiation therapy for cancer at Kingston and
Hamilton. In Porphyrin Photosensitization (Edited by D. Kessel and
T.J. Dougherty). Plenum Press, New York, pp. 53-62.

Kennedy, J. and Oswald, K. (1984) Hematoporphyrin derivative photoradia-
tion therapy, in theory and practice. In Porphyrins in Tumor
Therapy. (Edited by A. Andreoni and R. Cubeddu). Plenum Press, New
York, pp. 365-371.

Kessel, D. (1976) Effects of photoactivated porphyrins at the cell surface of leukemia L1210 cells. Biochem. 16, 3443-3449.

Kessel, D. (1981) Transport and binding of hematoporphyrin derivative and
related porphyrins by murine leukemia L1210 cells. Cancer Res. 41,
1318-1323.

Kessel, D. (1982) Determinants of hematoporphyrin-catalyzed photosensitization. Photochem. Photobiol. 36, 99-101.

Kessel, D. (1982) Components of hematoporphyrin derivative and their
tumor-localizing ability. Cancer Res. 42, 1703-1706.

Kessel, D. (1984) Chemical and biochemical determinants of porphyrin
localization. In Porphyrin Localization and Treatment of
Tumors. (Edited by D.R. Doiron and C.J. Gomer). Alan R. Liss, New
York, pp. 405-418.

Kessel, D. (1984) Hematoporphyrin and HPD: Photophysics, photochemistry
and Phototherapy. Photochem Photobiol. 39,851-859.

Kessel, D. (1984) Porphyrin localization: a new modality for detection
and therapy of tumors. Biochem Pharmacol 33, 1389-1393.

Kessel D. and Cheng, M-L. (1985) On the preparation and properties of
dihematoporphyrin ether, the tumor-localizing component of HPD.
Photochem Photobiol 41, 277-282.

Kessel, D. and Cheng, M-L. (1985) Biological and biophysical properties
of the tumor-localizing component of hematoporphyrin derivative.
Cancer Res. 45, 3053-3057.

Kessel, D. and Chou, T. (1983) Tumor-localizing components of the porphyrin preparation hematoporphyrin derivative. Cancer Res. 43, 1994-
1999.

Kessel, D. and Dutton, C. (1984) Photodynamic effects: porphyrin vs.
chlorin. Photochem. Photobiol. 40, 403-406.

Kessel D. and Kohn, K. (1980). Modes of transport and binding of
mesoporphyrin IX by leukemia L1210 cells. Cancer Res 40, 303-307.

Kessel, D. and Rossi, E. (1982) Determinants of porphyrin-sensitized
photooxidation characterized by fluorescence and absorption spectra.
Photochem. Photobiol. 35, 37-41.

Kessel, D. and Sykes, E. (1984) Porphyrin accumulation by atheromatous
plaques of the aorta. Photochem. Photobiol. 40, 59-61.

King, E., Doiron, D., Man, G., Profio, A.E. and Huth, G. (1982) Hematoporphyrin derivative as a tumor marker in the detection and localization of pulmonary malignancy. (1982) Recent Results Cancer Res. 82,
90-96.

King, E., Man, G.,J. LeRiche, J., Amy, R., Profio, A.E. and Doiron,
D. (1982) Fluorescence bronchoscopy in the localization of

bronchogenic carcinoma. Cancer 49, 777-782.

Kinsey, J., Cortese, D., Moses, H., Ryan, R. and Branum, E. (1981) Photo-dynamic effect of hematoporphyrin derivative as a function of optical spectrum and incident energy density. Cancer Res. 41, 5020-5026.

Kinsey, J., Cortese, D. and Neel, H. (1983) Thermal considerations in murine tumor killing using hematoporphyrin derivative phototherapy. Cancer Res. 43, 1562-1567.

Kinsey, J., Cortese, D. and Sanderson, D. (1978) Detection of hematopor-phyrin fluorescence during fiberoptic bronchoscopy to localize early bronchogenic carcinoma. Mayo Clin. Proc., 53, 594-600.

Klaunig, J., Selman, S., Shulok, J., Schafer, P., Britton, S and Goldblatt, P. (1985) Morphologic studies of bladder tumors treated with hematoporphyrin derivative phototherapy. Am. J. Path. 119, 236-243.

Kohn, K. and Kessel, D. (1980) On the mode of cytotoxic action of photo-activated porphyrins. Biochem. Pharmacol. 28, 2465-2470.

Koller, M-E. and Romslo, I. (1980) Uptake of protoporphyrin IX by iso-lated rat liver mitochondria. Biochem. J. 188, 329-335.

Kontos, H. and Hess, M. (1983) Oxygen radicals and vascular damage. In Myocardial Injury, (Edited by J.J. Spitzer) Plenum Press, New York, pp. 365-375.

Kreimer-Birnbaum, M., Klaunig, J., Keck, R., Goldblat, P.J., Britton, S.L. and Selman, J. (1984) Studies with hematoporphyrin derivative in transplantable urothelial tumors. In Porphyrins in Tumor Therapy. (Edited by A. Andreoni and R. Cubeddu). Plenum Press, New York, pp. 235-242.

Krieg, M. and Whitten, D. (1984) Self-sensitized photo-oxidation of pro-toporphyrin IX and related porphyrins in erythrocyte ghosts and microemulsions: a novel photo-oxidation pathway involving sinlget oxygen. J. Photochem. 25, 235-252.

Kyriazis, G., Balin, H. and Lipson, R. (1973) Hematoporphyrin derivative fluorescence test colposcopy and colpophotography in the diagnosis of atypical metaplasia, dysplasia and carcinoma in situ of the cervix uteri. Am. J. Obstet. Gynecol. 117, 372-380.

Lamola, A. (1981) Fluorescence methods in the diagnosis and management of diseases of tetrapyrrole metabolism. J. Invest. Dermatol. 77, 114-121.

Lamola, A. (1982) Fluorescence studies of protoporphyrin transport and clearance. Acta. Dermat. Suppl. 100, 57-66.

Lamola, A., Asher, I., Muller-Eberhard, U., and Poh-Fitzpatrick, M. (1981) Fluorimetric study of the binding of protoporphyrin to hemopexin and albumin. Biochem. J. 196, 693-698.

Lamola, A. and Doleiden, F. (1980) Cross-linking of membrane proteins and protoporphyrin-sensitized photohemolysis. Photochem. Photobiol. 31, 597-601.

Land, E.J. (1984) Porphyrin phototherapy of human cancer. Int. J. Rad. Biol. 46, 219-223.

Latham, P. and Bloomer, J. (1983) Protoporphyrin-induced photodamage: studies using cultured fibroblasts. Photochem. Photobiol. 37, 553-557.

Laws, E., Cortese, D., Kinsey, J., Eagan, R. and Anderson, R. (1981) Photoradiation therapy in the treatment of malignant brain tumors: a phase I (feasibility) study. Neurosurg. 9, 672-678.

Leonard, J. and Beck, W. (1971) Hematoporphyrin fluorescence: an aid in diagnosis of malignant neoplasms. Laryngoscope 81, 365-372.

Lim, H., Young, L., Hagan, M. and Gigli, I. (1985) Delayed phase of hematoporphyrin-induced phototoxicity: modulation by complement, leukocytes and antihistamines. J. Invest. Dermatol. 84, 114-117.

Lin, C-W., Bellnier, D., Prout Jr., G., Andrus. W. and Prescott, R. (1984) Cystoscopic fluorescence detector for photodetection of bladder carcinoma with hematoporphyrin derivative. J. Urol. 131, 587-590.

Lipson, R. and Baldes, E. (1961) Hematoporphyrin derivative: a new aid
for endoscopic detection of malignant disease. J. Thorac. Cardiovasc. Surg. 42, 623-629.

Lipson, R., Baldes, E. and Gray, M. (1967) Hematoporphyrin derivative for
detection and management of cancer. Cancer, 20, 2255-2257.

Lipson, R., Baldes, E., and Olsen, A. (1961) The use of a derivative of
hematoporphyrin in tumor detection. J. Natl. Cancer Inst. 26, 1-8.

Lipson, R., Baldes, E. and Olsen, A. (1964) Further evaluation of the use
of hematoporphyrin derivative as a new aid for endoscopic detection
of malignant disease. Dis. Chest. 46, 676-679.

Lipson, R., Baldes, E. and Olsen, A. (1970) Hematoporphyrin derivative: a
new aid for endoscopic detection of malignant disease. J. Thorac.
Cardiovasc. Surg. 42, 623-629.

Lipson, R., Pratt, J., Baldes, E. and Dockerty, M. (1964) Hematoporphyrin
derivative for detection of cervical cancer. Obstet. Gynecol. 24,
78-84.

Longas, M. and Poh-Fitzpatrick, M. (1980) High-pressure chromatography of
plasma free acid porphyrins. Analyt. Biochem. 104, 268-276.

Longas, M. and Poh-Fitzpatrick, M. (1982) A tightly-bound protein-
porphyrin complex isolated from the plasma of a patient with
variegate porphyria. Clin. Chim. Acta 118, 219-228.

Malik, Z. and Breitbart, H. (1980) Cross-linking of hemoglobin and inhibition of globin synthesis in reticulocytes induced by photoactivated protoporphyrin. Acta. Haematol. 64, 304-309.

Malik, Z., Creter, D., Cohen, A. and Djaldelli, M. (1983) Haemin affects
platelet aggregation and lymphocyte mitogenicity in while blood incubations. Cytobios 38, 33-38.

Malik, Z. and Djaldetti, M. (1980) Destruction of erythroleukemia,
myelocytic leukemia and Burkitt lymphoma cells by photoactivated protoporphyrin. Int. J. Cancer 26, 495-500.

Malik, Z., Lejbkowicz, F. and Salzberg, S. (1983) Cholesterol impregna-
tion into erythroleukemioa cell membrane induces rsistance to hematoporphyrin photodynamic effect. In Porphyrins in Tumor Phototherapy.
A. Andreoni and R. Cubeddu, eds., Plenum Press, N.Y., 1984. pp. 185-
191.

Margalit, R. and Cohen, S. (1984) Studies on hematoporphyrin and hematoporphyrin derivative equilibria in heterogeneous systems: porphyrin-
liposome binding and porphyrin aqueous dimerization. Biochim.
Biophys. Acta, 736, 163-170.

Margalit, R. and Rotenberg, M. (1984) Thermodynamics of porphyrin
dimerization in aqueous solutions. Biochem. J. 219, 445-450.

Margalit, R., Shaklai, N. and Cohen, S. (1983) Fluorometric studies on
the dimerization equilibrium of protoporphyrin IX and its haemato
derivative. Biochem. J. 209, 547-552.

Matthews-Roth, M. (1982) Photosensitization by porphyrins and prevention
of photosensitization by carotenoids. J. Natl. Cancer Inst. 69, 279-
285.

Matthews-Roth, M. (1984) Porphyrin photosensitization and carotenoid
protection in mice: in vitro and in vivo studies. Photochem.
Photobiol. 40, 63-67.

McCaughan, J., Guy, J., Hawley, P., Hicks, W., Inglis, W., Laufman, L.,
May, E., Nims, T. and Sherman, R. (1983) Hematoporphyrin derivative
and photoradiation therapy of malignant tumors. Lasers in Surg. Med.
3, 199-209.

McCaughan, J., Hicks, W., Laufman, L., May, E. and Roach, R. (1984) Palliation of esophageal malignancy with photoradiation therapy. Cancer
54, 2905-2910.

McCulloch, G. Forbes, I., See, K., Cowled, P., Jacka, F. and Ward,
A.D. (1984) Phototherapy in malignant brain tumors. In Porphyrin
Localization and Treatment of Tumors. (Edited by D.R. Doiron and

C.J. Gomer). Alan R. Liss, New York, pp. 709-718.

McCullough, J., Weinstein, G., Lemus, L., Rampone, W. and Jenkins, J.J. (1983) Development of a topical hematoporphyrin derivative formulation: characterization of photosensitizing effects in vivo. J. Invest. Dermatol. 81, 528-532.

McGhie, J., Wold, E., Pettersen, E. and Moan, J. (1983) Combined electron radiation and hyperthermia. Repair of DNA strand breaks in NHIK cells irradiated and incubated at 37, 42.5 or 45°. Rad. Res. 96, 31-40.

Medhi, S., Brisbin, D. and McBryde, W. (1976) The stability of porphyrin and metalloporphyrin molecular complexes in solution. Biochim. Biophys. Acta 444, 407-415.

Meffert, H., Bohm, F. and Bauer, E. (1983) Cellular membranes as possible main target of combined photo-chemotherapy. Studia Biophys. 94, 41-44.

Melloni, E., Marchesi, R., Emuanelli, H., Fava, G., Locati, L., Pezzoni, G., Savi, G. and Zunino, F. (1984) Hyperthermal effects in phototherapy with hematoporphyrin derivative photosensitization. Tumori 70, 321-325.

Mew, D., Wat, C., Towers, G. and Levy, J. (1983) Photoimmunotherapy: treatment of animal tumors with tumor-specific monoclonal antibody-hematoporphyrin conjugates. J. Immunol. 130, 1473-1477.

Mitchell, J., McPherson, S., DeGraff, W., Gamson, J., Zabell, A. and Russo, A. (1985) Oxygen dependence of hematoporphyrin derivative-induced photoinactivation of Chinese hamster cells. Cancer Res. 45, 2009-2011.

Miyoshi, N., Hisazumi, H., Ueka, O. and Nakajima. K. (1984) Cellular binding of hematoporphyrin derivative in human bladder cancer cell lines: KK-47. Photochem. Photobiol. 39, 359-363.

Miyoshi, N., Hisazumi, Nakajima, K., H., Ueka, O. and Nakajima. K. (1984) The similarity between the fluorescence spectra of hematoporphyrin derivative incorporated by rat bladder cancer tissues and trimethyl ammonium bromide micelles. Photobiochem Photobiophys 8, 115-121.

Moan, J. (1980) A new method to detect porphyrin radicals in aqueous solution. Acta. Chem. Scand. B 34, 519-521.

Moan, J. (1984) Fluorescence of porphyrins in cells. In Porphyrins in Tumor Therapy. (Edited by A. Andreoni and R. Cubeddu). Plenum Press, New York, pp. 109-124.

Moan, J. (1984) The photochemical yield of singlet oxygen from porphyrins in different states of aggregation. Photochem. Photobiol. 4, 445-449.

Moan, J. and Boye, E. (1981) Photodynamic effect of DNA and cell survival of human cells sensitized by hematoporphyrin. Photobiochem. Photobiophys. 2, 301-307.

Moan, J. and Christensen, T. (1980) Photodynamic effects of human cells exposed to light in the presence of hematoporphyrin. Localization of the active dye. Cancer Lett. 11, 209-214.

Moan, J. and Christensen, T. (1980) Porphyrins as tumor localizing agents and their possible use in photochemotherapy of cancer. Tumor Res. 15, 1-10.

Moan, J. and Christensen, T. (1981) Cellular uptake and photodynamic effect of hematoporphyrin. Photobiochem. Photobiophys. 2, 291-299.

Moan, J. and Christensen, T. (1981) Photodynamic effects on human cells exposed to light in the presence of hematoporphyrin: localization of the active dye. Cancer Lett., 11, 209-214.

Moan, J. and Christensen, T. (1979) Photodynamic inactivation of cancer cells in vitro. Effect of irradiation temperature and dose fractionation. Cancer Lett. 6, 331-335.

Moan, J., Christensen, T. and Jacobsen, P. (1984) Porphyrin-sensitized photoinactivation of cells in vitro. In Porphyrin Localization and Treatment of Tumors. (Edited by D.R. Doiron and C.J. Gomer). Alan R. Liss, New York, pp. 419-442.

Moan, J., Christensen, T., and Jacobsen, P.B. (1984) Photodynamic effects of cells in vitro labeled with hematoporphyrin derivative. Photobiochem. Photobiophys. 7, 349-358.

Moan, J., Christensen, T. and Sommer, S. (1982) The main photosensitizing components of hematoporphyrin derivative. Cancer Lett. 15, 161-166.

Moan, J., Høvik, B., and Sommer, S. (1984) A device to determine fluence-response curves for photoinactivation of cells in vitro. Photobiochem. Photobiophys. 8, 11-17.

Moan, J., Johanssen, J., Christensen, T., Espevik, T. and McGhie, J. (1982) Porphyrin-sensitized photoinactivation of human cells in vitro. Am. J. Path. 109, 184-192.

Moan, J., McGhie, J. and Christensen, T. (1982) Hematoporphyrin derivative: photosensitizing efficiency and cellular uptake of its components. Photobiochem. Photobiophys. 4, 337-345. Photobiol.

Moan, J., McGhie, J. and Jacobsen, P. (1983) Photodynamic effects on cells in vitro exposed to hematoporphyrin derivative and light. Photochem. Photobiol. 37, 599-604.

Moan. J. and Pettersen, E. (1981) X-irradiation of human cells in culture in the presence of haematoporphyrin. Int. J. Rad. Biol. 40, 107-109.

Moan, J., Pettersen, O. and Christensen, T. (1979) The mechanism of photodynamic inactivation of human cells in vitro in the presence of haematoporphyrin. Br. J. Cancer 39, 398-407.

Moan, J., Rimington, C. and Western, A. (1985) The binding of dihematoporphyrin ether (Photofrin II) to human serum albumin. Clin. Chim. Acta 145, 227-236.

Moan, J., Smedshammer, L. and Christensen, T. (1980) Photodynamic effects of human cells exposed to light in the presence of hematoporphyrin. pH effects. Cancer Lett. 9, 327-332.

Moan, J. and Sommer, S. (1981) Fluorescence and absorption properties of the components of hematoporphyrin derivative. Photobiochem. Photobiophys. 3, 93-103.

Moan, J. and Sommer, S. (1983) Uptake of the components of hematoporphyrin derivative by cells and tumours. Cancer Lett. 21, 167-174.

Moan, J. and Sommer, S. (1984) Action spectra for hematoporphyrin derivative and photofrin II with respect to sensitization of human cells in vitro to photoinactivation. Photochem. Photobiol. 40, 631-634.

Moan, J. and Sommer, S. (1985) Oxygen dependence of the photosensitizing effect of hematoporphyrin derivative in NHIK 3025 cells. Cancer Res 45, 1608-1610.

Moan, J., Steen, H., Feren, K. and Christensen, T. (1981) Uptake of hematoporphyrin derivative and sensitized photoinactivation of C3H cells with different oncogenic potential. Cancer Lett. 14, 291-295.

Moan, J., Waksvik, H. and Christensen, T. (1980) DNA single-strand breaks and sister chromatid exchanges induced by treatment with hematoporphyrin and light or by X-rays in human NHIK 3025 cells. Cancer Res. 40, 2915-2918.

Morgan, W., Smith, A. and Koskelo, P. (1980) The interaction of human serum albumin and hemopexin with porphyrins. Biochim. Biophys. Acta 624, 271-285.

Mossoba, M., Rosenthal, I., Carmichael, A. and Riesz, P. (1984) Photochemistry of porphyrins as studied by spin trapping and electron spin resonance. Photochem. Photobiol. 39, 731-734.

Munson, B. (1979) Photodynamic inactivation of mammalian DNA-dependent RNA polymerase by hematoporphyrin and visible light. Int. J. Biochem., 10, 957-960.

Murgia, S., Pasqua, A. and Poletti, A. (1982) A flash photolysis study of hematoporphyrin IX in neutral, aqueous and ethanolic solutions. Photobiochem. Photobiophys. 4, 329-336.

Musser, D., Datta-Gupta, N. and Fiel, R. (1980) Inhibition of DNA dependent RNA synthesis by porphyrin photosensitization. Biochem. Biophys. Res. Commun. 97, 918-925.

Musser, D. and Datta-Gupta, N. (1984) Inability to elicit rapid cytocidal
    effects on L1210 cells derived from Porphyrin-injected mice followed
    by in vitro photoirradiation. J. Natl. Cancer Inst., 72, 427-434.
Musser, D., Wagner, J. and Datta-Gupta, N. (1982) The interaction of
    tumor localizing porphyrins with collagen and elastin. Res. Commun.
    Chem. Path. Pharmacol. 36, 251-259.
Musser, D., Wagner, J., Weber, F. and Datta-Gupta, N. (1979) The effect
    of tumor localizing porphyrins on the conversion of fibrinogen to
    fibrin. Res. Commun. Chem. Path. and Pharmacol. 26, 357-380.
Musser, D., Wagner, J., Weber, F. and Datta-Gupta, N. (1980) The binding
    of tumor localizing porphyrins to a fibrin matrix and their effects
    following photoirradiation. Res. Commun. Chem. Path. and Pharmacol.
    505-525.
Packer, A., Tse, D., Guu, X. and Hayreh, S. (1984) Hematoporphyrin
    photoradiation therapy for iris neovascularization. Arch. Opthalmol.
    102, 1193-1197.
Parker, J. and Stanbro, W. (1984) Dependence of photosensitized singlet
    oxygen production on porphyrin structure and solvent. In Porphyrin
    Localization and Treatment of Tumors. (Edited by D.R. Doiron and
    C.J. Gomer). Alan R. Liss, New York, pp. 259-284.
Paine, J., Dolphin, D. and Gouterman, M. (1978) Exiton and electron in-
    teraction in covalently-linked porphyrins. Can. J. Chem. 56, 1712-
    1715.
Parrish, J. (1982) Therapeutic in vivo photochemistry: photochemical
    toxicity studies in humans. J. Natl. Cancer Inst. 69, 273-278.
Pasqua, A., Poletti, A. and Murgia, S.M. (1982) Ultrafiltration techni-
    ques as a tool for the investigation of hematoporphyrin aggregates in
    aqueous solution. Med. Biol. Environ. 10, 287-291.
Pasternack, R. (1973) Aggregation properties of water-soluble porphyrins.
    Ann. N.Y. Acad. Sci., 206, 614-630.
Pasternak, R., Gibbs, E. and Villafranca, J. (1983) Interactions of por-
    phyrins with nucleic acids. Biochem. 22, 5409-5417.
Patrice, T., Le Bodic, M., Le Bodic, L., Spreux, T., Dabouis, G. and Her-
    vouet, L. (1983) Neodymium-yttrium aluminum garnet laser destruction
    of non-sensitized and hematoporphyrin derivative-sensitized tumors.
    Cancer Res. 43, 2876-2879.
Perlin, D., Murant, R., Gibson, S. and Hilf, R. (1985) Effects of photo-
    sensitization by hematoporphyrin derivative on mitochondrial
    adenosine triphosphatase-mediated proton transport and membrane in-
    tegrity of R3230AC mammary adenocarcinoma. Cancer Res. 45, 653-658.
Perria, C., Delitala, G., Francaviglia, N. and Altomonte. M. (1981)
    Anterior pituitary fluorescence with a hematoporphyrin derivative.
    ICRS Med. Sci. 9, 705-706.
Perria, C., Delitala, G., Francaviglia, N. and Altomonte. M. (1983) The
    uptake of hematoporphyrin derivative by cells of human gliomas:
    determination by fluorescence-microscopy. IRCS
Peterson, D., McKelvey, S. and Edmonson, P. (1981) A hypothesis for the
    molecular mechanism of tumor killing by porphyrins and light. Med.
    Hypoth. 7, 201-205. Med. Sci. 11, 46-47.
Pimstone, N., Horner, I., Shaylor-Billings, J. and Gandhi, S. (1982)
    Hematoporphyrin-augmented phototherapy: dosimetric studies in ex-
    perimental liver cancer in the rat. Proc. Int. Soc. Optical
    Eng. 357, 60-67.
Poletti, A., Murgia, S. and Cannistraro, S. (1981) Laser flash photolysis
    of the hematoporphyrin-β-carotene system. Photobiochem.
    Photobiophys. 2, 167-172.
Pooler, J. and Valenzeno, D. (1981) Dye-sensitized photodynamic inactiva-
    tion of cells. Med. Phys. 8, 614-628.
Pooler, J. and Valenzeno, D. (1982) A method to quantify the potency of
    photosensitizers that modify cell membranes. J. Natl. Cancer Inst.
    69, 211-215.

Profio, A.E. (1984) Laser excited diagnosis of hematoporphyrin derivative for diagnosis of cancer. IEEE J. Quantum Electronics QE-20, 1502-1507.

Profio, A.E., Carvlin, M.,Sarnaik, J. and Wudl, L. (1984) Fluorescence of hematoporphyrin derivative for detection and characterization of tumors. In Porphyrins in Tumor Therapy. (Edited by A. Andreoni and R. Cubeddu). Plenum Press, New York, pp. 321-338.

Profio, A.E. and Doiron, D. (1981) Dosimetry considerations in photo-therapy. Am. Assoc. Phys. Med. 8, 190-196.

Profio, A.E., D.R. Doiron, O.J. Balchum and G.C. Huth (1983) Fluorescence bronchoscopy for localization of carcinoma in situ. Am. Assoc. Phys. Med. 10, 35-39.

Profio, A.E. and Sarnaik, J. (1984) Fluorescence of HPD for tumor detection and dosimetry in photoradiation therapy. In Porphyrin Localization and Treatment of Tumors. (Edited by D.R. Doiron and C.J. Gomer). Alan R. Liss, New York, pp. 163-176.

Rasmussen-Taxdal, D., Ward, G. and Figge, F. (1955) Fluorescence of human lymphatic and cancer tissue following high doses of intravenous hema-toporphyrin. Cancer 8, 78-81.

Reddi, E., Jori, G. and Rodgers, M. (1983) Flash photolysis studies of hemato- and copro-porphyrins in homogenous and microheterogeneous dispersions. Studia. Biophys. 94, 13-18.

Reddi, E., Jori, G., Rodgers, M. and Spikes, J. (1983) Flash photolysis studies of hemato- and copro-porphyrins in homogeneous and microheterogeneous aqueous dispersions. Photochem. Photobiol. 38, 639-646.

Reddi, E., Rodgers, M., Spikes, J. and Jori, G. (1984) The effect of medium polarity on the hematoporphyrin-sensitized photooxidation of L-tryptophan. Photochem. Photobiol. 40, 415-421.

Reddi, E., Rossi, E. and Jori, G. (1981) Factors controlling the ef-ficiency of porphyrins as photosensitizers of biological systems to damage by visible light. Med. Biol. Environ., 9, 337-351.

Rettenmaier, M., Berman, M., DiSaia, P., Burns, R. and Berns, M. (1984) Photoradiation of gynecologic malignancies. Gynecol. Oncol. 17, 200-206.

Rettenmaier, M., Berman, M., DiSaia, P., Burns, R., Weinstein, G., McCul-lough, J. and Berns, M. (1984) Gynecologic uses of photoradiation therapy. In Porphyrin Localization and Treatment of Tumors. (Edited by D.R. Doiron and C.J. Gomer). Alan R. Liss, New York, pp. 767-775.

Reyftmann, J., Morliere, P., Goldstein, S., Santus, R., Dubertret, L. and Lagrange, D. (1984) Interactions of human serum low density lipo-proteins with porphyrins: a spectroscopic and photochemical study. Photochem. Photobiol. 40, 721-729.

Richard, P., Blum, A. and Grossweiner, L. (1983) Hematoporphyrin photo-sensitization of serum albumin and subtilisin BPN. Photochem. Photobiol. 37, 287-291.

Richelli, F. and Grossweiner, L. (1984) Properties of a new state of hematoporphyrin in dilute aqueous solution. Photochem. Photobiol. 40, 599-606.

Riopelle, R. and Kennedy, J. (1982) Some aspects of porphyrin neurotoxicity in vitro. Can. J. Physiol. Pharmacol. 60, 707-714.

Roberts, J. (1984) The photodynamic effect of chlorpromazine, promazine and hematoporphyrin on lens protein. Invest. Ophthalmol. Vis. Sci. 25, 748-750.

Rossi, E., Van de Vorst, A. and Jori, G. (1981) Competition between the singlet oxygen and electron transfer mechanisms in the porphyrin-sensitized photooxidation of L-tryptophan and tryptamine in aqueous micellar dispersions. Photochem. Photobiol. 34, 447-454.

Salet, C. and Moreno, G. (1981) Photodynamic effects of haematoporphyrin on respiration and calcium uptake in isolated mitochondria. Int. J. Rad. Biol. 39, 227-230.

Salet, C., Moreno, G., Vever-Bizet, C. and Brault, D. (1984) Anoxic photodamage in the presence of porphyrins: evidence for the lack of effects on mitochondrial membranes. Photochem. Photobiol. 40, 145-147.

Salet, C., Moreno, G. and Vinzens, F. (1983) Effects of photodynamic action on energy coupling of Ca²⁺ uptake in liver mitochondria. Biochem. Biophys. Res. Commun. 115, 76-81.

Salzberg, S., Lejbkowicz, F., Ehrenberg, B. and Malik, Z. (1985) Protective effect of cholesterol on Friend leukemic cells against photosensitization by hematoporphyrin derivative. Cancer Res. 45, 3305-3310.

Sandberg, S. and Romslo, I. (1980) Porphyrin-sensitized photodynamic damage of isolated rat liver mitochondria. Biochim. Biophys. Acta 593, 187-195.

Sandberg, S. and Romslo, I. (1981) Porphyrin-induced photodamage at the cellular and the sub-cellular level as related to the solubility of the porphyrin. Clin. Chim. Acta 109, 193-201.

Sandberg, S., Romslo, I., Høvding, G. and Bjørndal, T. (1982) Porphyrin-induced photodamage as related to the subcellular localization of the porphyrins. Acta Dermat. Suppl. 100, 75-80.

Schothorst, A., Van Steveninck, J., Went, L. and Suurmond, D. (1972) Photodynamic damage of the erythrocyte membrane caused by protoporphyrin in protoporphyria and in normal red blood cells. Clin. Chim Acta 39, 161-170.

Schnipper, L., Lewin, A., Schwartz, M. and Crumpacker, C. (1980) Mechanisms of photodynamic inactivation of herpes simplex viruses. Comparison between methylene blue, light plus electricity, and hematoporphyrin plus light. J. Clin. Invest. 65, 432-438.

Schuller, D., McCaughan, J. and Rock, R. (1985) Photodynamic therapy in head and neck cancer. Arch. Otolaryngol. 111, 351-355.

Schwartz, S., Berg, M., Bossenmaier, I. and Dinsmore, H. (1960) Detection of porphyrins in biological materials. Meth. Biochem. Anal. 8, 221-287.

See, K., Forbes, I. and Betts, W. (1984) Oxygen dependency of photocytotoxicity with haematoporphyrin derivative. Photochem. Photobiol. 39, 631-634.

Selensky, R., Holten, D., Windsor, M., Paine III, J., Dolphin, D., Gouterman, M. and Thomas, J. (1981) Exitonic interactions in covalently-linked porphyrin dimers. Chem. Phys. 60, 33-46.

Selman, S., Keck, R., Klaunig, J., Kreimer-Birnbaum, M., Goldblat, P. and Britton, S. (1983) Acute blood flow changes in transplantable FANFT-induced urothelial tumors treated with hematoporphyrin derivative and light. Surg. Forum 34, 676-678.

Selman, S., Kreimer-Birnbaum, M., Klaunig, J., Goldblatt, P., Keck, R. and Britton, S. (1984). Blood flow in transplantable bladder tumors treated with hematoporphyrin derivative and light. Cancer Res. 44, 1924-1927.

Sery, T. (1979) Photodynamic killing of retinoblastoma cells with hematoporphyrin and light. Cancer Res. 39, 96-100.

Sery, T. and Dougherty, T. (1984) Photoradiation of rabbit ocular malignant melanoma sensitized with hematoporphyrin derivative. Current Eye Res. 3, 519-528.

Seybold, P. and Gouterman, M. (1969) Porphyrins. XIII Fluorescence spectra and quantum yields. J. Mol. Spectros. 31, 1-13.

Sima, A., Kennedy, J., Blakeslee, D. and Robertson, D. (1981) Experimental porphyric neuropathy: a preliminary report. J. Can. Sci. Neurol. 8, 105-114.

Slater, T. and Riley, P. (1966) Photosensitization and lysosomal damage. Nature 209, 151-154.

Smith, K. (1979) Protoporphyrin IX: some recent research. Acc. Chem. Res., 12, 374-381.

Smith, R., Doran, D., Mazur, M. and Bush, B. (1980) High-performance liquid chromatographic determination of protoporphyrin and zinc protoporphyrin in blood. J. Chromatog. 181, 319-327.

Soma, H. and Nutahara, S. (1983) Cancer of the female genitalia. In Lasers and Hematoporphyrin Derivative in Cancer Therapy. Y Hayata and TJ Dougherty, eds, Igako-Shoin, Tokyo, New York. pp 97-109.

Sommer, S., Moan, J., Christensen, T. and Evensen, J. A chromatographic study of hematoporphyrin derivatives. In Porphyrins in Tumor Phototherapy. A. Andreoni and R. Cubeddu, eds., Plenum Press, N.Y., 1984. pp. 81-91.

Sommer, S., Rimington, C. and Moan, J. (1984) Formation of metal complexes of tuomor-localizing porphyrins. FEBS Lett. 172, 267-271.

Sorata, Y., Takahama, U. and Kimura, M. (1984) Protective effect of quercetin and rutin on photosensitized lysis of human erythrocytes in the presence of hematoporphyrin. Biochim. Biophys. Acta 799, 313-317.

Spears, J., Serur, J., Shropshire, D. and Paulin, S. (1983) Fluorescence of experimental atheromatous plaques with hematoporphyrin derivative. J. Clin. Invest. 71, 395-399.

Spikes, J. (1975) Porphyrins and related compounds as photodynamic sensitizers. Ann. N.Y. Acad. Sci. 244, 496-508.

Spikes, J. (1982) Photodynamic reactions in photomedicine. In The Science of Photomedicine (Edited by J.D. Regan and J. Parrish. Plenum Publ. Co., New York, pp. 113-144.

Spikes, J. (1983) Photosensitization in mammalian cells. In Photoimmunology (Edited by J.A. Parrish, M.L. Kripke and W.L. Morrison). Plenum Publ. Co., New York, pp. 23-49.

Spikes, J., Matis, W. and Rodgers, M. (1983) The photochemical behavior of porphyrins in solution and incorporated into liposomal membranes. Stud. Biophys. 94, 19-24.

Starr, W., Marijinssen, J., van den Berg-Blok, A. and Reinhold, H. (1984) Destructive effect of photoradiation on the microcirculation of a rat mammary tumor growing in 'sandwich' observation chambers. In Porphyrin Localization and Treatment of Tumors. (Edited by D.R. Doiron and C.J. Gomer). Alan R. Liss, New York, pp. 637-646.

Stenstrøm, A., Moan, J., Brunborg, G. and Eklund, T. (1980) Photodynamic inactivation of yeast cells sensitized by hematoporphyrin. Photochem. Photobiol. 32, 349-352.

Strom, R., Crifo, C., Simonetta, M., Federici, G., Mavelli, I. and Agro, A.F. (1977) Protoporphyrin IX sensitized photohemolysis: stoichiometry of the reaction and repair by reduced glutathione. Physiol. Chem. Phys. 9, 63-74.

Suwa, K., Kimura, T. and Schaap, A. (1977) Reactivity of singlet molecular oxygen with cholesterol in a phospholipid membrane matrix. A model for oxidative damage of membranes. Biochem. Biophys. Res. Commun. 75, 785-792.

Svaasand, L. (1982) Properties of thermal waves in vascular media. Med. Phys. 9, 711-714.

Svaasand, L. (1984) Optical dosimetry for direct and interstitial photoradiation therapy of malignant tumors. In Porphyrin Localization and Treatment of Tumors. (Edited by D.R. Doiron and C.J. Gomer). Alan R. Liss, New York, pp. 91-114.

Svaasand, L., Doiron, D. and Dougherty, T. (1983) Temperature rise during photoradiation therapy of malignant tumors. Med. Phys. 10, 10-17.

Svaasand, L. and Ellingsen, R. (1985) Optical penetration in human intracranial tumors. Photochem. Photobiol. 41, 73-76.

Swincer, A., Trenerry, V. and Ward, A.D. (1984) The analysis and some chemistry of haematoporphyrin derivative. In Porphyrin Localization and Treatment of Tumors. (Edited by D.R. Doiron and C.J. Gomer). Alan R. Liss, New York, pp 285-300.

Swincer, A., Ward, A.D. and Howlett, G. (1985) The molecular weight of haematoporphyrin derivative, its gel column fractions and some of its components in aqueous solution. Photochem. Photobiol. 41, 47-50.

Tangeras, A. and Flatmark, T. (1979) In vitro binding of protoheme IX and protoporphyrin IX to components in the matrix of rat liver mitochondria. Biochim. Biophys. Acta 588, 201-210.

Tomio, L., Zorat, P.L., Corti, L., Calzavara, F., Cozzani, I., Reddi, E., Salvato, G. and Jori, G. (1982) Cancer chemotherapy: biochemical bases and experimental results. Med. Biol. Environ. 10, 303-307.

Tomio, L., Zorat, P., Corti, L., Calzavara, F., Reddi, E. and Jori, G. (1983) Effect of hematoporphyrin and red light on AH-130 solid tumors in rats. Acta. Radiol. Onc. 22, 49-53.

Tomio, L., Zorat, P., Jori, G., Reddi, E., Salvato, B., Corti, L. and Calzavara, F. (1982) Elimination pathway of hematoporphyrin from normal and tumor-bearing rats. Tumori 68, 283-286.

Tomio, L., Zorat, P., Corti, L., Calzavara, F., Cozzani, I., Reddi, E., Salvato, B. and Jori, G. (1982) Cancer phototherapy: biochemical basis and experimental results. Med. Biol. Enviorn. 10, 303-307.

Truscott, T.G. (1980) Laser flash photolysis of molecules of medical relevance. In Lasers in Biology and Medicine (F. Hillenkamp, R. Pratesi and C.A. Sacchi, eds.). Plenum Publishing Corp., New York, pp. 235-250.

Tse, D., Dutton, J., Weingeist, T., Hermsen, V. and Kersten, R. (1984) Hematoporphyrin photoradiation therapy for intraocular and orbital malignant melanoma. Arch. Opthalmol. 102, 833-838.

Tsuchiya, A., Obara, N.,Miwa, M., Ohi, T., Kato, H. and Hayata, Y. (1983) Hematoporphyrin derivative and laser photoradiation in the diagnosis and treatment of bladder cancer. J. Urology 130, 79-82.

Van der Putten, W. and van Gemert, M. (1983) A modelling approach to the detection of subcutaneous tumours by haematoporphyrin-derivative fluorescence. Phys. Med. Biol. 28, 639-645.

Van der Putten, W. and Van Gemert, M. (1983) Haematoporphyrin-derivative fluorescence in vitro and in an animal tumor. Phys. Med. Biol. 28, 633-638.

Van Steveninck, J., Kogeler, J. and Dubbelman, T. (1984) Involvement of imidazole-imidazole and imidazole-sulfhydryl interactions in photo-dynamic crosslinkling of proteins. Biochim. Biophys. Acta 788,35-40.

Verweij, H., Dubbelman, T. and Van Steveninck, J. (1981) Photodynamic protein cross-linking. Biochim. Biophys. Acta 647, 87-94.

Verweij, H. and Van Steveninck, J. (1982) Model studies on photodynamic cross-linking. Photochem. Photobiol. 35, 265-267.

Volden, G., Christensen, T. and Moan, J. (1981) Photodynamic membrane damage of hematoporphyrin derivative-treated NHIK 3025 cells in vitro. Photobiochem. Photobiophys. 3, 105-111.

Wakulchik, S., Schiltz, J. and Bickers, D. (1980) Photolysis of protoporphyrin-treated human fibroblasts in vitro: studies on the mechanism. J. Lab. Clin. Med. 96, 158-167.

Waldow, S. and Dougherty, T. (1984) Interaction of hyperthermia and photoradiation therapy. Radiation Res. 97, 380-385.

Waldow, S., Henderson, B. and Dougherty, T. (1984) Enhanced tumor control following sequential treatments of photodynamic therapy (PDT) and localized microwave hyperthermia in vivo. Lasers in Surg. Med. 4, 79-85.

Ward. B., Forbes, I., Cowled, P., McEvoy, M. and Cox, L. (1982) The treatment of vaginal recurrences of gynecologic malignancies with phototherapy following hematoporphyrin derivative pretreatment. Am. J. Obstet, Gynecol. 142, 356-357.

Wat, C., Mew, D., Levy, J. and Towers, G. (1984) Photosensitizer-protein conjugate: potential use as photoimmunotherapeutic agents. In Porphyrin Localization and Treatment of Tumors. (Edited by D.R. Doiron and C.J. Gomer). Alan R. Liss, New York,pp. 351-360.

Weishaupt, K., Gomer, C. and Dougherty, T. (1976) Identification of singlet oxygen as the cytotoxic agent in photo-inactivation of a murine tumor. Cancer Res. 36, 2326-2329.

Wile, A., Dahlman, A., Burns, R. and Berns, M. (1982) Laser photoradiation therapy of cancer following hematoporphyrin sensitization. Lasers in Med. Surg. 2, 163-168.

Winkelman, J. (1961) Intracellular localization of 'hematoporphyrin' in a transplanted tumor. J. Natl. Cancer Inst. 27, 1369-1377.

Wise, B. and Taxdal, D. (1967) Studies of the blood-brain barrier using hematoporphyrin. Brain Res. 4, 387-389.

Xian-wen, H., Xin-min, S., Jian-guo, X., Xian-jun, F., Yu-hai, Z., Qi-chu, M., Hong, S., Shi-lin, X. and Ru-gang, Z. (1983) Clinical use of hematoporphyrin derivative in malignant tumors. Chinese Med. 96, 754-758.

Zorat, P., Corti, L., Tomio, L., Maluta, S., Rigon, A., Mandotti, G. Jori, G., Reddi, E. and Calzavara, F. (1983) Photoradiation therapy of cancer using hematoporphyrin. Med. Biol. Environ. 11, 511-516.